Constructing the Future

Also available from Taylor & Francis

Agents and Multi-Agent Systems in Construction
**Edited by: Chimay J. Anumba,
Onuegbu Ugwu and Zhaomin Ren**   Hb: 0–415–35904–X

Understanding I.T. in Construction
Ming Sung   Pb: 0–415–23190–6

Concurrent Engineering in Construction Projects
Edited by: Chimay J. Anumba et al.   Hb: 0–415–39488–0

People and Culture in Construction
**Andrew Dainty Barbara Bagilhole and
Stuart Green**   Hb: 0–415–34870–6

Information and ordering details

For price availability and ordering visit our website **www.tandf.co.uk/builtenvironment**

Alternatively our books are available from all good bookshops.

Constructing the Future

nD Modelling

Edited by
Ghassan Aouad,
Angela Lee and Song Wu

Routledge
Taylor & Francis Group

LONDON AND NEW YORK

First published 2007
by Taylor & Francis
This edition published 2013
Routledge
2 Park Square, Milton Park, Abingdon, Oxfordshire OX14 4RN

Simultaneously published in the USA and Canada
by Routledge
711 Third Avenue, New York, NY 10017, USA

First issued in paperback 2016

Taylor & Francis is an imprint of the Taylor & Francis Group,
an informa business

© 2007 Taylor and Francis

Typeset in Times New Roman by
Newgen Imaging Systems (P) Ltd, Chennai, India

British Library Cataloguing in Publication Data
A catalogue record for this book is available from the British Library

Library of Congress Cataloging in Publication Data
    Constructing the future : nD modelling / edited by Ghassan Aouad,
Angela Lee & Song Wu.
        p. cm.
    Includes bibliographical references and index.
    1. Building – Data processing. 2. Architectural design – Data
processing. 3. Buildings – Computer-aided design. 4. Multidisciplinary
design optimization. I. Aouad, Ghassan. II. Lee, Angela, Ph. D.
III. Wu, Song, Dr.

    TH437.C656 2006
    690.0285–dc22   2006010379

ISBN 13: 978-1-138-97161-5 (pbk)
ISBN 13: 978-0-415-39171-9 (hbk)

Contents

Contributors

Ghassan Aouad (*School of Construction and Property Management, University of Salford, UK*) Professor Aouad is Dean of the faculty of Business and Informatics in the UK, and director of the £3M EPSRC IMRC Centre (Salford Centre for Research and Innovation in the Built and Human Environment: SCRI). He led the prestigious £443k EPSRC platform grant (from 3D to nD modelling). Professor Aouad's research interests are in modelling and visualisation, development of information standards, process mapping and improvement and virtual organisations. He has published and presented internationally on his work.

Yusuf Arayici (*School of Construction and Property Management, University of Salford, UK*) Dr Arayici is a research assistant working in Virtual Planning at the University of Salford. He was responsible for building information capture and processing for the Building Data Integration System in the IntelCities project. Yusuf has been investigating the use of laser scanner technology for building surveys. He now works on 3D scanner technology in the VEPs project. He previously worked on the DIVERCITY project (Distributed Virtual Workspace for enhancing Communication within the Construction Industry).

Matthew Bacon (*Director of ARK e-management Ltd, UK Visiting Professor, University of Salford, UK*) Professor Bacon co-founded ARK e-management Ltd in 2000, a company providing management consultancy and technology development services. ARK (Assets Resources and Knowledge) believes that new management practices are required if businesses are to effectively exploit the power of the Internet, and that the effective management of digital information is a key requirement in this regard. Bacon originally trained as an architect. In the 1980s he worked with RUCAPS, one of the first commercial Object-orientated CAD technologies, and then later he with AutoCAD. In 1995 he joined BAA plc to work with the core team developing the BAA Project Process. He was responsible for leading the development of the Design Management Process, which embraced briefing for major projects. At BAA he developed a number of technology 'tools' to support this work. As Head of

Process and Technology a key part of his role was to support innovation in project teams in terms of the management of information and process, using leading edge technologies. Whilst at BAA, Bacon represented the business within the International Alliance for Interoperability, and served on the Board of Directors. He established the Client Briefing Group, which comprised some major UK companies and public bodies.

Richard J. Coble (*CEC-RJC Consulting, USA Shiloh Consulting, USA Visiting Professor, University of Salford, UK*) Professor Coble's work life extends into both Academic and Construction Engineering Management. He owns SHILOH Consulting, with offices in Orlando, Miami Gainesville, Phoenix and a new office set to open in Rio de Janiero, Brazil. His work in most cases ties to Llitigation Support, Visualization Implementation and Risk Management. Dr Coble is a Research Professor at Florida International University, where SHILOH also sponsors the SHILOH Lecture Series on Construction Risk Management. This is where Professor Coble brings in a scholar or industry professionals to speak and write a paper on construction risk management. Professor Coble has the AIC designation with the American Institute of Constructors and holds their highest qualification status of Certified Professional Constructor (CPC). He is an active member of ASCE and has been for 37 years. He often makes professional presentations and has authored over 100 scholarly publications. In his spare time he is a State of Florida, Department of Juvenile Justice, Certified Volunteer Juvenile Parole Officer.

Rachel Cooper (*Director of the Adelphi Research Institute for Creative Arts and Sciences, University of Salford, UK*) Rachel is a Professor of Design Management, and undertakes research in the areas of design management, new product development, design in the built environment, design against crime and socially responsible design. All her projects have been in collaboration with industry, working both nationally and internationally. Currently, Professor Cooper is Principal Investigator of 'Vivacity 2020', a five-year study of urban sustainability for the 24-hour city, and Co-Director of SCRI. Professor Cooper was Founding Chair of the European Academy of Design, and is also Founding Editor of The Design Journal. She is currently a member of the Strategic Advisory Team for Environment and Infrastructure division of the UK's Engineering & Physical Sciences Research Council (EPSRC) and is Panel Convenor for the Postgraduate Awards in Visual Arts and Media for the Arts and Humanities Research Board (AHRB), where she is also currently serving on the Management Board. Professor Cooper has written over 100 papers and 6 books.

Steve Curwell (*School of Construction and Property Management, University of Salford, UK*) Steve (BSc, MSc, RIBA, MILT) is Professor of Sustainable Urban Development (SUD) and a leading European researcher in this area through his key role in 14 EU and UK national research projects to a total value of circa €15M over the last 15 years. He was the Scientific and Technical Director of the IntelCities IP and has extensive research collaboration experience with over 350 research groups, city authorities and IST companies via four EU projects BEQUEST (FP4), CRISP, INTELCITY and LUDA (FP5). He is the author or co-author of 75 publications, 5 books, 50+ research papers, learning packages and research web sites.

Nashwan Dawood (*Centre for Construction Innovation and Research, University of Teesside, UK*) Professor Dawood's research lies within the field of project and construction management, including the application of VR in the construction process, 4D visual planning, risk management, intelligent decision support systems, cost forecasting and control business processes. This has resulted in over 140 published papers in refereed international journals and conferences, and research grants from the British Council, Construction Industry, Engineering Academy, EPSRC, DTI and EU. He has been a visiting professor to five overseas universities.

Robin Drogemuller (*CSIRO & Co-operative Research Centre for Construction Innovation, Australia Visiting Professor, University of Salford, UK*) Professor Drogemuller holds qualifications in architecture, mathematics and computing, and worked as an architect and construction manager in both temperate and tropical Australia before entering academia at the Northern Territory University and then James Cook University of North Queensland. At CSIRO, Professor Drogemuller leads a team of 12 people in the area of IT support to design and manufacturing with a major focus on AEC applications. This group has developed and is continuing to develop commercial software to support construction innovation.

Thomas Froese (*Department of Civil Engineering, University of British Columbia, Canada*) Thomas is an Associate Professor at the University of British Columbia. Professor Froese's research focus concentrates on computer applications and information technology to support construction management, particularly information models and standards of construction process data for computer-integrated construction. Professor Froese originally studied Civil Engineering at the University of British Columbia before obtaining his PhD from Stanford University in 1992.

Dennis Fukai (*Insitebuilders, USA*) Dr Fukai focuses his research on construction communications, specifically the interaction of architects, engineers and constructors in the pre-construction phases of project development after the general design for a building has been completed. The goal of his research is to facilitate the exchange of technical information between these professionals in order to reduce the potential for confusion and conflicts during the construction process. The objective is to demonstrate and apply practical solutions to the industry of multi-dimensional graphical information systems as a visual approach to design and construction management. Dr Fukai's work currently centres on design and construction information delivered in the form of interactive models, 3D photo overlays and 4D information models. Current projects include the implementation of an augmented graphical interface in the form of animated sequence assemblies for construction, assembly and maintenance of complex structures. This work is supported by the use of graphic narratives and illustrative 'cartoons' used to deliver construction information quickly and effectively to owners, managers and field employees.

Andy Hamilton (*School of Construction and Property Management, University of Salford, UK*) Andy is Director of the Virtual Planning in the 6* rated Research Institute for the Built and Human Environment (BuHu) at the University of Salford. He leads the system design work in the Virtual Environmental Planning systems EU INTERREG project (4 million Euro, 2004–07). He was the Deputy Scientific and technical Director of the IntelCities project (2004–05). He worked as an EU recognised expert consultant on the EU funded Lithuanian eCities project in 2004. Andy is the author of 50+ publications.

David Hands (*School of Art and Design, University of Salford, UK*) Dr Hands is the Associate Director of the Design and Innovation Research Group, and Programme Leader for BA (Hons) Design Management at the University of Salford. His areas of expertise include design policy analysis, implementing technology transfer systems through design and recently completed doctoral research into embedding 'crime resistant' thinking and tools into the briefing process for both product and spatial designers. He has written and published over 60 international academic papers and articles on a variety of design management topics, and has published two major books – one focusing on Design Management (Routledge, 2002) and *Manchester: A Guide to Recent Architecture* (BT Batsford, 2000). His forthcoming publication *Design Management Fieldwork* (with Prof. Robert Jerrard) will be published in autumn 2006 focusing on Design Management Audits. He is also the co-founder of the Design Management Network which has an international membership dedicated to supporting and promoting the value of design and design management to both industry and academia. Dr Hands is External Examiner to the BA [Hons] Programme Design Management for the Creative Industries at the University of Nottingham; and MA Design Management at the London College of Communication, University of the Arts London. He is also a visiting lecturer to the MSc Design Management Programme, L'Institut Européen de Design, Toulon, France.

Tarek M. Hassan (*Department of Civil & Building Engineering, Loughborough University, UK*) Dr Hassan is a Senior Lecturer and Director of the European Union Research Group at Loughborough University. His academic experience is complimented by 10 years of industrial experience with several international construction organisations. Dr Hassan has been leading and partnered in 10 projects funded by the European Commission under the IST (Information Society Technologies) programme. His research interests include

Advanced Information and Communication Technologies (ICT), applications of e-commerce, virtual enterprises business relationships, legal aspects in ICT environments, information modelling and simulation, design management, management information systems, strategic management and dynamic brief development. Dr Hassan has over 100 publications.

Margaret Horne (*School of the Built Environment, University of Northumbria, UK*) Margaret researches and teaches in the field of 3D computer modelling and visualisation for architecture, and Virtual Reality for the built environment. She is co-author of a recent book, *Artists' Impressions in Architectural Design*, which analyses the ways in which architects have presented their designs for clients and public, both historically and contemporarily, thus analysing the correlation between architectural representation and eventual buildings. She is a Chartered Information Technology Professional and has lectured in universities around the world, introducing applications of IT to built environment staff and students. She has helped practising architects evaluate the role of computers in architecture, and recently researched into the visualisation of photovoltaic clad buildings in collaboration with Ove Arup and Partners. Currently researching into Virtual Reality for the Built Environment, within a broad interdisciplinary context, Margaret is helping to implement a Virtual Reality Facility for the School of the Built Environment, Northumbria University.

Kalle Kähkönen (*VTT Building & Transport, Finland*) Dr Kähkönen works as a chief research scientist at the VTT Technical Research Centre of Finland. He holds degrees from two universities: MSc (Civil Engineering) from the Helsinki University of Technology and PhD from the University of Reading (UK). He has carried out research and development in numerous areas also covering the implementation of research results in companies as a consultant. Dr Kähkönen's main interests are in advanced technology and solutions for modern project management and project business.

Jan Karlshøj (*Rambøll Danmark, Denmark*) Dr Karlshøj has a Master of Science and PhD in data modelling in the construction industry. He has been working at consulting engineer companies for 13 years, where he has been giving tasks within the structural domain and a number of ICT development-related tasks. He is working as a Chief Consultant at Rambøll. Dr Karlshøj has been active in a number of national ICT development projects for the construction industry, and has the project lead in the '3D visualisation and simulation' project under the national IT implementation programme: Digital Construction. He has had a long interest in IFC, and has held several positions within the International Alliance for Interoperability.

Andreas Kohlhaas (*GIStec GmbH, Germany*) Andreas is Business Development and Project Manager for 3D GIS applications and CAD-GIS Integration at GIStec GmbH, a spin off from the Fraunhofer Institute GD, Darmstadt. Besides leading many 3D-City-Model projects, he is an active member of the Special Interest Group 3D of the Geo-Data initiative of North Rhine Westphalia since 2002. The necessity for CAD-GIS integration in 3D city models and the combination of IFCs and CityGML is a continuing focus of his professional activities. Formerly Andreas was R&D Manager, Graphisoft Deutschland, and General Manager of Graphisoft Nord.

Jarmo Laitinen (*JT Innovators Ltd, Finland Acting Professor, Tampere University of Technology, Finland*) Professor Laitinen (Dr, MSc, Eur.Ing) has a background in structural design, production processes and their technologies. He has wide expertise of European RTD programmes and knowledge of many different R&D methodologies. Professor Laitinen has participated in and co-ordinated several EC-research projects, and has evaluated for the EC Framework programmes. He was the chairman of the steering committee of the national VERA programme launched by Tekes (IT enabled business networking in construction), member Nordic IAI chapter, and has coordinated the IST-1999-60002-GLOBEMEN (IMS) project.

Yiu Wai Lam (*University of Salford, UK*) Yiu Wai is Professor of Acoustics and the Head of the Acoustics Research Centre at the University of Salford. Professor Lam has over 20 year's experience of working in environmental acoustics, industrial noise and building acoustics and is internationally recognised for his research and expertise in these fields. He is known particular for his expertise in the development and application of computer modelling techniques in these areas. He has a long track record of successful EPSRC, European, and industrial funded researches in building acoustics, noise control, and outdoor sound propagation. During the 1990's, he developed a computer model for the prediction of noise transmission through metal cladding building elements, which is still being used in the UK metal cladding manufacturing industry. In the field of room acoustics, he has developed an adaptive beam tracing technique for the prediction of internal room acoustics and is a leading propellant for diffuse reflection modelling in such models. Currently he is working closely with the industry to develop environmental noise propagation models that take into account of the complex effects of the atmosphere. He is also the co-ordinator of an Engineering Network in Building and Environmental Acoustics, which promotes the understanding and application of acoustics in the built environment sector. He was awarded the Institute of Acoustics' Tyndall Medal in 2000 for his contribution to acoustics. He is currently a member of the Institute of Acoustics' Research Coordination Committee, a member of the BSI EH/1/3 committee on industrial and residential noise, and a member of the Acoustics Technical Committee of the American Helicopter Society. Professor Lam is an Associate Editor of the international journal Applied Acoustics.

Angela Lee (*School of Construction and Property Management, University of Salford, UK*) Dr Lee is the Programme Director for BSc (Hons) in

Architectural Design and Technology at the University of Salford. She is/has worked on various EU, CIB and EPSRC-funded projects including 'Revaluing Construction', 'PeBBu', '3D to nD Modelling', 'Process Protocol', 'Salford: China research tour', 'nD game' and 'Women and Construction'. Her research interests include design management, performance measurement, process management, nD modelling, product and process modelling and requirements capture. She has published extensively in both journal and construction papers in these fields. Dr Lee completed a BA (Hons) in Architecture at the University of Sheffield, and her PhD at the University of Salford.

Jun Lee (*School of Construction and Property Management, University of Salford, UK*) Jun trained as an architect in Korea and was awarded an MSc in Architecture in 2002. He is currently a doctoral candidate at the University of Salford. His PhD aims to develop an integrated building information modelling that will aid visualisation of the interaction between various design criteria, in the form of a computer game for school kids.

Amanda Jane Marshall-Ponting (*School of Construction and Property Management, University of Salford, UK*) Amanda obtained a BSc (Hons) in Applied Psychology from John Moores University in 1998, and an MRes in Informatics at Manchester University. Her introduction to the University of Salford came through involvement with the BEQUEST network, an EU concerted action concerned with Sustainable Urban Development. She has been involved with a number of Salford projects including VULCAN (Virtual Urban Laboratories for computer aided network), Intelcity and 3D to nD Modelling; the latter project is being used as the main PhD case study to develop her interests in multi-disciplinary collaboration, organisational working and information systems.

John Mitchell (*Faculty of the Built Environment, University of New South Wales, Australia CQR Pty Ltd*) John is an architect by profession, and principal of the consulting practice CQR Pty Ltd. He is Chairman of the IAI Australia and New Zealand Chapters, as well as a Deputy Vice-Chairman of the International body of the International Alliance for Interoperability. Concurrently, in the role of Adjunct Associate Professor at the Faculty of the Built Environment, University of New South Wales, he leads a course in multi-disciplinary design collaboration, using an IFC model server, and a suite of IFC compliant applications. John also supports general Object-Based CAD courses as well as applying computer-based city modelling expertise to the UNSW City Futures Research Centre as part of a focus on advanced urban scale building model technology development.

Wafaa Nadim (*School of Construction and Property Management, University of Salford, UK*) Wafaa trained as an architect and has an MSc in Construction Information Technology from the University of Salford. She is currently working on the ManuBuild EU-funded project at Salford.

Khalid Naji (*Department of Civil Engineering, University of Qatar, Qatar*) Dr Naji holds a PhD in Civil Engineering from the University of Florida in the area of Public Works Planning and Management. He received his MSc degree in Civil Engineering from the University of Texas at Austin, and his BSc degree in Civil Engineering from the University of Qatar. Dr Naji's area of interest is computer applications in construction engineering, project planning and scheduling and applications of virtual reality in construction engineering. He served as Head of the Civil Engineering Department at the University of Qatar from 2001 until 2005, where he succeeded with his colleagues in receiving International ABET accreditation recognition for the Civil Engineering Programme. He is a member of ASCE and a number of committees in both the university and national levels at the State of Qatar. Dr Naji is also a registered Grade-A consulting engineer in the State of Qatar, and owns an Architectural and Engineering consulting firm in Doha.

Zhaomin Ren (*School of Technology Glamorgan University, UK*) Dr Ren is a Senior Lecturer at the School of Technology at Glamorgan University. He has years of research experience in different countries and more than 16 years industrial experience. His research interests are in the fields of computer-aided engineering, advanced IT, collaborative engineering, trust and legal issues of eBusiness and eEnginering and project management. He has over 30 publications in these fields.

Leonardo Rischmoller (*Director, School of Construction, Faculty of Engineering Universidad de Talca, Chile*) Dr Rischmoller obtained his Civil Engineer degree at Ricardo Palma University in Lima, Perú and then his PhD and MSc degrees at the Pontificia Universidad Católica de Chile. As engineer and researcher, he worked for Bechtel where he developed technology, methods and procedures by which to conduct 4D planning; he pioneered the first 4D application on a live project that rendered significant and demonstrable savings in man-hours and materials, giving solid credence to his assertion that 4D technology is viable and cost effective. Prior to this, he worked on several small to medium-sized construction projects as construction engineer. Currently, he is the Director of the School of Construction at Universidad de Talca, Chile, Construction Management consultant and also works as part-time professor at the Construction and Engineering Management program at Pontificia Universidad Católica de Chile.

Jukka Rönkkö (*VTT Digital Information Systems, Finland*) Jukka works as a team leader at the VTT Technical Research Centre of Finland. His team concentrates on interactive computer graphics, 3D user interface research and virtual reality applications. He has worked on simulator user interfaces and construction industry visualisations. He has an MSc (Computer Science) from the Helsinki University of Technology.

Martin Sexton (*School of Construction and Property Management, University of Salford, UK*) Dr Sexton is Associate Head for Research in the RAE 2001 6* rated School of Construction and Property Management at the University of Salford.

He has researched in the organisation and management of construction and property field for a number of years, with particular focus on the creation, management of exploitation of innovation in firm and project contexts. Dr Sexton has over 120 publications. He is a member of the EPSRC-funded Innovative Manufacturing Research Centre (IMRC) at the University of Salford. Dr Sexton is Joint Director of the CIB Working Commission in the Organisation and Management of Construction.

Yonghui Song (*School of Construction and Property Management, University of Salford, UK*) Dr Song is a research fellow working in Virtual Planning at the University of Salford. He played a major role in the design and implementation of the Building Data Integration System in the IntelCities project. He now works on the architecture of planning systems in the VEPS project. Before wining a scholarship to do his PhD at Salford, he won the Excellent System Development award for the Railway 114 system, implemented by the Ministry of Railways of the People's Republic of China.

Joseph H. M. Tah (*School of Construction and Property Management, University of Salford, UK*) Joseph is Professor of Construction Information Technology and Associate Head of School of Construction and Property Management at the University of Salford. His current research focus is on the development of nD virtual prototyping environments for the performance modelling and simulation of sustainable urban environments and constructed facilities, the underlying processes and the underpinning adaptive supply networks involved in the procurement of such facilities using Distributed Computing and Artificial Intelligence techniques.

Ali Tanyer (*Middle East Technical University, Turkey*) Dr Tanyer studied architecture at the Middle East Technical University, Ankara, Turkey. After obtaining his MSc in Building Science in 1999 he joined the University of Salford to undertake a PhD. His main research interests are in the area of Construction Information Technology, particularly, integrated computer environments, 4D simulation and visualisation. Dr Tanyer is currently working as an Assistant Professor at the Middle East Technical University.

Walid Tizani (*School of Civil Engineering, University of Nottingham, UK*) Dr Tizani's main research areas are the application of novel information technologies to engineering design and the investigation of structural performance of steel frames and connections. His experience includes the application of virtual prototyping and virtual reality techniques for the design of buildings; the application of knowledge-based systems and object-orientated technology in the provision of advice and decision support for structural design, construction-led and AEC integrated design, collaborative design, the economic appraisal of structural steelwork and blind-bolt connection to tubular hollow sections. Dr Tizani is a Senior Lecturer and lectures in Steel Structures, IT for Engineers, and Engineering Communication.

Hongxia Wang (*School of Construction and Property Management, University of Salford, UK*) Hongxia is a part-time research fellow working in Virtual Planning, University of Salford. She played a major role in the data modelling of the Building Data Integration System in the IntelCities project. She now works on data processing in the VEPS project. Before Hongxia became a PhD candidate in Salford in 2003, she worked as a lecturer in computing in China for nine years and has more than 20 research and technical publications. Her PhD research topic is nD modelling for urban planning.

Jeffrey Wix (*AEC3 Ltd, UK Visiting Professor, University of Salford, UK*) Professor Wix has been active in research and development of IT for building construction since 1980. In 1988, he developed low-cost CAD for construction and building services applications. He returned to research consultancy in 1993 and is active in developing next generation software applications, information sharing capabilities between software and strategies for information development within major organisations. Current key work involves developing ICT-supported regulation representation and checking systems, creating new forms of cost and environmental impact representation, and defining shared data links between the GIS and building construction worlds. He leads work in development of the 'Information Delivery Manual' for delivery of discrete sets of information in specific business processes, focussing on the definition of business objects within a process framework and how they can be supported by information sharing and ontology. He is engaged in standards development for building information modelling, data exchange and dictionary frameworks within ISO, CEN and BSI. Professor Wix is a member of the Model Support Group of the IAI, acts as Technical Coordinator for the UK Chapter, and led development of the IFC 2x information model.

Song Wu (*School of Construction and Property Management, University of Salford, UK*) Dr Wu is a BSc (Hons) Quantity Surveying lecturer and a Research Fellow at the University of Salford. He has worked on many projects, including 3D to nD Modelling and Process Protocol II. He was a Quantity Surveyor in Singapore and China for three years. He was awarded an MSc in Information Technology in Construction and a PhD at the University of Salford. Dr Wu's research interests include product and process modelling, construction IT and knowledge management.

Foreword

The success of any significant construction undertaking rises and falls with the ability of the project's principal stakeholders to consider and communicate multi-disciplinary concerns, constraints, goals and perspectives. This has to happen in a timely, economical, accurate, effective and transparent way.

Multi-dimensional or nD modelling is possibly the most promising tool and method available to address this challenge. Research and development in the last decade has brought about a dramatic improvement in the ability of the individual disciplines involved in creating a facility (architects, various engineers, builders, etc.) to model, simulate, analyse and visualise the concerns and project elements of their disciplines in the computer. Many disciplines now routinely use these computational tools for their work. Over time the accuracy, economy and speed of these discipline-specific models and their performance predictions has increased significantly and have – in many cases – undoubtedly contributed to safer structures, lower lifecycle costs or more beautiful buildings.

However, by their nature, these advancements have not yet enabled breakthrough performance improvements on a widespread scale in the construction industry. Reports of wasted efforts and materials, facilities that don't meet client needs, facility-development processes that are opaque to most of the stakeholders, sub-optimisation of a facility design for a particular objective, and so on are still too common.

Multi-dimensional or nD modelling extends the reach of these discipline-specific models to enable facility stakeholders to consider the multi-disciplinary nature of construction projects more holistically and effectively. nD modelling relates and integrates the information and processes required for the definition, design, procurement, construction, start-up and handover and operations of facilities. It provides a shared and transparent basis for decision making and documentation of project goals, designs and outcomes.

While significant efforts are still needed to make the nD vision common practice, this book provides a wide-ranging overview of the many issues that impact the development and uptake of the nD concept, ranging from risks and legal issues to technology and technology transfer. A major contribution of the book lies in the accumulation of the world's leading researchers and practitioners in this field. The presented work may at times be parallel or orthogonal and build upon or conflict with each other, but will enable the reader to assess the status and value of the emerging field of nD modelling.

Professor Martin Fischer
Department of Civil and Environmental Engineering
and (by Courtesy) Computer Science, Stanford University
Center for Integrated Facility Engineering, Stanford University
Visiting Professor, University of Salford

Preface

Recent years have seen a major change in the approach to construction innovation and research. There has been a huge concentration, from both the academic and industrial communities, on the development of a single building/product model and/or on the expansion of 3D CAD modelling with other design attributes (such as time, cost, accessibility, acoustics, crime etc). Thus forming an integrated nD model that is shared by all participants in the design and construction process. The information in the model is linked, so that when the design is changed, for example, the cost of the project will also change to reflect the new design.

The terms 'nD modelling' and 'nD CAD' are now gaining increased usage in the field of information communication technologies (ICTs) and building design, and is strongly considered to be the future of the AEC/FM (Architectural, Engineering, Construction and Facilities Management) industry.

This book brings together leading researchers from around the globe who are working on integrated or disparate aspects of nD modelling, but who are all working to construct the same shared vision of an improved future of design and construction. This book is divided into four parts. The first part introduces the concept of nD. This is followed by 'the scope' which defines different 'dimensional' aspects of the nD phenomenon from around the world. The third part looks at specific application aspects, such as visualisation, risk and technology diffusion issues involved in its widespread application. The final section, part four, takes a step forward into the future, of where research and strategy in this field is heading.

The targeted audiences of this book are construction policy makers, professional bodies, leading industry professionals, academics and all those who are interested in construction of the future. It will be valuable to those organisations that are interested in nD modelling in that they can benefit from the experience of others, and those who have already implemented the concept, as there may be findings that could be applicable to their own practice. The Editors welcome future collaboration with those who are interested in driving the nD agenda forward.

Acknowledgements

The Editors would like to, first and foremost, thank all the contributors to this book for their valuable contribution.

All the participants of the five nD modelling workshops, and all those who are actively engaged in the nD modelling discussion group. Their intellectual insights have helped to develop the 3D–nD modelling tool.

Last, and by no means least, the Engineering and Physical Sciences Research Council who funded the 3D–nD modelling project which helped to set up an international nD modelling network.

Microsoft product screen shot(s) reprinted with permission from Microsoft Corporation.

Abbreviations

1D	1 Dimensional
2D	2 Dimensional
3D	3 Dimensional
4D	4 Dimensional
4D-PS	4D Planning and Scheduling
A_T	Total absorption area
AEC/ FM	Architecture, Engineering, Construction and Facilities Management Industry
AHRB	Arts and Humanities Research Board
AIC	American Institute of Constructors
AIGA	American Institute of Graphic Arts
API	Application Programming Interface
ARK e-management	Asset Resources and Knowledge e-management Limited, a UK company
ASCE	American Society of Civil Engineers
A-weighting, C-weighting	Noise analysis weightings
BAA	British Airports Authority
BCA	Building Code of Australia
BCS	British Crime Survey
BEM	Boundary element method
BIM	Building Information Modelling
BLIS	Building Lifecycle Interoperable software
BLIS-XML	Building Lifecycle Interoperable software eXtendable Mark-up Language
BOQ	Bill of Quantities
BRep	Boundary Representation
BSF	Building Schools for the Future, UK initiative
BSI	British Standards Institute
BuHu	Research Institute for the Built and Human Environment, University of Salford
CAD	Computer Aided Design
CAVI	Centre for Advanced Visualization and Interaction, Aarhus University

CAVT	Computer Advanced Visualisation Tools
CAWS	Common Arrangement of Work Sections
CCRM	Centre for Construction Risk Management, Florida International University
CE	Concurrent Engineering
CFD	Computerised Fluid Dynamics
CGM	Computer Graphics Metafile
CIS	CIMsteel integration standard
CITB – ConstructionSkills	Construction Industry Training Board – Construction Skills, UK
CityGML	City Geographical Mark-up Language
CMM	Capability Maturity Model
CNC	Computer Numerically Controlled
CPC	Certified Professional Constructor
CPIC	Construction Industry Project Information Committee
CPR	Construction Process Reengineering
CPTED	Crime Prevention through Environmental Design
CPWR	Centre to Protect Workers Rights, USA
CRC-CI	Cooperative Research Centre for Construction Innovation, Australia
CSG	Constructive Solid Geometry
$D_{n,e}$	Standardised level difference
D_{nT}	Standardised level difference
DAC	Design Against Crime
dB	Decibel
DfES	Department for Education and Skills, UK
DGN	Microstation Design File Format
DIVERCITY	Distributed Virtual Workspace for enhancing Communication within the Construction Industry, research project at the University of Salford
DKK	Danish Krone (currency for Denmark)
DOF	Degrees of freedom
DOPs	Digital Ortho-Photos
DR	Digital Realities
DSM	Digital Surface Model
DTM	Digital Terrain Model
DWG	Drawing file format
DXF	Autodesk Drawing eXchange Format
E	Young's modulus
e-CP	e-City Platform
EDM	Express Data Manager
EPIC	Electronic Product Information Co-operation
EPSRC	Engineering & Physical Sciences Research Council, UK

ESRI	Environmental & Systems Research Institute
EU	European Union
EXPRESS	A formal language for information modelling (ISO 10303–11).
f_c	Critical or coincidence frequency
f_0	Mass-spring-mass resonance frequency
FDTD	Finite different time domain
FEM	Finite element method
GA	(structural) General Arrangement (of drawings)
GEMINI	Geospatial Metadata Interoperability Initiative
GEPUC	Production Management Center, Chile
GI	Geospatial Information
GIS	Geographical Information System
GML	Geographical Mark-up Language
GMP	Guaranteed Maximum Price
GPbS	General Problem Solving
Grid	GIS raster data format
GSM	Global System for Mobile Communications
GUID	Globally Unique Identifier
Hz	Hertz
I3D	Interactive 3D
IAI	International Alliance for Interoperability
IANL	Internal Ambient Noise Level
ICT	Information and Communication Technologies
ID	IDentification
IFC	Industry Foundation Classes
IFCXML	Industry Foundation Classes eXtendable Mark-up Language
IFG	Industry Foundation Classes for Geographical Information System
IP	Intellectual Property
IPR	Intellectual Property Rights
IRMI	International Risk Management Institute
ISO	International Standards Organisation
IST	Information Society Technologies
IT	Information Technology
JCT	Joint Contracts Tribunal
$L_{eq,T}$	Continuous Sound Pressure Level over a time period T
$L'_{n(Tmf,max),w}$	Impact sound insulation of floors
L'_{nT}	Standardised impact sound pressure level
$L'_{nT,w}$	Weighted standardised impact sound pressure level
L_p	Sound Pressure Level

LAS	American Society for Photogrammetry and Remote Sensing Lidar Data Exchange Format Standard
LF	Lateral energy fraction
LSE	Large Scale Engineering industry
M&E	Mechanical & Electrical
m/s	Metre per Second
MOU	Memorandum of Understanding
MS	Microsoft
NBC	National Building Codes, Canada
NBS	National Building Specification
nD	n Dimensional
NECs	Noise Exposure Categories
NES	National Engineering Specification
NIST	National Institute of Standards and Technology, USA
Ns/m^4	Flow Resisitivity
ODBC	Open Data Base Connectivity
OGC	Open Geospatial Consortium
OS	Ordnance Survey
P	Sound Pressure Amplitude
p_o	Reference Sound Pressure Value
Pa	Pascal
P&S	Planning & Scheduling
PBS	Product Breakdown Structure
PC	Personal Computer
PCM	Process Control Module
PIO	Project Information Officer
POI	Points of Interest
PPC	Percentage of Planned Activities Completed
QS	Quantity Surveyor
R	Reduction index
R_w	Weighted sound reduction index
RATAS	JARMO
RECC	Finnish Real Estate and Construction Cluster
RFD	Resource Definition Framework
RFID	Radio Frequency Identification Device
SBD	Secured by Design
SCP	Situational Crime Prevention
SCRI	Salford Centre for Research and Innovation in the Built and Human Environment, University of Salford
SDAI	Standard Data Access Interface
SEA	Statistical energy analysis
SHP	ESRI Shape File

SIAS	Spatial Information and Access System, a product of Geobyte Software GmbH, Germany
SI-weighting	Assessment of speech interference
SM	Simulation Manager
SMM	Standard Method of Measurement
SMM7	Standard Method of Measurement, seventh edition
SMS	Short Messaging Service
SOA	Service Oriented Architecture
SOR	Statement of Requirements
SQL	Structured Query Language
STI	Speech Transmission Index
SOTR	Statement of Technical Requirements
SUD	Sustainable Urban Development
T	Reverberation Time
T_o	Reference Reverberation Time
T_{mf}	Mid-Frequency Reverberation Time
TIN	Triangulated Irregular Network
UK	United Kingdom
UKSGB	UK Standard Geographic Base
Uniclass	Unified Classification for the Construction Industry
USA	United States of America
VCE	Virtual Construction Environment
VDE	Virtual Design Environment
VE	Virtual Environment
VR	Virtual Reality
VRML	Virtual Reality Modelling Language, an ISO standard for 3D modelling
VULCAN	Virtual Urban Laboratories for Computer Aided Network
VUP	Virtual Urban Planning
W3C	World Wide Web Consortium
WBS	Work Breakdown Structure
WP	Work Package
WPs	Work Packages
XML	eXtendable Mark-up Language
ν	Poisson's ratio
ρ	Density

Part I

nD modelling
The concept

nD modelling is profoundly becoming a synergetic idiom associated with information communication technologies in the AEC/FM industry. Part 1 of *Constructing the Future: nD Modelling* defines the concept of nD modelling. It is anticipated that the chapters presented here will give readers a good insight into the need and potential for nD modelling. In the first chapter, Lee *et al.* discuss the need for nD before detailing a staged development approach, starting from 2D, 3D, 4D and thus through to nD modelling. Finally, the 3D to nD modelling project is introduced – the impetus and pivotal project that generated worldwide research into the nD concept and thus formed the stimulus for this book.

Tizani's chapter on engineering design compliments and supplements the introductory chapter by providing a more detailed account of nD modelling. It confers that much of the work carried out by the building designers involves simple and relatively deterministic tasks. However, design automation is typically not possible in traditional building design due to a lack of formalisation of the necessary information, and the perceived lack of control that automation would introduce to the design process. Tizani demonstrates how rapid and virtual prototyping can be applied to effect automation using multi-storey buildings as an example, and concludes by identifying the greater precision and speed benefits of automation. The complexity of the traditional design process is clearly defined with its many shortfalls highlighted, and subsequently outlays the requirements for such an integrated IT system.

The final chapter of the first part of this book specifically describes how main contractors' data can be improved to provide better client value and more cost-efficient production. Laitinen provides an overview of the conceptualisation, development, introduction and use of nD BIMs (Building Information Models) over a period of ten years. The chapter is based on research findings that adopt the IDEF0 modelling methodology to describe the information management process throughout the building process lifecycle from both as-is and to-be perspectives. An integration of design and production planning based on the product model approach is presented.

nD modelling

The background

Angela Lee, Song Wu and Ghassan Aouad

Introduction

One of the main challenges facing the construction industry today is how to improve the efficiency and effectiveness of the integrated design and construction process. Moreover, what contribution can the effective use of information technology make to this?

Designing a building is part art and part alchemy. It is no longer simply a question of organising a range of facilities on a single site; the needs of a whole host of project stakeholders have to be satisfied. Thus, the way in which the building will fulfil the expectations of all those parties who form the spectrum of building stakeholders is increasingly becoming the measure of its success. The stakeholders include not only the organisations and individuals who occupy the building, but also those who have provided it, those who manage it and those who live with it – the community in general.

The design not only has to be buildable (in terms of cost and time), but stakeholders are increasingly enquiring about its maintainability, sustainability, accessibility, crime deterrent features and its acoustic and energy performance. Often, a whole host of construction specialists are involved in instigating these aspects of design, such as accessibility auditors, FM specialists and acoustic consultants. With so much information and from so many experts, it becomes very difficult for the client to visualise the design, any changes applied and the subsequent impact on the time and cost of the construction project. Changing and adapting the design, planning schedules and cost estimates to aid client decision making can be laborious, time consuming and costly. Each of these design parameters that the stakeholders seek to consider will have a host of social, economic and legislative constraints that may be in conflict with one another. Furthermore, as each of these factors vary – in the amount and type of demands they make – they will have a direct impact on the time and cost of the construction project. The criteria for successful design therefore will include a measure of the extent to which all these factors can be co-ordinated and mutually satisfied to meet the expectations of all the parties involved.

Traditionally, specialist input of each of these design criteria is usually undertaken in a sequential step-by-step fashion whereby the design succeeds a number

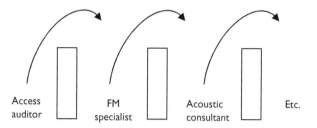

| Access auditor | FM specialist | Acoustic consultant | Etc. |

Figure 1.1 Sequential 'over-the-brick-wall' approach.

of changes after satisfying the legal requirements of a specialist consultant, and then proceeds to the next consultant who in turn again makes a number of design recommendation changes. In this sense, specialist design changes are made in isolation from each other creating an over-the-brick-wall effect where each discrete change by one consultant plays little or no regard to the next (see Figure 1.1). Therefore, it is often difficult to balance the design between aesthetics, ecology and economism – a 3D view of design that acknowledges its social, environmental and economic roles – in order to satisfy the needs of all the stakeholders.

These problems are mainly attributed to the vast amount of information and knowledge that is required to bring about good design and construction co-ordination and communication between a traditionally fragmented supply chain. The complexity of the problem increases with the fact that this information is produced by a number of construction professionals of different backgrounds. Therefore, without an effective implementation of IT and processes to control and manage this information, the problem will only expound as construction projects become more and more complex, and as stakeholders increasingly enquire about the performance of buildings (sustainability, accessibility, acoustic, energy, maintainability, crime, etc.).

Current developments that aim to address this problem have been led by two factors, namely the technology push and strategic pull. The technological push has led to the development and implementation of hardware and software to improve a number of functions, including the development of building information models (e.g. integration of 3D + time/4D modelling), and information standardisation. The strategic pull can be gauged by the growing number of workshops, seminars and conferences on the subject.

nD modelling

An nD model is an extension of the building information model made by incorporating all the design information required at each stage of the lifecycle of a building facility (Lee *et al.*, 2003). Thus, a building information model (BIM) is a computer model database of building design information, which may also contain information about the building's construction, management, operations and maintenance (Graphisoft, 2003). From this database, different views of the information can be generated automatically, views that correspond to traditional design documents such as plans, sections, elevations and schedules. As the documents are derived from the same database, they are all co-ordinated and accurate – any design changes made in the model will automatically be reflected in the resulting drawings, ensuring a complete and consistent set of documentation (Graphisoft, 2003). It builds upon the concept of 2D, 3D and 3D.

2D and 3D modelling in the construction industry takes its precedence from the laws governing the positioning and dimensions of a point or object in physics whereby a three-number vector represents a point in space, the x and y axes describing the planar state and the z axis depicts the height (Lee *et al.*, 2003). Further, 3D modelling in construction goes beyond the object's geometric dimensions and replicates visual attributes such as colour and texture. This visualisation is a common attribute of many AEC design packages, such as 3D Studio Max and ArchiCAD, which enable the simulation of reality in all its aspects or allow a rehearsal medium for strategic planning.

Combining time sequencing in visual environments with the 3D geometric model (x, y, z) is commonly referred to as 4D CAD/modelling (Rischmoller *et al.*, 2000). Using 4D CAD, the processes of building construction can be demonstrated before any real construction activities occur (Kunz *et al.*, 2002). This will help users to find the possible mistakes and conflicts at the early stage of a construction project, and to enable stakeholders to predict the construction schedule. Research projects around the world have taken the concept, developed it further and software prototypes and commercial packages have begun to emerge. In the United States, the Center of Integrated Facility Engineering (CIFE) at Stanford University has implemented the concept of the 4D model on the Walt Disney Concert Hall project. In the United Kingdom, the University of Teesside's VIRCON project integrates a comprehensive core database designed with Standard Classification Methods (Uniclass) with a CAD package (AutoCAD, 2000), a Project Management Package (MS Project) and Graphical User Interfaces as a 4D/VR model to simulate construction processes of an £8-million, 3-storey development for the University's Health School (Dawood *et al.*, 2002). Commercial packages are also now available, such as 4D Simulation from VirtualSTEP, Schedule Simulator from Bentley and 4D CAD System from JGC Corporation.

nD modelling develops the concept of 4D modelling and aims to integrate an *n*th number of design dimensions into a holistic model which would enable users to portray and visually project the building design over its complete lifecycle. nD modelling is based upon the BIM, a concept first introduced in the 1970s and the basis of considerable research in construction IT ever since. The idea evolved with the introduction of object-oriented CAD; the 'objects' in these CAD systems (e.g. doors, walls, windows, roofs) can also store non-graphical data about the building in a logical structure. The BIM is a repository that stores all the data 'objects' with each object being described only once. Both graphical and non-graphical documents, such as drawings and specifications, schedules and other data respectively are included. Changes to each item are made in only one place and so each project participant sees the same information in the repository. By handling project documentation in this way, communication problems that slow down projects and increase costs can be greatly reduced (Cyon Research, 2003).

Leading CAD vendors such as AutoDesk, Bentley and Graphisoft have promoted BIM heavily with their own BIM solutions and demonstrated the benefits of the concept. However, as these solutions are based on different, non-compatible standards, an open and neutral data format is required to ensure data compatibility across the different applications. Industry Foundation Classes (IFC), developed by the International Alliance for Interoperability (IAI), provide such capabilities. IFCs provide a set of rules and protocols that determine how the data representing the building in the model are defined and the agreed specification of classes of components enables the development of a common language for construction. IFC-based objects allow project models to be shared whilst allowing each profession to define its own view of the objects contained in that model. This leads to improved efficiency in cost estimating, building services design, construction and facility management: IFCs enable interoperability between the various AEC/FM software applications allowing software developers to use IFCs to create applications that use universal objects based on the IFC specification. Furthermore, this shared data can continue to evolve after the design phase and throughout the construction and occupation of the building.

3D to nD modelling project

The 3D to nD research project, at the University of Salford, had developed a multi-dimensional computer model that will portray and visually project the entire design and construction process, enabling users to 'see' and simulate the whole life of the project. This, it is anticipated, will help to improve the decision-making process and construction performance by enabling true 'what-if' analysis to be performed to demonstrate the real cost in terms of the variables of the design issues (see Figure 1.2). Therefore, the trade-offs between the parameters can be clearly envisaged:

- Predict and plan the construction process
- Determine cost options
- Maximise sustainability
- Investigate energy requirements
- Examine people's accessibility
- Determine maintenance needs
- Incorporate crime deterrent features
- Examine the building's acoustics

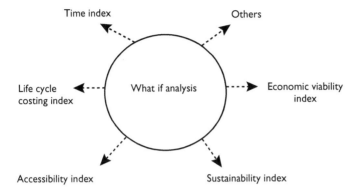

Figure 1.2 What-if analysis indexes of the 3D to nD modelling project.

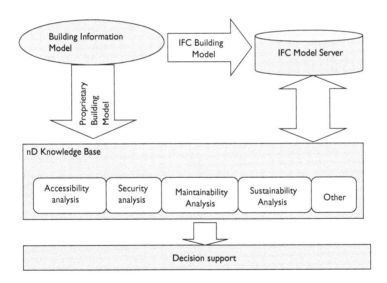

Figure 1.3 System architecture of the nD modelling prototype tool.

The project aimed to develop the infrastructure, methodologies and technologies that will facilitate the integration of time, cost, accessibility, sustainability, maintainability, acoustics, crime and thermal requirements. It assembled and combined the leading advances that had been made in discrete information communication technologies (ICTs) and process improvement to produce an integrated prototyping platform for the construction and engineering industries. This output will allow seamless communication, simulation and visualisation,

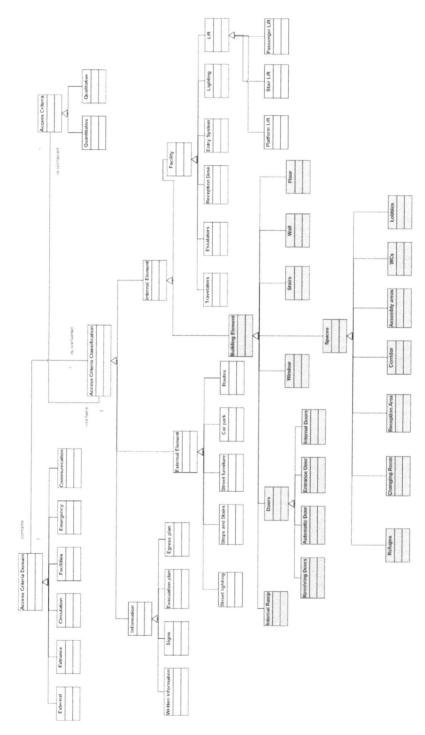

Figure 1.4 Section of the accessibility data model.

and intelligent and dynamic interaction of emerging building design prototypes so that their fitness of purpose for economic, environmental, building performance and human usability will be considered in an integrated manner. Conceptually, this will involve taking 3D modelling in the built environment to nth number of dimensions. The project was funded by the EPSRC (Engineering and Physical Sciences Research Council) under a Platform grant to the tune of £0.5 million for 4 years. The unique nature of this grant encouraged blue-sky innovative research, international collaboration and supports future funding opportunities.

The developed nD tool builds upon the concept of BIM and is IFC based: the system architecture is illustrated in Figure 1.3.

- nD knowledge base: platform that provides information analysis services for the design knowledge related to the various design perspective constraints of the nD modelling (i.e. accessibility requirements, crime deterrent measures, sustainability requirements etc.). Information from various design handbooks and guidelines on the legislative specifications of building component will be used together with physical building data from building information model to perform individual analysis (see Figure 1.4).

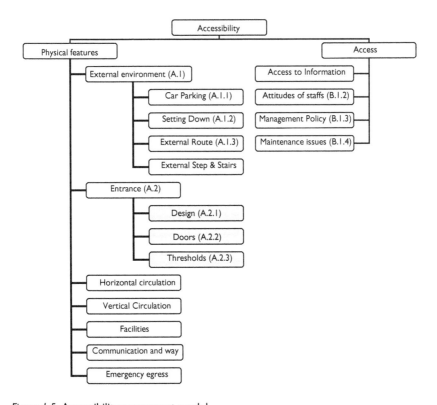

Figure 1.5 Accessibility assessment model.

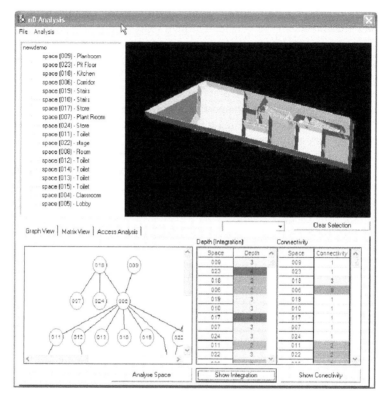

Figure 1.6 Screenshot of nD prototype tool displaying accessibility analysis.

- Decision support: multi-criterion decision analysis (MCDA) techniques have been adopted for the combined assessment of qualitative criteria (i.e. criteria from the Building Regulations and British Standard documents that cannot be directly measured against in their present form) and quantitative criteria (e.g. expressed in geometric dimensions, monetary units, etc.). Analytic Hierarchy Process (AHP) is used to assess both qualitative criteria (i.e. criteria that cannot be directly measured) and quantitative criteria (e.g. expressed in dimensions, monetary units, etc.). The accessibility assessment model based on the AHP methodology is developed to support the decision making on accessible design (Figure 1.5).

So far, the nD prototype tool incorporates whole-lifecycle costing (using data generated by Salford's Life-Cycle costing research project), acoustics (using the Rw weighted sound reduction index), environmental impact data (using BRE's

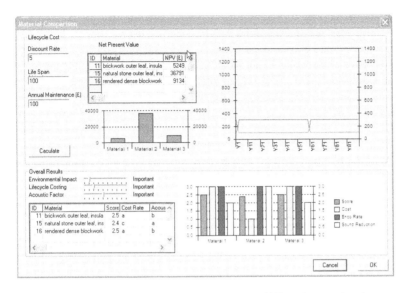

Figure 1.7 Screenshot of nD prototype tool displaying AHP technique for determining suitable building material in accordance to lifecycle costing, acoustics and environmental impact factors.

 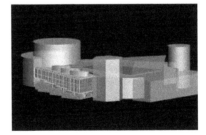

Figure 1.8 The Lowry extension.

'Green Guide to Specification' data), crime (using the Secured by Design Scheme standards) and accessibility (using BSI:83001). Technology for space analysis is also developed to support the accessibility analysis. A screenshot of the tool can be found in Figures 1.6 and 1.7; further information can be sought by visiting http://nDmodelling.scpm.salford.ac.uk. Figure 1.8 demonstrates the prototype on a live case study – the extension of The Lowry, Salford, UK. The Lowry is a national landmark millennium building. The Lowry is an architectural flagship with a unique and dynamic identity, designed by Michael Wilford and Partners.

Opened on 28 April 2000, it brings together a wide variety of performing and visual arts under one roof. The Lowry houses two theatres for performing arts (1,730 and 466 seats) presenting a full range of drama, opera, ballet, dance, musicals, children's shows, popular music, jazz, folk and comedy; gallery spaces (1,610 metres of floor space) showing the works of LS Lowry alongside contemporary exhibitions; and ArtWorks, an interactive attraction designed to encourage individual creativity. The success of the building has led to the need for an extension of the existing building. The extension of the Lowry is used only to test and validate the nD modelling prototype tool and technologies we developed, that is IFC-based space analysis, and not to assess the feasibility of the design itself: the results of the nD modelling prototype tool will not be used to inform the design of the building.

Bibliography

Adachi, Y. (2002) Overview of IFC Model Server Framework. In proceedings of the International ECPPM Conference – eWork and eBusiness in Architecture, Engineering and Construction, Portoroz, Slovenia.

Alshawi, M. and Ingirige, B. (2003) Web-enabled project management: an emerging paradigm in construction. *Automation in Construction*, 12, 349–364.

Alshawi, M., Faraj, I., Aouad, G., Child, T. and Underwood, J. (1999) An IFC Web-based Collaborative Construction Computer Environment. Invited paper, in proceedings of the International Construction IT conference, Construction Industry Development Board, Malaysia, 8–33.

Aouad, G., Sun, M. and Faraj, I. (2002) Automatic generation of data representations for construction application. *Construction Innovation*, 2, 151–165.

AutoCAD (2000) www.AutoCAD.com (Data accessed December).

Christiansson, P., Dalto, L., Skjaerbaek, J., Soubra, S. and Marache, M. (2002) Virtual Environments for the AEC Sector: The Divercity Experience. In proceedings of the International ECPPM Conference – eWork and eBusiness in Architecture, Engineering and Construction, Portoroz, Slovenia.

Cyon Research (2003) www.cyon.com (Data accessed June)

Dawood, N., Sriprasert, E., Mallasi, Z. and Hobbs, B. (2002) Development of an integrated information resource base for 4D/VR construction processes simulation. *Automation in Construction*, 12, 123–131.

Drogemuller, R. (2002) CSIRO & CRC-CI IFC Development Projects, ITM Meeting, Tokyo.

Durst, R. and Kabel, K. (2001) Cross-functional teams in a concurrent engineering environment – principles, model, and methods, in M.M. Beyerlein, D.A. Johnson and S.T. Beyerlein (eds), *Virtual Teams*, JAI, Oxford, pp. 167–214.

Eastman, C. (1999) *Building Product Models: Computer Environments Support Design and Construction*, CRC Press LLC, Florida.

Emmerson, H. (1962) *Studies of Problems before the Construction Industries*, HMSO, London.

Fischer, M. (2000) Construction Planning and Management using 3D & 4D CAD Models, *Construction IT* 2000, Sydney, Australia.

Graphisoft (2003) The Graphisoft Virtual Building: Bridging the Building Information Model from Concept into Reality. Graphisoft Whitepaper.

Kunz, J., Fischer, M., Haymaker, J. and Levitt, R. (2002) Integrated and Automated Project Processes in Civil Engineering: Experiences of the Centre for Integrated Facility Engineering at Stanford University, Computing in Civil Engineering Proceedings, ASCE, Reston, VA, 96–105, January 2002.

Latham, M. (1994) Constructing the Team: Joint Review of Procurement and Contractual Arrangements in the UK Construction Industry. Department of the Environment, HMSO, 1994.

Lee, A., Marshall-Ponting, A.J., Aouad, G., Wu, S., Koh, I., Fu, C., Cooper, R., Betts, M., Kagioglou, M. and Fischer, M. (2003) Developing a Vision of nD-enabled Construction, Construct IT Report, ISBN: 1-900491-92-3.

Leibich, L., Wix, J., Forester, J. and Qi, Z. (2002) Speeding-up the Building Plan Approval – The Singapore e-Plan Checking Project Offers Automatic Plan Checking Based on IFC, The international conference of ECPPM 2002 – eWork and eBusiness in Architecture, Engineering and Construction, Portoroz, Slovenia, 2002.

Rischmoller, L., Fisher, M., Fox, R. and Alarcon, L. (2000) 4D Planning & Scheduling: Grounding Construction IT Research in Industry Practice. In Proceedings of CIB W78 conference on Construction IT, June, Iceland.

Yu, K., Froese, T. and Grobler, F. (2000) A development framework for data models for computer-integrated facilities management. *Automation in Construction*, 9, 145–167.

Chapter 2

Engineering design

Walid Tizani

Introduction

Construction projects broadly undergo four phases: design, construction, maintenance and demolition. Most interactions occur within the design and construction phases that, consequently, have the highest potential for improvement. Improving the efficiency of the design and construction processes must involve facilitating the interactions and communications between inter-disciplinary parties. Information technology, although commonly used in the communication between design processes, is not yet fully exploited and is an obvious area to target for delivering such an improvement.

The design phase is primarily concerned with specifying the 'product' that best fulfils the client's brief, ensures safety during construction and use and achieves minimum overall cost. This process requires interactions between the disciplines of architecture, building services, structural engineering and site construction. Each specialist within these disciplines has specific understanding of the design problem. The decisions made by these specialists can interact and conflict in complex ways that cannot be readily predicted. This causes compartmentalised decision making that can cause problems with downstream design activities, and conflict with upstream decisions and intentions. So, although the design process itself constitutes just 5 per cent of the costs associated with a typical construction project, its success affects the build cost and the quality of the remaining 95 per cent of the project (Latham, 1994; Egan, 1998). The design process is, therefore, a critical part of any project and supporting it is an important factor in improving the overall life cycle of construction projects.

This chapter is concerned with the design phase, and concentrates more specifically on the engineering design process. It provides an overview of the problems within the current engineering design process and outlines a strategy for improving it.

Complexity of design process

Design is a creative process typically involving a cycle of proposition and assessment activities. The process is driven by the goal to satisfy the intentions of the client and the resulting user requirements. These requirements impose constraints on potential design solutions. The proposition activity generates possible solutions to the design problem. It is a creative activity and is usually heavily influenced by experience. It is also open-ended as there is, usually, no unique solution to the problem. The assessment activity checks whether a proposed solution complies with the constraints and measures its fitness in how well it satisfies a set of objectives resulting from the requirements. It also has the goal of appraising the relative merit of one solution compared with others. The assessment activity either results in the communication of the design solution in the form of specification, having deemed it to satisfy the constraints, or leads to another cycle of design involving the proposition of a new solution which can be a refinement of the last.

Addressing design requirements form the point of view of a single discipline is a major creative task given the open-ended nature of it. The design for a construction facility is further complicated by the multi-disciplinary nature of it and the fact that each discipline adds its own sub-set of requirements and constraints. The multi-disciplinary design process is thus a fluent and complex process that attempts to satisfy discipline-based, cross-discipline-based as well global constraints and requirements. A successful design process should address the multi-disciplinarily nature of the problem so as to minimise cross-discipline conflicts while at the same time not losing sight of the original overall requirements.

The design process is highly reliant on the management, flow and usage of information. Boujut and Lawreillard (2002) estimates that 75 per cent of an engineer's design work consists of seeking, organising, modifying and translating information, and that this information is often unrelated to his own personal discipline. Only 25 per cent of his time remains for specific engineering efforts. It is therefore quite reasonable to assume that improving the management and standardisation of building design information is a key area in which to improve the efficiency of the design process (Smith, 2005).

Much communications between the disciplines is normally necessary in such a process. The communications deal with preliminary specification for the various parts of the overall design with much iteration. The process carries its own momentum in that each specification set is built upon by others in such a way that it is difficult to roll back the process and start again and each different initial start might result in a different final design. It is not unknown that such a final design might not adequately fulfil the original intentions set by the client. One of the main causes of such variation from the original intentions is the inability to easily assess the consequences of design decisions of one discipline onto the others. The specification provided by one specialism tends to focus primarily on its own imposed constraints with the minimum cross-checking with other affected specialisms.

Strategy for improvement

The integration of all aspects of the design of a facility and allowing for the ability to concurrently carry out the multi-disciplinary and cross-dependent design activities has been the subject of much research. The researches have attempted to devise an integrated design process supported by a standardised format for specifying the complete design solution. Most of these researches have concentrated on either part of the process and have chosen a typical building type for the study area (such as multi-storey steel or concrete building, portal frame industrial buildings, etc.). This is understandable given the widely varying design requirements for the various types of buildings and the enormous amount of data specification required to be integrated.

There has been notable success in the area of standardised design data specification judging from the infiltration of their use in practice. Two notable standardised data models or product models are the CIMsteel integration standard (CIS) (SCI, 2002) and the International Foundation classes (IFC) (IAI, 2002). The concept of product modelling has the ultimate goal of consolidating the information involved in the building design process into a single coherent document format, so that the building can be designed holistically, and without information boundaries. The CIS is specialised in modelling the engineering design and fabrication of steel buildings while the IFC is broader but less detailed. The CIS is currently the standard format for the exchange of design specification for the steel industry especially in the United States. The IFC on the other hand, tend to be favoured within the architectural and construction disciplines.

Product models have on the whole been used for the purpose of data exchanges. With such standards readily available, most engineering software, although they operate their own internal processes and data models, provide translators to import and export standard models. Therefore, process integration has been partially successful by exchanging the model between specialised disciplinary software. So, it is possible for a structural engineer to provide the 'model' electronically (CIS in this case) to be imported into a specialised steel detailing and fabrication software.

With regards to process integration and concurrency, successes on the implementation side have been less obvious. Software within the same disciplines have integrated more aspects of the specialised design (such as the seamless analysis, design and detailing in structural engineering). However, there have been no noticeable advances on multi-disciplinary design processes. Unlike product modelling, there is no agreed process model yet. Modelling processes is far more complex than modelling products, which are essentially static and can have agreed logical descriptions. Given the complexity of design, a process model needs to allow for a flexible and fluent process that allows for the designer's creativity.

An integrated process model, to be effective, must make use of an integrated product model. However, product models tend to look at the problem from the product end, that is attempting to describe the physical model with its various entities and logical inter-relationships. It does not, therefore, provide a flexible means of describing the outcome of a design process that might require the description of an incomplete or imprecise state of design. Additionally, they do not provide for the formalisation or description of the layer of the non-product set of information capturing design requirements, intents and constraints. As these models form gateways through which IT systems may interact, the sophistication and completeness of these models has a fundamental effect on the effectiveness of the building design process as a whole (Smith, 2005).

A strategy for improving the design process should therefore not only model the process itself but also model appropriately the necessary information as well as the means to specify its outcome in a product model. The process must be carried out in an integrated and concurrent fashion. This is so that all design decisions can be checked for their adherence to the original design intentions and can be immediately assessed in terms of their compliance with the constraints imposed by a particular specialism and in terms of their effects on others.

The implementation of such a radical improvement to the process might require a complete overhaul of the way designs are currently conducted. This includes cultural and technological changes. The cultural issues deal with the working practices and involve fundamental changes to the organisational structure of the industry. The technological issues deal with the systems, methods used in the evolution of the product specification and the development and communication of design information.

The cultural issues are difficult to resolve directly and, it can be argued, a fundamental change would not be desirable. This is because construction projects are complex in nature and involve the interaction of deeply specialised disciplines that cannot be fully integrated. An improvement strategy should therefore concentrate on improving the interactions between these disciplines. This can be done through providing better support for the interactions between the disciplines by improving the technologies used. This should be done without forcing major change affecting the main organisational structure of the industry and without forcing fundamental changes to the interaction points between the disciplines by improving the way the design information is represented and used throughout the building design process.

Improving the representation and usage of information must be made to encompass all design phases. This is so as to allow the consideration of wider aspects than what is currently possible. Within the nD context, this means that the proposed design solution must take into account all major aspects considered by all disciplines involved.

Kiviniemi and Fisher (2003) showed that the requirements and solutions for the design should be linked together so that the designers can access all the requirements related to the issues they are working with. This can make it possible to check whether the new solutions are in accordance with the requirements. Kiviniemi goes on to suggest that information and communication technologies (ICTs) could provide valuable tools to identify conflicts, but that it was most suitable for these conflicts to be then resolved using human-based decisions. Tizani and Smith (2003) proposed that such a system could be realised through the creation of a heavily inter-connected product and process model that highlighted conflicts within the design, enabling building designers to work in a co-ordinated way. Smith (2005) suggested a formal way of representing design intentions and the use of these in order to assess the compliance of the design solution with original intentions.

This chapter proposes an approach to integrated design using 'integrated process and information model', hence an nD model. The model attempts to address the points raised earlier. This model was built to address problems identified in the current design process for multi-story steel buildings. At this stage the implementation of this approach is at the proof of concept stage and does not consider all involved disciplines, but concentrates on the architecture-structural-services and fabrication, thus, for now, leaving out the site construction phase. The following sections describe a high-level model of the current design process and identify a number of generic problems that should be addressed in an integrated IT system aimed at process improvement.

Typical design process

The current building design practice is outlined, and a series of key problems relating to the flow and usage of information are identified. These problems primarily relate to the nonintegrated nature of the existing design processes, the lack of consistent data formats and the inability to describe formally, the motivations behind the designers' decisions.

Analysis of the current process

To highlight the inefficiencies within the current design process it is necessary to develop a typical model of the current design processes. This process model demonstrates typical information flow for the design of a multi-storey steel-framed structure. The description of the current design process is presented as a series of IDEF0 diagrams (NIST, 1991). The focus has been on the information flow between the designers involved in the process. The model focuses on projects which employ specialist consultants for the design activities in the building, for example architects and structural designers.

A high-level representation of the design process is described in Figure 2.1 (Ruikar *et al.*, 2005). The Figure describes the process of 'Managing the building design, fabrication and construction process'. There are five design and construction activities from defining the client's brief (A21) to handover to the user (A25). The major inputs to the process are the client requirement, materials and products. The major outputs are a building ready for use and the documentation about the building, which according to today's practice is in the form of paper documents, such as drawings. There are several intermediate outputs, such as the room schedule resulting from the briefing activity, which are used as controls or input for later activities.

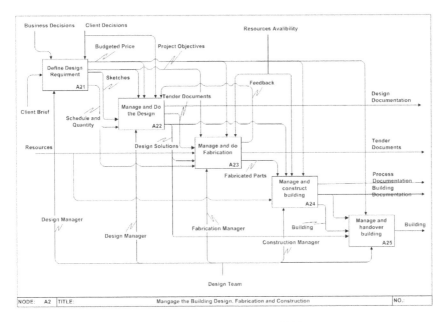

Business Decisions   Client Decisions

Resources Avalibility

Budgeted Price   Project Objectives

Define Design
Requirment
Sketches
A21

Feedback

Tender Documents

Design
Documentation

Client Brief

Manage and Do
the Design
A22

Schedule and
Quantity

Resources

Design Solutions

Manage and do
Fabrication
A23

Tender
Documents

Fabricated Parts

Manage and
construct
building
A24

Process
Documentation

Building
Documentation

Design Manager

Fabrication Manager

Building

Manage and
handover
building
A25

Building

Design Manager

Construction Manager

Design Team

NODE: A2 TITLE: Mangage the Building Design, Fabrication and Construction   NO.:

Figure 2.1 Typical building design process (Ruikar *et al.*, 2005).

Typical information flow within the current process

Information flow of a typical design process is shown in Figure 2.2 (Ruikar *et al.*, 2005). The process illustrates the flow of information and data within a typical design process of a multi-storey steel-framed building. The work starts with the development of the building concept. This involves the architect and the client. The architect prepares the client brief and specifies the total space requirements that conform to relevant standards and spaces as per expected occupancy. The circulation areas, number and size of cores are detailed by the architect. Conceptual floor layout drawings are prepared and the design is then delivered to the structural designer. Typically the designer makes 2D floor layouts for the structural system; this is done using the architectural layout plans. Structural 'General Arrangement' drawings (GA) for the floor and then the whole structure are considered. This includes the position of the columns, primary and secondary beams. Cost models for different structural grid arrangements are typically calculated using rough cost estimating tools (spreadsheets and general rules of thumb) and the optimum system both architecturally and structurally is finalised in discussion with the architect. After the floor plans have been finalised, the building cores and/or bracing systems are designed to provide lateral stability to the structure. This completes the conceptual design of the building.

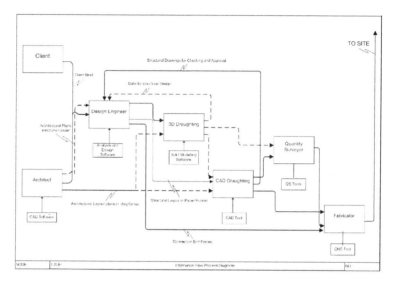

Figure 2.2 Typical information flow: client to fabricator (Ruikar *et al.*, 2005).

After the completion of conceptual design, work progresses to the detailed design stage. The process starts with the development of the architectural working drawings, which are completed using CAD (Computer Aided Design) software. Once these are finalised they are passed on to the design engineer both in paper form and in electronic format, for example the dwg file format. Electronic copies may also, for prestigious projects, be submitted for creating a 3D model of the building. At this stage, the design engineer starts by formulating the overall framing arrangement on the architectural flooring plans and the structural layout plans are finalised. The appropriate loads are calculated and appropriate structural members sizes are selected. It is worth noting here that connections are not normally designed at this stage. It is typical to only specify the assumed behaviour used in the structural analysis (e.g. moment resisting or not). The design is completed with the help of structural analysis and design software or by selecting typical beams and columns per floor layout within the structure and using simple Excel spreadsheets for design. Once the design is completed, the information is submitted (in paper form) by the design engineer to the draughtsman for developing a 3D model of the building. It should be noted that 3D models are prepared mainly for major projects and this step is usually completely skipped for day-to-day projects. For major projects, visualisation packages, such as AutoCAD 3D and 3D Studio Max (AutoDesk, 2005) are used to create the 3D model. The main purpose of this 3D model is for visualisation and client benefit. The model may also be used for clash detection checks, to ensure that structural design models and the architectural models match. Any discrepancies or clashes that are detected in the architectural and structural models are noted by the 3D draughtsman and this information is relayed to the design engineer usually in the form of memos. The structural design is refined and amended to accommodate the changes. For example, the structural layouts may be changed and the size of the initially designed core for structural stability is refined and finalised. After several iterations that involve input from the architect, design engineer and the draughting engineer, the design is finalised and information is relayed to the structural draughtsman and the site working drawings are prepared.

The structural draughting process is completed using CAD software such as AutoCAD. After several iterations in which the design is refined and drawings finalised, the completed drawings are submitted for costing to the quantity surveyors (QS) and to the fabricator for fabrication. The fabricator then receives information from the design engineer, the CAD draughtsman and the QS. This information is relayed both in electronic and paper form as illustrated in Figure 2.3. Typically a fabrication model (usually a 3D model) is created using software such as Xsteel or StruCAD (AceCAD, 2005; Tekla, 2005). The detailed connection design is carried out, usually using pre-defined macros embedded in the 3D modeller. It is not unknown that suitable and cost-effective connection details cannot be provided within the constraints, of using specified section sizes and connection behaviour, imposed on the fabricator. Such detailing should have been considered at the member sizing stage.

The fabrication model can then be used to generate CNC (Computer Numerically Controlled) instructions for automated fabrication. The difference between the two 3D models, that created by the consultants (architect or structural engineer) and that by the specialist steel fabricator, is that the latter adheres to tight fabrication tolerances while the earlier is mainly used to visualise the concept and is not as precise.

As seen from the process description and Figure 2.2, there is a heavy reliance on IT systems in the design process. There is a considerable amount of data transfer between each of the design stages. Also, different software packages are used at each stage of the design process. These software packages do not necessarily allow for the seamless transfer of data between processes and disciplines. For example, in the design process illustrated, the design engineer uses an in-house design spreadsheet alongside analysis software for the structural design process. The 3D modelling process uses specialised solid modelling software. The design layout finalised by the design engineer is relayed to the 3D draughtsman in the form of design spreadsheets. This format of information is not compatible with the solid modelling software requiring manual re-entry of building information on the part of the 3D draughting engineer. The 3D draughting process alters the structural design of the building as per how it synchronises with the architectural layout. These alterations in design information are communicated to the design engineer for re-evaluation. The design engineer interprets this information and makes the appropriate change to the building's design for re-assessment.

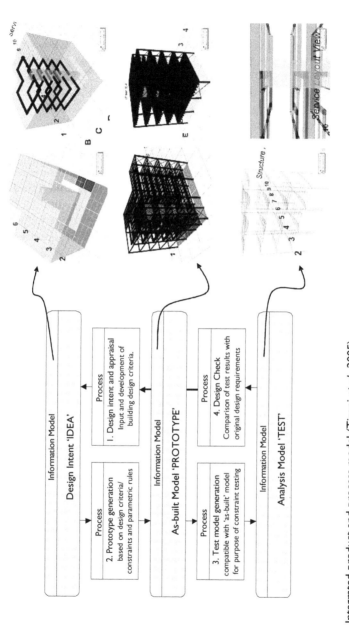

Figure 2.3 Integrated product and process model (Tizani *et al.*, 2005).

This generates an iterative process that involves a high level of re-working of data. Such information/ data flow leads to a lack of integration between different stages in the design processes and hinders collaborative design. It also creates a compartmentalised design process that only allows rigid or mostly paper-based information to be transferred across boundaries. Such data transfers result in degradation of the data integrity as documented by Eastman (1999).

Shortcoming of current design process

The design process described in the previous section illustrates the multi-disciplinary nature of the design. It also highlighted a number of shortcomings despite only describing the major information flow points. With the purpose of formulating the possible means of addressing these shortcomings, an analysis of the process and the information flow has been conducted (Smith, 2005). This has led to the identification of a number of higher-level problems:

- lack of formalised building requirements;
- rigid design flow;
- manual re-entry of data;
- compartmentalised decision making;
- lack of fidelity;
- unaccommodating of design experimentation;
- unaccommodating of late design changes;
- lack of design automation;
- each of these general problems is described in turn;
- lack of formalised building requirements.

The required functionality of the building and design intent of the architect in form and space are not readily available in a formal and declarative manner. The type of information that might represent such concepts is not present in existing product models, and so this information is lost. Therefore, there is a need to

- capture the ideas behind the building instead of just design decisions; and
- provide a set of measurements that the performance of the transient or ulti-mately completed design solution can be measured against.

Rigid design flow

The organisation of the building design process varies considerably between pro-jects due to differences in contractual arrangements and the particular preferences of building design teams. However, within any given project, the building design process is very rigid and inflexible. The order in which design activities must be carried out is decided upon at an early stage in the design. This 'rigidity' stems primarily from the poor integration of building design documentation, rather than from the building designers themselves.

Manual re-entry of data

The prevalent use of IT systems at every stage of the building design process makes these tools an integral, and critical, part of the design process. For example, and as illustrated in Figure 2.2 several different structural design, structural visualisation and costing software are used in the process of structural design. However, the data files that these systems use are rarely compatible. Additionally, the transfer of design data between the various partners is currently based on 2D drawings, which predominantly are exchanged in paper format even though they are increasingly produced using CAD. This requires a great deal of manual reentry when data is being taken from one IT system and being transferred into another. This has a three-fold effect:

- The design process is delayed.
- The manual data interpretation costs man hours.
- A significant source of error is introduced.

At many stages in the design of a building, documents will be created, like hand-drawn floor layouts, which are not in an IT compatible format. The information in these documents must similarly be manually imported into an IT system at some downstream design activity.

Compartmentalised decision making

Design decisions are made without being able to appreciate fully the impact that they have upon the rest of the building design. For example, at the structural design stage the designer is not able to fully account for the influence of other building elements such as the building services layout and cladding arrangements and how they interact with the structural system. The cost implication of such building elements is much higher than the actual structural steelwork and their integration with the structural solution have a significant influence on the overall construction process. Also the extent to which practical fabrication and construction issues are considered at this design stage depends mainly on work experiences of those responsible for the design. This is most apparent in the following generalised cases:

- The building designer must make a design decision within his or her own design discipline, which is likely to cause detrimental effects within another design discipline, of which the designer has little knowledge.
- The building designer is unable to consider the entirety of the building design within his or her own design specialty, and so must break the design down into components that can be individually designed, or face the possibility

of allowing local design decisions to have effects that propagate throughout the design.

The former problem can only be remedied by integrating the design of the building between design disciplines so that these knock on effects can be immediately understood by the building designers, and so that it is not necessary to simplify the design of the building, or design the building in ways that make such concessions to simplification.

Lack of fidelity

Building design fidelity means that design assumptions and any analysis based on them are compatible with the physical or as-built building. One such fidelity issue is ensuring that the structural analysis of the building correctly represents the physical design of the building, or that the calculated cost of the building project correctly represents the design. Current practice does not achieve such fidelity due to many simplifications being made to the analysis and design models. Designers tend to simplify and modify the models to fit with their own assumptions rather than modelling what is going to be built. For example, current methods of analysis and design tend to simplify or even ignore the member/connection interface.

Unaccommodating of design experimentation

The multi-disciplinary nature of the building design process means that most design activities require the input from more than one actor before their impact can be fully appreciated. Transferring the building design from one party to another entails a great deal of overhead in terms of time and man-power, and so the traditional building design process enforces the requirement to reach final design decisions as soon as possible.

This reduces the amount of design iteration that may be carried out, which effectively stifles creativity. This in turn leads to the development of potentially less efficient designs.

Unaccommodating of late design changes

In the traditional design process, late changes to the specifications of a building design during the detailed design stage are common. It is common for floor loading to alter due to a late change in the client's floor area requirements, or for a column to be removed. When these late changes occur, the building design process is stressed and two key issues are highlighted:

- A vast amount of design checking must be carried out to find out if the building design is still satisfactory, and to locate the parts of the building design that need to be changed.

- The building designer's motivations behind the original design are not obvious from looking at the existing design documents. This means that previously made and rectified mistakes may be remade.

It is also common for changes to the design to occur during the construction phase, though these problems are principally due to logical mistakes in the design that had not been previously realised. These problems principally stem from the rigid design flow that does not operate in a way that can easily accommodate the necessary re-design activities.

Lack of design automation

Much of the work carried out by the building designers involves simple and relatively deterministic tasks. This is particularly apparent in structural analysis and design conformance checking, but also in carrying out simple design tasks such as sizing members.

However, automation is typically not possible in traditional building design due to a lack of the formalisation of the necessary information, and the perceived lack of control that automation would introduce to the design process. If not for the design information's fragmentation and lack of formalisation, these types of tasks could be carried out by IT systems directly. In doing so, they would also be carried out with greater precision and speed.

For such automated decision making to be made possible, the boundary between the building designers' decisions and the automated decisions must be formalised. For it to be practicable, the implementation must ensure that the building designers can ultimately override the automated decisions in all cases.

Requirements for an integrated IT system

An integrated IT system aimed at enabling an 'nD' design process should address a number of high-level requirements. Below is a list of such requirements set to satisfy the strategy outlined earlier and the problems of the current design process:

- Design requirements, constraints and intents must be formalised to form part of the design information. Such information will be used to assess fitness and compliance of design solutions.
- The decisions made by the building designers should be captured in a suitable way so as to allow the logic behind the decisions to be maintained and then re-used at later stages in the building's design.
- Information should be available concerning the whole building project in a compatible manner in order to allow all building design aspects to be considered concurrently. The information should maintain fidelity between assumptions used for analysis and design solution.

- The effect of any changes to the building's design should be apparent so that collaborating non-specialists designers can appreciate the effect these changes have upon the rest of the building design.
- The design process should be flexible enough to allow for experimentation and for the variation in design methodologies and domains of responsibility that are required by different partnerships.
- Low-level design processes should be automated where accuracy and speed are more relevant than creativity.
- The process should accommodate large-scale changes made late in the building's design by reducing the amount of manual re-working required.
- The building design should be conducted through a unified interface based around the integrated building design process in order to make sure that all design aspects can benefit from the advantages of such an integration.
- Fidelity problems are largely caused by the lack of consistent modelling techniques. Integrating the design of the building with its analytical representations, and developing automated systems for the creation and usage of these analytical models can overcome these issues.

An implementation of an integrated design system

This section describes an integrated system designed to overcome some of the problems that exist in the design process that stem from the nature and the flow of building design information and adhere to the system requirements mentioned in the previous section.

An integrated process and information model has evolved from earlier work (Ruikar *et al.*, 2001; Smith, 2005; Tizani *et al.*, 2005). Within the term 'process and information model', the 'information' holds the design specification (product model) and information used in the design process (design requirements, constraints, intents, analyses, etc.). The information is described in terms of the logical format of information required to produce a formalised model suitable for implementation in an IT solution. The 'process' consists of the activities involved in the development and creation of the 'product'. The processes are described in terms of the activities carried out by building designers alongside the operation and implementation of automated processes. The concept of this integrated model is schematically shown in Figure 2.3 (Tizani *et al.*, 2005).

Figure 2.3 shows four high-level processes: (1) design intent and appraisal, (2) prototype generation, (3) test model generation and (4) design checking. These processes operate upon three tiers of information: design intent, as-built model and analysis model (or IDEA, PROTOTYPE and TEST). The interaction within the process and information model is described here in a simplified manner:

- The Design Intent process allows the designer to input or modify their design intents, which will be stored in the Design Intent model constituting the 'IDEA'. The IDEA is composed of the decisions and choices made by each of the building designers. As such, this tier contains the most valuable information in the building's design, and the other two tiers are ultimately a logical development of the ideas expressed within.

- The Prototype Generation process uses the information stored in the Design Intent Model to generate the physical model, constituting the 'PROTO-TYPE'. The PROTOTYPE is the outcome of the total design process and principally includes the physical product. This is the description of actual products that make up the building, such as for a steel building: the steel frame components, floor system and cladding system and so on. The generation process relies heavily on automated processes to generate suitable design solutions bounded by the constraints set in the Design Intent Model.

- The Test Model Generation process applies analysis processes onto the prototype and store the results in the Analysis Model. These analysis processes test the conformance of the building to the set constraints and the general engineering principles. They are run for each of the conformance tests required. For example, to test the structural engineering requirement, a structural analysis model is created, and based on the response of the structure a safety check is carried out (e.g. member sizing check), or to estimate the cost of the building a cost analysis model is created and checked against the cost targets. Other conformance tests can be carried out in a similar fashion. For example, to ascertain the fire safety or fire escape requirements a suitable analysis model is created and checked against the regulations.

- The Design Checking process applies checking processes onto the stored Analysis Model and checks the compliance of the Prototype with the requirements and constraints stored in the Design Intent Model.

This sequence outlined a cycle of generating, from the IDEA, a PROTOTYPE and then TEST to gauge its performance. Further cycles can be carried out as required by refining the IDEA.

This integrated process and information model allows for the conduct of an integrated design covering more than one aspect or one discipline (nD approach). The level of integration of the implemented software system will depend on the coverage provided in the various processes and information model tiers. In effect the integrated model constitutes a framework that is scalable to cover wider aspects of design.

Within the current prototype implementation, the system covers the architectural–structural and services integration. The software has been developed using the object-oriented methodology and written in the C++ programming language. All processes and product models are implemented in terms of objects logically inter-connected. A modular approach is taken to the addition of agents used to carry out the processes of structural analysis, checking and design. The system uses an interactive realistic 3D modelling interface, akin to virtual reality. This was developed in OpenGL (Open Graphic Library) to create a real-time dynamic system where all actions upon the model can be carried out through the 3D interface.

This is a significant advance over the use of static VRML models whose main use is visualisation.

The interface has the ability to simultaneously visualise all aspects of the building's design. It includes the ability to have specialised views for each discipline that combine several different visualisation options alongside the tools appropriate for that discipline. Default views are provided for each of the traditional roles including the architect, structural designer and services designer. Model manipulation is specialised for each of the views so that no information over-load will occur. However, each view can be customised to visualise any of the available options by overriding the default set. So, the structural engineering view might be customised to super-impose any of the architect details over the structural details.

All of the design ideas are visualised in terms of their corresponding PROTO-TYPE model (which is at all times synchronised with the design intent) so that the interactions between the building components can be better understood. The results from the analysis tier are also visualised within the same 3D interface, which makes manual design checking possible and provides the building's designers with an intuitive way of visualising the effect that changes have upon the model.

The use of the system to follow an integrated design process is best explained through running a typical design scenario.

Design scenario

The process model makes no assumptions concerning the roles of actors within the building design process, though it is helpful to consider the design activities in terms of the actors traditionally involved in those roles. As such, the activities within the processes developed in this model can be divided between the client, architect, structural engineer and services engineer. A typical design scenario that may be followed using the software is provided here.

The client is responsible for providing the building's global requirements. This includes the building's target cost that may be subdivided between the structure, services, flooring and cladding. The client is also responsible for setting the total floor space required within the building as a whole (Figure 2.4). At all times during the building's design, the costs and areas are automatically calculated and compared with the client's requirements. Any failure for the actual figures to meet the targets will be flagged so that all building design participants will be aware of the issue.

The architect is responsible for specifying the layout of the building's perimeter, the number of floors and the internal height requirements for each floor (Figure 2.5). The height of any given storey of the building can be given in terms of the non-structural floor depth, clear depth and the ceiling depth, which includes the services and structural depths. This provides a number of requirements for the structural and services design to conform to, and sufficient information to specify a general model of the building in terms of size and spacing.

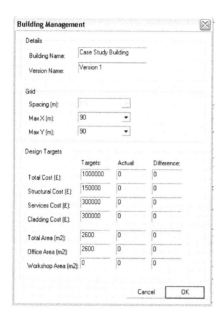

Figure 2.4 Input of global targets.

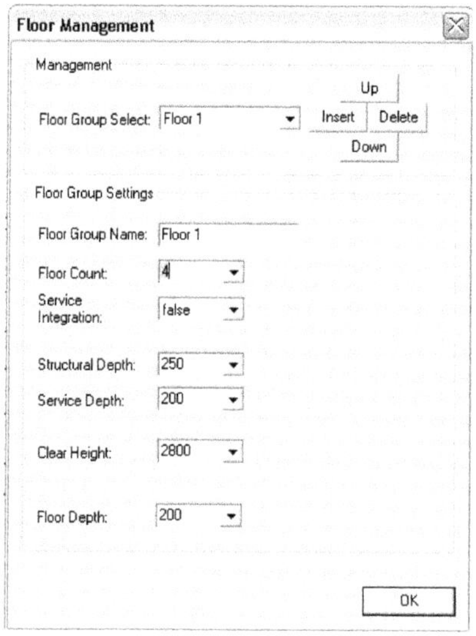

Figure 2.5 Input of constraints of floor zone.

The architect is also responsible for specifying the cladding types assigned to each part of the building's perimeter. This provides sufficient information to calculate the claddings cost, which will then be compared with the clients brief. The cladding information also provides an early indication as to the external look of the building, which may be useful for assessment of early prototypes by the client. In terms of the building's structure, the cladding information also provides data regarding its loading requirements that will then be incorporated into the structural analysis. The architect is responsible for specifying the floor usage for each part of the building (Figure 2.6). This provides information for the cost modelling process, loading information for the structural analysis process, and allowable structural response information for the building design checking process. The architect may also provide information regarding extra loading areas within the building, such as around building cores so that the appropriate loads will be added to the building's structural analysis model. The architect may also designate column spacing area within the building in order to prevent the structural engineer from positioning columns inappropriately. The architect is also responsible for specifying cores in terms of their location and designated purpose (Figure 2.7). This information will be important for use by the structural engineer and the services engineer.

The architect's design input, therefore, has a far-reaching effect on the building's overall design. It must meet the client's requirements in terms of the spacing

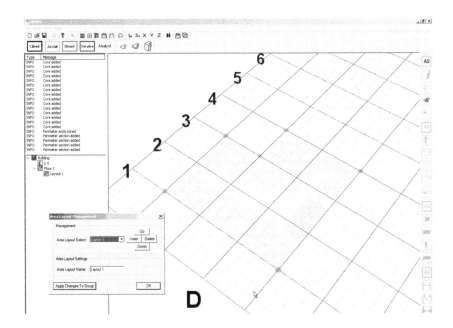

Figure 2.6 Definition of floor plan and usage.

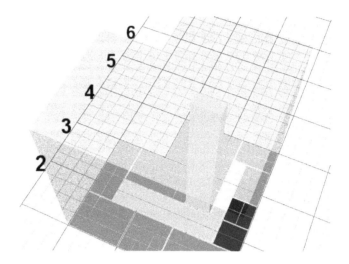

Figure 2.7 Definition of building cores.

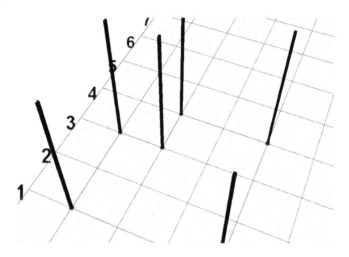

Figure 2.8 Positioning of column grids to suit architectural constraints.

requirements while setting forth the requirements of the structural and building services engineers. All the input so far will form part of the Design Intent Model.

The structural engineer's first task is to specify the column positions within the building (Figure 2.8). The height is not specified, as this is automatically calculated based on the architect's building information. It is also not necessary to specify the dimensions of the columns at this stage, as they may be calculated

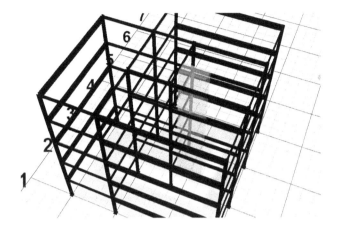

Figure 2.9 Structural: propose framing system.

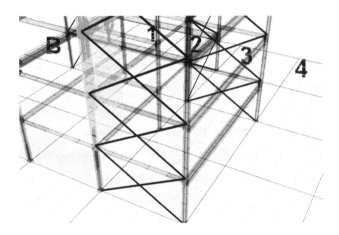

Figure 2.10 Structural: devise structural bracing system.

automatically from parametric rules and then later refined based on the columns structural response. The next step in the structural design is to locate the primary beams and either specify their dimensions or, as with the columns, allow the software to automate this process (Figure 2.9). The connection detailing between the primary beams and columns is optionally generated automatically, and may be redesigned later when structural analysis information is available. Bracing is similarly specified using simple connections (Figure 2.10). 'Floor areas' are then specified, which allow for the designation of the secondary beams and structural

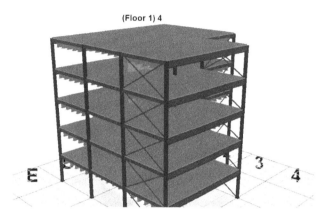

Figure 2.11 Structural: complete structural system prototype.

floor in one action (Figure 2.11). As with previous structural design stages, any or all aspects of the floor may be specified by the engineer including the secondary beam dimensions, spacing, orientation and the floor type and floor dimensions. The detailing for the remaining design aspects, if any, will then be calculated automatically.

The structural engineer is therefore able to commit to any level of detail that he feels necessary, and this allows him to leave the less globally influential design issues to the software that is better suited to this task. As a result of these actions, data is generated concerning the self-weight of the structure and the cost of the building elements for use later in the design process.

The structural engineer has so far provided a set of intentions by providing location and types of elements but without selecting specified structural specification. The automatic generation of the PROTOTYPE solution is done using processes that operate general rules of thumb using the loading information (from the input area usage) and spanning of structural elements from the given dimensional constraints. The cost information is generated by the Test Model generation process.

At this stage a PROTOTYPE structural framing of the building does exist that would comply with the architectural constraints.

The services engineer is able to specify the location of service entry points to the building, the location of service cores and the main service routing in plan around each floor (Figures 2.12 and 2.13). The height of the actual service ducts is based on the service strategy chosen, and will either be positioned beneath the structural layer or within the structural layer. In the case of the latter, the beams will be appropriately notched, and this will be taken into account in the building's structural analysis and costing.

With the information provided through these processes, there is sufficient data to produce a 3D structural analysis model of the building (Figure 2.14). This is

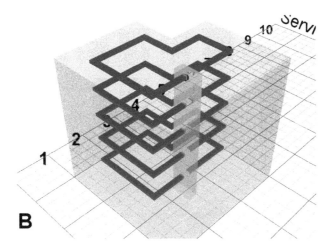

Figure 2.12 Services view: adding services duct connected to services core.

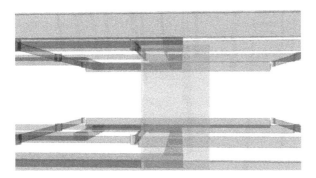

Figure 2.13 Services/structural: checking structural floors – services integration.

generated automatically to accurately model the members and connections with the appropriate use of analysis members and dummy members to model connection offsets, hence maintaining compatibility between the as-built structure and the analysis model. Loading information is automatically generated from the information previously input by the architect, structural engineer and service engineer in terms of factored dead and live loads. The analysis is carried out to produce the structural analysis response model (Figure 2.15). This information is used to perform a design check on the structural members, which may then be used by the structural engineer or automated processes to iteratively improve the building's design.

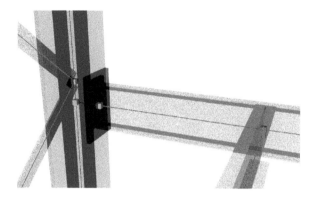

Figure 2.14 Structural general structural analysis model compatible with as-built details.

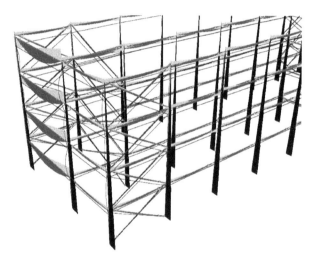

Figure 2.15 Structural: structural analysis results.

The information here has given an overview of a sequential design process. At this stage a complete building prototype is available for assessment. The client or the designer (of any of the involved disciplines) can inspect the compliance of the proposed solution with their own constraints. For example, assessments will be possible of the target costs, floor area, storey clear heights, building height, structural member checking, service ducts zone and so on.

The information describing the building together with all relevant input is stored in the appropriate tier of the information model. Subsequent changes or refinements, by any of the designers to the model, are concurrently available to all other disciplines. Effecting concurrency is facilitated by the use of the single information model. In addition, partial building designs may still be carried by concentrating the work on a single discipline. For example, a structural engineer can input a steel frame and carry out analysis and design without the availability of architectural constraints.

nD decision support

Decision support is provided through the ability to carry out rapid prototyping within an integrated system that facilitates concurrency between a number of inter-related design disciplines. The process modelled within the experimental software can also be described as intuitive since it is modelled on a best-practice approach that concentrates on the satisfaction of what is judged to be the critical goals required from the design process. With the ability to quickly generate workable solutions that meet the requirement of architectural and services constraints, the designer is left to concentrate on high-level decisions.

In addition to this, the system includes a parametric cost model that is based on the quantity surveying principles. The cost model adopted is judged to provide 'rough' estimates accurate enough to allow for comparative costs sufficient for the purpose of economic appraisal. Using the model, the designer can input a target budget figure that can then be compared with the cost estimate of the proposed design solution.

The experimental software is a so-called 'multi-document project' allows the user to generate 'what if' scenarios that allow several design solutions to be compared by cost and structural performance. For example, the designer might produce a number of solutions differing by aspects such as architectural layout with various floor usage, services–structural integration, column spacing, floor systems and so on. The overall effect of alternative design options can then be readily appreciated.

Summary and conclusions

This chapter has introduced the area of engineering design. The process of engineering design for construction projects is complicated due to its multi-disciplinary nature and the fact that most buildings require unique solutions.

The domain of the design of multi-storey steel buildings was chosen as the area into which an investigation of a possible nD approach to the design is possible. A typical design process for such building types was modelled and the design information flow was presented and analysed. The analysis of the typical design process has highlighted a number of generic problems that were described. A list of requirements for such an approach to design was then provided based on these identified problems.

An implementation of a prototype-integrated design software system for buildings made with steel frames was then introduced. The implementation uses an integrated process and information model. The information model includes, in addition to modelling the 'product' in a PROTOTYPE, the modelling of loose data such as requirements and constraints. This type of data was formalised and stored in a 'Design Intent' Model. The information model also includes an 'Analysis Model', which stores within it the results of test analyses conduct on the PROTOTYPE. The process model is composed of a number of high-level processes to acquire the design intent, generate the PROTOTYPE and the Analysis Model and check the compliance of the proposed design solution.

A typical sequential design scenario was provided. Non-sequential design processes are also possible. The scenario demonstrated the rapid prototyping style of design where minimum amount of information is required to produce a complete building PROTOTYPE. The scenario also illustrated how the multi-disciplinary approach to design can be followed in the software system.

This process and information model enables an integrated and concurrent approach by allowing the simultaneous consideration of all significant building design aspects. This is managed through a coherent interface where all modifications to the building's design are made and where the wide-scale consequences of designers' actions can be examined. A sequential process is allowed for, to the extent of providing a general order in which design activities may be carried out, though with flexibility to allow for varied working styles formed by various working arrangements.

The research and implementation of the described system demonstrated that multi-discipline building design can be supported through the adoption of a domain-specific level of conceptual design information that operates at a level above the types of information traditionally modelled in building design systems. A method is proposed by which a software system can be developed that incorporates this type of model. It is shown how this system promotes a parallel design over the traditional iterative design approach and thereby facilitates a more efficient and natural design process that requires less re-working and maintenance.

It is worth noting however, that the current domain of testing has been limited to the client and architect interfaces with structural and services engineering. Future work will include testing the scalability of the system by introducing more disciplines, for example the area of construction management.

References

AceCAD (2005) StruCAD software, www.acecad.com

AutoDesk (2005) AutoCAD software, www.autodesk.com

Boujut, J.F. and Laureillard, P. (2002) 'A Co-operation Framework for Product-Process Integration in Engineering Design', *Design Studies*, 23, 497–513.

Eastman, C. (1999) *Building Product Models: Computer Environments Supporting Design and Construction*, CRC Press, Boca Raton FL, USA.

Egan, Sir J. (1998) 'Rethinking Construction, Report of the Construction Task Force on the Scope for Improving the Quality and Efficiency of the UK Construction Industry', Department of Environment, Transport and the Regions (DETR), London, UK.

IAI, International Alliance for Interoperability (2002) 'Industry Foundation Classes'. Online. Available HTTTP: <http://cig.bre.co.uk/iai_uk/>.

Kiviniemi, A. and Fischer, M. (2003) 'Requirements Management with Product Models', in *Construct IT: Developing a Vision of nD-enabled Construction*, G. Aouad (ed.), pp. 77–81, University of Salford, Sanlord.

Latham, Sir M. (1994) 'Constructing the Team, Final Report of the Government/Industry Review of Procurement and Contractual Arrangements in the UK Construction Industry', HMSO, London.

NIST (National Institute of Standards and Technology) (1991) 'Integration Definition for Function Modeling (IDEF0)'. Online. Available HTTP: <http://www.erc.cims.edu.cn/erc_e/topic/arture/modelingmethod/idef0.htm> (accessed March 2005).

Ruikar, D., Smith, R., Tizani, W. and Nethercot, D. (2001) 'Best Practice in the Design of Multi Storey Steel Framed Structures', Proceedings of the First Innovation in AEC Conference at Loughborough.

Ruikar, D., Tizani, W. and Smith, R.A. (2005) 'Design Process Improvement Using a Single Model Environment'. Proceedings of the Tenth International Conference on Civil, Structural and Environmental Engineering Computing. Paper 57, B.H.V. Topping, (ed.), Civil-Comp Press, Stirling, Scotland.

SCI (2002) 'CIMsteel Integration Standards 2, Volume 4, The Logical Product Model', Online and paperform. Available HTTP: <http://www.cis2.org>. The Steel Construction Institute, Ascot, Berkshire, UK.

Smith, R.A. (2005) 'Design Centric Information and Process Modelling for Integrated Building Design', PhD thesis under preparation, University of Nottingham, UK.

Tekla (2005) XSteel software, www.xsteel.com

Tizani, W. and Smith, R. (2003) 'Overview of an IT Vision for Engineering Design', *Construct IT: Developing a Vision of nD-enabled Construction*, G. Aouad (ed.), pp. 108–111, University of Salford, Salford.

Tizani, W., Smith, R.A. and Ruikar, D. (2005) 'Virtual Prototyping for Engineering Design', in Proceedings of the 5th Conference on Construction Applications of Virtual Reality, CONVR 2005, On CD, 12–13 September, Durham, UK.

Chapter 3

Lessons learned from ten years of conceptualising, developing, introducing and using nD BIMs

Jarmo Laitinen

Introduction

In this chapter, an overview of the conceptualization, development, introduction and use of nD BIMs (Building Information Models) over a period of 10 years is provided. Initially, the conceptualization and development of nD BIMs by a principal contractor in design and construct projects is described. Then, the broader introduction and use of nD BIMs is discussed. In conclusion, recommendations from lessons learned over 10 years of nD BIM innovation are stated.

When the work described in this chapter began, nD BIM was science fiction. Moreover, it was a science fiction which only very few people had any interest in. These people were scattered around the world in a few institutions and in even fewer companies. Today, nD BIM is a technological reality which more and more companies are exploiting. In Finland, for example, one company alone has already created and exploited over 500 BIMs, and now more and more companies are improving their business performance by exchanging and sharing nD BIMs. Back in the early 1990s, things were very different. Since then, many thousands of small steps have been taken towards a goal that has moved ever closer: the integration of building design, building production and building operation. We have not yet arrived at this goal, but we are much closer. Now, the attractions of the goal are even more sharply in focus: built environments with lower life-cycle costs and higher life-cycle performance.

Background

Steps towards the use of nD BIMs began with an aim of improving the data management of a principal contractor who carries out design and construct projects. The improvement of data management was seen as being essential if clients were to be provided with better value through more cost-efficient production. On a more detailed level, aims were to

- identify the ways in which design information can be made amenable to computer interpretation;

- define the structure and content for a cost and value engineering management system; and
- demonstrate the working parameters of a cost and value engineering management system.

Work began with an analysis of the construction process chain throughout the whole life cycle from the briefing phase to the use and maintenance phases. The analysed phases were as follows: client briefing, design, production planning, construction and building use, and maintenance. There were three processes which were considered from the information management point of view:

- the customer service process – how to provide decision support and services to the customer throughout the building project,
- teamwork between designers and contractors in design and construct type projects and
- the construction materials and subcontracting supply process.

Consideration of these processes led to the conclusion that the principal contractor should have the ability to create a production model from the designers' data (which could be available either as paper drawings, CAD-files or in the future as BIM objects). Further, it was concluded that after its creation, the model should be integrated with the contractor's applications for cost estimation, scheduling and so on and can finally be used to provide services for the customer.

The description of the principal contractors' building process is presented below as IDEF0 diagrams. Since the process description is based on the principal contractor's viewpoint, the activities of owners and designers are in a minor role. A more general analysis of Finnish practice at the time can be found in the generic present-state systematization by VTT. The focus in the model presented here is in the co-operation between designers and contractors (in Finland at the time, few contractors had in-house design departments) and on information flows and their contents. The model is focused on design and construct type of projects, where the contractor has the responsibility for the design activities.

The top level of the IDEF0 model consists of one single box (Figure 3.1) which illustrates the whole construction project. The major outputs are a building ready for use by the client or other end users and the documentation about the building, which according to today's practice is in the form of paper documents, such as drawings.

Interviews and analyses of the IDEF0 models suggested that the support for the decision making of the client, as well as that of the contractor, was needed most in the briefing phase. Further, it became apparent that the biggest needs and problems in the designer–contractor collaboration was in the tendering phase. It was concluded that the principal contractor should be able to use information coming from the designers as input data, and to merge with its own construction know-how. Another issue identified as important was cost and value engineering, that is

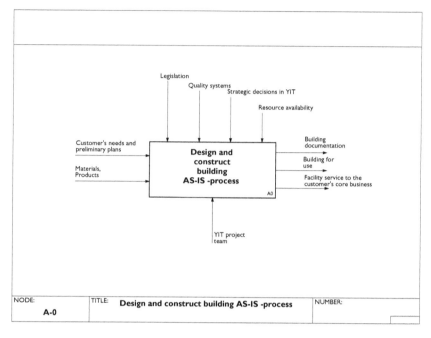

Figure 3.1 A-0, Design and construct building, AS-IS process.

the ability to analyse design solutions from the perspective of accomplishing the construction, as well as to examine alternatives and the impact they will have on the overall costs.

Decision support for the customer was considered to be most important. In the principal contractor's practice, the customer's needs and demands were not in general specified in sufficient detail. Moreover, they were not presented in the form of measurable attributes. The checking of quality requirement fulfilment was based on the design supervisor's own viewpoint and judgement. Design solutions that fulfil the customer's needs and demands, being yet cost effective and easy to construct, were very difficult to find.

In residential construction projects, the customer was presented with little choice. The principal contractor had difficulties in knowing about the alternatives or their impact on the total costs or on the project process. Use and maintenance cost assessment (life-cycle costing) was hardly carried out in the design stage. Figure 3.2 illustrates the discrepancy which existed between the client's initial expectations and the end result that the client eventually got. The different factors that contributed to this discrepancy are further explained below.

1 Customers were not able to express their expectations in a formal or verbal way.

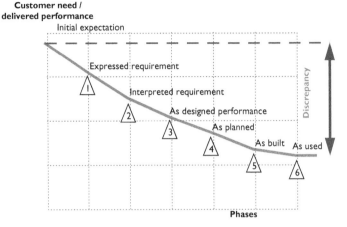

Figure 3.2 An illustration of the origins of the discrepancy between initial customer expectations and the final result. The figure is based on answers to questionnaires from owners and users.

2   The designer interprets requirements in a different way to that which the customer assumes.
3   The technical solution chosen did not fulfil the requirements.
4   The contractor took into account constructability issues in planning the execution, which was in conflict with the client's or designer's intentions.
5   The actual work was carried out differently from the plans, due to poor quality assurance and other such factors.
6   The building was not operated and maintained according to the instructions.

The overall conclusion was that there were two main causes of this discrepancy: inadequate information management and the inherent conflict between customer and supplier.

Towards nD BIM

At the time when the prototyping began, there existed no fully developed product data models or standards which, alone, were sufficient for the data exchange needs of the principal contractor. Nonetheless, it was recognized that on-going research and standardization could offer the general paradigm of a core model, supported by aspect models, as a framework for how the data exchange could be achieved. For integration with the production know-how of the contractor, the concept of a production model emerged. Some work had already been done by a number of researchers, as well as within the STEP and IFC

(Industry Foundation Classes) core models, to incorporate some data structures which would be useful in such production models.

Despite these limitations, a new approach for the principal contractor was proposed that was based on the application of product modelling technology, especially for the purpose of cost and value engineering. It was envisaged that the utilization of the product model would cover all phases of the building project. Moreover, it was envisaged that product model and performance driven construction would develop together. The proposed new approach was seen as a 'target' process. When technologies are at an early stage of development, it is important to stress that an envisaged process is a 'target' process. The target may be hit or exceeded. However, before that the target may be reached at the first attempts.

The target process which was defined is described here. The description focuses particularly on the design process and on controlling it, with cost and value engineering used as the basis for this. This is due to results of the analysis of the construction process information flow which as carried out. The aim was to integrate the information of the designers and the contractor in such a way that the data from the design work could be used as source data directly, without manual input, for calculating the tender as well as for production planning (i.e. the information produced by the designers would be integrated with the contractor's know-how). At the time, designers did not have appropriate modelling tools, so the contractor needed to have an IT-system which allowed the building up a of product model which contained production information – what at the time was referred to as the so-called 'production model'. Another objective was regular communication between the various partners, for example, to analyse design solutions or to generate alternative solutions to support decision making in the earlier phases of the process. The basis for this new approach consisted of enterprise-specific information which had been stored beforehand, such as accepted good practice structural details as well as typical building cases.

There were three main requirements for the target process:

1   Provide more information and alternative solutions to the customer and the other participants within the building construction process.
2   Provide more accurate information in earlier phases of the process.
3   Utilize information created beforehand (cases) and classified technical solutions embodying the contractor's knowledge.

The target process is illustrated by Figure 3.3 from a decision support point of view, which is based on the approach that key activities are shifting to earlier phases.

It was considered that by using product modelling technology it would be possible to reach at least some of these targets. It was also considered that the model should be enterprise based, in terms of using the contractor's production knowledge, but yet open for other disciplines within the construction process. This meant that one would first define the minimum 'core' data to be shared by

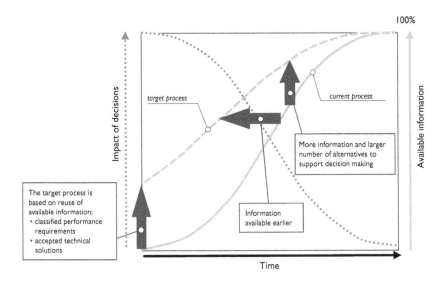

Figure 3.3 Decision-making versus available information.

partners involved so that all the parties could get the data needed for their own usage and are able to share with the others the data that they need. This forms the basic framework for production model development. The traditional design process was divided into different phases of design, starting with briefing and ending with the production of working drawings. It was anticipated that in integrated, model-based design, there would be no such clear distinction between the phases; instead, the phases of the design process constantly interact and complement each other. Further, it was anticipated that the data content of the artefact being designed would become more detailed and data would accumulate continuously as the process advances. It was also anticipated that all new data would be produced once only and would be used as input data for the next 'phase'. It was anticipated that the traditional demarcations between phases of design would vanish and be replaced by new 'decision points'. In particular, decisions would be made in smaller parts and earlier, so that design and decision making would be interactive. Figure 3.4 illustrates the target process in IDEF0 format.

Figure 3.5 illustrates the target process from the customer's, contractor's and building user's points of view. This shows that information management is organized in a systematic manner with discussion, feedback and self-learning being enabled.

Subsequently it was determined that 'as built' models would be a logical continuation for the production model. It was considered that in such models there would be information of designers (why and based on what the technical solution was chosen) and of the contractor (materials, systems and maintenance information). The target situation is shown in Figure 3.6.

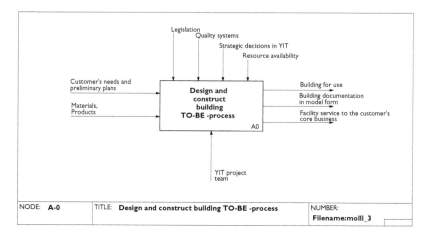

Figure 3.4 A-0, Design and construct, TO-BE process.

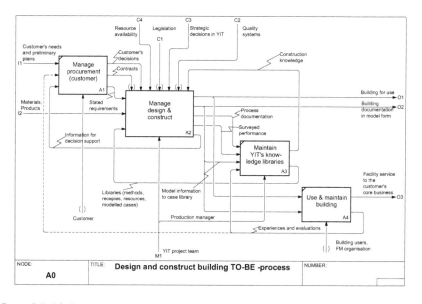

Figure 3.5 A0, Design and construct, TO-BE process.

Using nD BIMs

Today, the principal contractor has the opportunity to use nD BIMs in their day-to-day business. Some of the software which was developed along the way has now been commercialized. In the meantime, the use of nD BIMs has spread out to many other organizations in the Finnish Real Estate and Construction

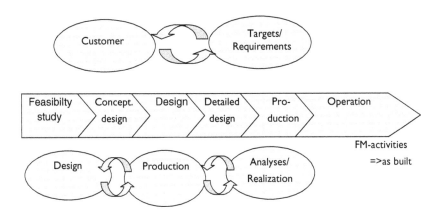

Figure 3.6 Target situation for the use of model support throughout the construction process life cycle.

Cluster (RECC). For example, design and construction of a Finnish university building has been carried out using nD BIM. Moreover, the design process has been much improved as a result. Significantly, the building owner has been able to base decisions on more comprehensive and accurate information at an early stage. Achieving this involved overcoming numerous technological challenges. In particular, during the project, BIM's have not had all the graphical details and visualization parameters needed to create sophisticated virtual reality (VR) simulations of a building. As a result, separately created VR models have been used alongside nD BIM. The various analyses facilitated by nD BIM revealed that the original design had a number of limitations that previously would not have been discovered until after they had become costly realities. For example, the construction cost was high, environmental impacts exceeded target and the energy consumption was also high. Informed by nD BIM-enabled analyses, the architect redesigned the building. The revised design retained the original aesthetic intent but with a better volume-to-surface ratio. As a result, construction, energy and environmental costs are now well on target. The process improvements facilitated by the use of nD BIMs are shown in Figures 3.7 and 3.8.

nD BIM recommendations

Many lessons have been learnt in 10 years that have involved the conceptualization, development, introduction and use of nD BIMs. These lessons learnt are summarized in the ten recommendations that follow:

1   Make **strategic** use of nD BIMs. The introduction of nD BIMs offers far-reaching opportunities. For example, building owners throughout the

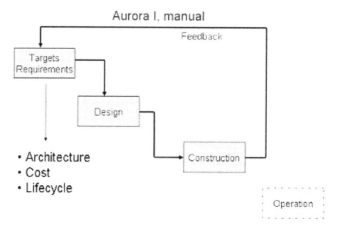

Figure 3.7 Pre-nD BIM process.

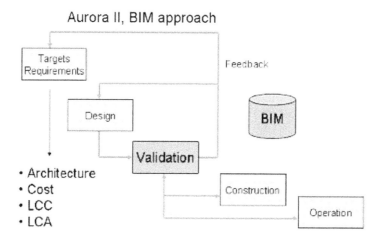

Figure 3.8 nD BIM-enabled process.

world may soon need nD BIMs for facilities management but lack local providers.

2   Make **selective** use of nD BIMs. Like most technologies, nD BIMs can bring major benefits in some applications, but only minor benefits in others. Before investing, forecast net benefits for alternative nD BIM applications.

3   Make **informed** use of nD BIMs. Much of the hardware, software and communications technologies required to set up nD BIMs is being continually improved. So, obtain the most up-to-date information.

4   Make **routine** business use of nD BIMs. Exploit opportunities to enhance existing business processes, for example, the exchange of design information between domains.

5   Make **unique** business use of nD BIMs. 'The rules of the AEC/FM game' will be changed by nD BIMs. This will enable organizations to offer new combinations of old services and to offer completely new services.

6   Make **targeted** use of nD BIMs. Use nD BIMs to work towards improvement targets, for example, a 10 per cent reduction in energy consumption or a 50 per cent reduction in the time required to carry out building cost calculations.

7   Make **phased** use of nD BIMs. It is important to take one step at a time when setting up and using nD BIMs, so make sure that targets are incremental and allow sufficient time for learning and for reflecting upon what has been learnt.

8   Make **integrated** use of nD BIMs. In order to maximize benefits, it is important to integrate nD BIMs with established technologies and practices that still add value, for example, physical mock-ups and samples.

9   Make **team** use of nD BIMs. Form business alliances that can best exploit the potential of nD BIMs, for example, alliances that enable one BIM to be used by multiple businesses across traditional domain boundaries.

10   Make **open** use of nD BIMs. The next few years will see dynamic development in nD BIMs. So, do not limit your business to a fixed selection of hardware, software and communications technologies.

Part 2

nD modelling

The scope

The concept and need for nD modelling was elucidated in Part 1. The second part of this book will define a number of key parameters of nD modelling. These include time – what is traditionally considered as 4D modelling – through to risk, safety, accessibility, acoustics and crime. These parameters are by no means exhaustive of the scope of nD modelling, but their inclusion in this book was through a select number of leading academe and practitioners from across the globe who have successfully integrated a parameter within an nD model. It is hoped that the following chapters will provide a snapshot of the scope of nD, and begin to advocate to readers how other parameters – such as facilities management/maintainability, sustainability and so on – can be included.

In the first chapter, Dawood presents current planning and scheduling practices in the United Kingdom before demonstrating the VIRCON project, a 4D modelling tool developed at the University of Teeside. He describes the development of the database, and how a Uniclass code structure was adopted. Finally, the Stockport case study is presented to demonstrate the benefits of the system.

Rischmoller's chapter continues the theme of construction scheduling with a Latin American perspective. The traditional construction scheduling process is detailed, covering the use of Gantt charts right through to construction safety. The need for improvement to address uncertainty and variability in the scheduling process is also highlighted. Finally, an nD construction scheduling process is proposed, which cumulates product and process modelling methodologies. The CAVT concept (Computer Advanced Visualisation Tools), developed in Chile specifically for industry, illustrates how the need for improved usability in 4D modelling is required, and the potential to include other dimensions, hence nD. It is proven that visualisation, planning, analysis and communication capabilities embedded in the nD construction scheduling development processes can improve the traditional scheduling process to the extent that it could finally be used as a real production control mechanism that could tackle problems proactively and re-schedule the project timely as necessary.

Synergistically to Dawood and Rischmoller, Coble reports that the use of Computer Aided Design (CAD) in risk management in construction is a new but not commonly utilised management technique. CAD has tremendous risk management potential in the management of construction projects. In this regard, Coble demonstrates how the benefits of hazard avoidance can be reaped through a visual context.

Naji's chapter covers the topical and timely subject of health and safety on construction sites. He extends the nD model to develop a dynamic tool that will investigate potential site issues in a virtual construction environment. The methodology and system architecture for such a tool is explicated and demonstrated using a case study in Qatar.

The ability to automate code checking to reduce problems of human interpretation of building codes is discussed by Drogemuller and Ding, using accessibility regulations in Australia as an example. The introduction of object-oriented CAD systems in the 1990s makes it now feasible to build automated code checking systems: Three state-of-the-art systems are described in this chapter: the Finnish 'Solibri Model Checker', Singapore's 'ePlanCheck' and finally, CSIRO's 'Design Check'.

Lam's chapter provides an informative insight into acoustics in the built environment. It begins by briefly introducing the concept of acoustics and discusses legislative conditions. It illustrates some of the basic modelling methods that can be used to estimate the acoustic performance of building elements and the building space, and gives an assessment of the usefulness in building design. A brief introduction to the more advanced modelling method is also given.

Finally, in Hands and Cooper's chapter, crime is pushed to the forefront in the design agenda. The chapter portrays how the public's increased perception of the fear of crime has led to new ways of design. Crime prevention features are most often considered retrospectively and applied to both products and built environments after the event of crime, rather than at the initial stages of their design development. It discusses how nD modelling will allow the multi-disciplinary design team to envision and anticipate potential crime misuse scenarios within the initial stages of the design process so that crime tensions and trade-offs can be fully explored, thus allowing decision makers to prioritise crime issues early.

Chapter 4

Planning and scheduling practices in the UK

Nashwan Dawood

Introduction

This chapter presents current planning and scheduling practices in the UK. It begins by identifying planning and scheduling processes before demonstrating the VIRCON project.

Figure 4.1 shows a generic planning process, thus identifying the relationship among the planning processes at all levels. These processes are subject to frequent iterations prior to completing the plan on a live construction project. For example, if the initial completion date does not meet the contractual finished date, project resources, cost or even scope may need to be re-defined. In addition, planning is not an exact science. Two different teams could generate very different plans for the same project.

The core processes and the facilitating processes are described as follows:

- Core processes: some planning processes have clear dependencies that require them to be performed in essentially the same order on most projects. For example, activities must be defined before they can be scheduled or budgeted. These core planning processes may be iterated several times during any one phase of a project. They includethe following:

    - Scope planning – developing a written scope statement as the basis for future project decisions.
    - Scope definition – subdividing the major project deliverables into smaller, more manageable components.
    - Activity definition – identifying the specific activities that must be performed to produce the various project deliverables.
    - Activity sequencing – identifying and documenting interactivity dependencies.
    - Activity duration Estimating – estimating the number of work periods which will be needed to complete individual activities.
    - Schedule development – analysing activity sequences, activity durations and resource requirements to create the project schedule.

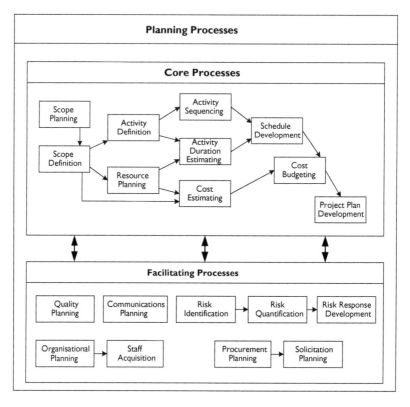

Figure 4.1 Relationships among the planning processes.

- Resource planning – determining what resources (people, equipment, materials) and what quantities of each should be used to perform project activities.
- Cost estimating – developing an approximation (estimate) of the costs of the resources needed to complete project activities.
- Cost budgeting – allocating the overall cost estimate to individual work items.
- Project plan development – taking the results of other planning processes and putting them into a consistent, coherent document.

• Facilitating processes: Interactions among the other planning processes are more dependent on the nature of the project. For example, on some projects there may be little or no identifiable risk until after most of the planning has been done and the team recognises that the cost and schedule targets are extremely aggressive and thus involve considerable risk. Although these facilitating processes are performed intermittently and as needed during project planning, they are not optional. They include the following:

- Quality planning – identifying which quality standards are relevant to the project and determining how to satisfy them.
- Organisational planning – identifying, documenting and assigning project roles, responsibilities and reporting relationships.
- Staff acquisition – getting the human resources needed assigned to and working on the project.
- Communications planning – determining the information and communications needs of the stakeholders: who needs what information, when will they need it and how will it be given to them.
- Risk identification – determining which risks are likely to affect the project and documenting the characteristics of each.
- Risk quantification – evaluating risks and risk interactions to asses the range of possible project outcomes.
- Risk response development – defining enhancement steps for opportunities and responses to threats.
- Procurement planning – determining what to procure and when.
- Solicitation planning – documenting product requirements and identifying potential sources.

• Controlling processes: project performance must be measured regularly to identify variances from the plan. Variances are fed into the control processes in the various knowledge areas. To the extent that significant variances are observed (i.e. those that jeopardise the project objectives), adjustments to the plan are made by repeating the appropriate project planning processes. For example, a missed activity finish date may require adjustments to the current staffing plan, reliance on overtime or trade-offs between budget and schedule objectives. Controlling also included taking preventive action in anticipation of possible problems. The controlling processes measure the performance of the executing processes and send feedback to the planning processes and the closing processes. Figure 4.2 illustrates how the following controlling processes interact:

- Overall change control – coordinating changes across the entire project.
- Scope change control – controlling changes to project scope.
- Schedule control – controlling changes to the project schedule.
- Cost control – controlling changes to the project budget.
- Quality control – monitoring specific project results to determine if they comply with relevant quality standards and identifying ways to eliminate causes of unsatisfactory performance.
- Performance reporting – collecting and disseminating performance information. This includes status reporting, progress measurement and forecasting.

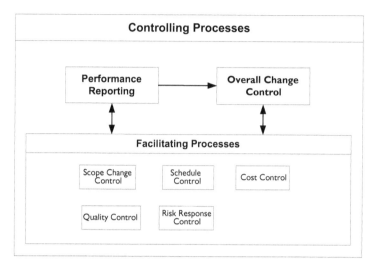

Figure 4.2 Relationships among the controlling processes.

  – Risk response control – responding to changes in risk over the course of
    the project.

With respect to the tools that planners use to develop the planning process
and control of site activities, a thorough review of the capabilities of these and
other Critical Path Analysis software packages can be found in Heesom and
Mahdjoubi (2002). Figure 4.3 shows the preferred planning software related to
the value of contracts handled by each planner. It is clear that Primavera is
preferred for larger value contracts and Power Project for smaller ones, see Winch
and Kelsey (2004).

   Recent advances in 4D planning (synchronised 3D with time) offer the opportunity to perform detailed planning due to its visualisation and coordination
capabilities. 4D planning facilities the integration of 3D products and time and
gives planners the opportunity to visualise progress before construction starts.
The reminder of this chapter discusses the information infrastructure for 4D planning and two real-life case studies in which 4D planning was used.

Data capture and database development

The objective of this section is to identify information required and needed to
run 4D modelling. A comprehensive database was designed, implemented and
populated with real live construction data. This was used as the base for the development of 4D and CSA models and the experimental work of the VIRCON
project. The database is composed of a core database of building components

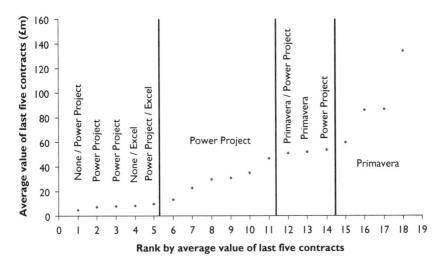

Figure 4.3 Decision support tools used by construction planners.

which are in turn, integrated with a CAD (Computer Aided Design) package (AutoCAD, 2000), a Project Management Package (MS Project) and Graphical User Interfaces.

The core database was designed using the Unified Classification for the Construction Industry (Uniclass). Uniclass is a new classification scheme for the construction industry and follows the international work set out by ISO Technical report 14177. One of the benefits of using the Uniclass method, apart from providing standards for structuring building information, is that it provides a media for integrating PBS (Product Breakdown Structure) with WBS (Work Breakdown Structure). This is an important aspect for delivering a meaningful 4D model.

The development of the core database was based on the relational database concept and implemented using MS Access 97. Integrated interfaces among MS Access Database, AutoCAD Drawings and MS Project Schedules were developed and implemented. Furthermore, the British Standards of layering convention (BS 1192-5) was adapted and implemented. The database was populated with a detailed 3D model (whole building and M&E), schedules of work and resources of the School of Health Project. The project also addresses object definition, structuring the data, and establishing the relationships and dependencies within the data set, the WBS and building objects as well as modelling the building in 3D in order to capture the essential space/time critical attributes of tasks. Practical application of database throughout the construction process has been highlighted and discussed. Figure 4.4 shows the structure of the information flow.

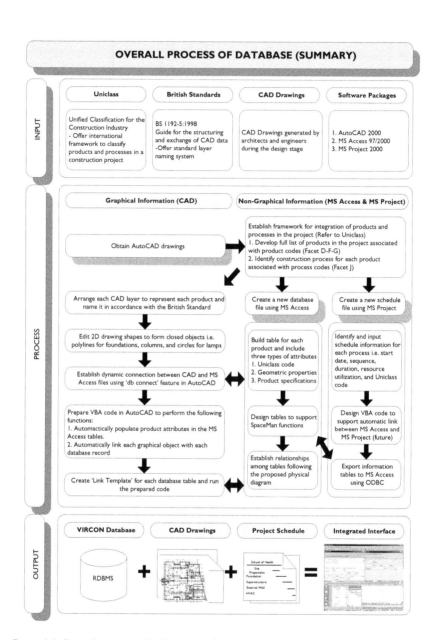

Figure 4.4 Overall process of information flow for 4D development.

The core database

The main objective of this section is to introduce the development of a project database for a construction project. Construction data and information were captured from many sources including the contractors, sub-contractors, Architects, clients and specialised contractors. Interviews and brain-storming sessions were followed to identify missing information and strategies for the project information. Substantial efforts were exerted to restructure and compile the data.

Preliminary structure

Preliminarily, building elements and components were being identified, classified and structured using standard classification methods (Uniclass and ISO STEP). The main output of this task was the class diagram as shown in Figure 4.5. A typical class diagram model is composed of the following:

- Class objects' name and grouping, attributes and methods (development of PBS); and
- Inheritance and aggregated relationships between and among the classes and objects.

Access physical class diagram

Based on the preliminary class diagram, MS Access can be used to deliver the physical diagram at the implementation stage. As shown in Figure 4.6, project and building tables are treated as the centre hub joining a variety of product tables on the left-hand side with the associated processes on the right-hand side. It is vital to notice that embedding the Uniclass code system in the database structure allows representation of classes and inheritances hence overcomes one of the limitations of relational database against object-oriented database.

Uniclass code structure

Uniclass is a relatively new classification scheme for the construction industry. The Construction Industry Project Information Committee (CPIC), representing the four major sponsor organisations (the Construction Confederation, the Royal Institute of British Architects, the Royal Institution of Chartered Surveyors and the Chartered Institution of Building Services Engineers), and the Department of the Environment Construction Sponsorship Directorate were responsible for commissioning and steering the project, which was developed by NBS (National Building Specification) Services on behalf of CPIC.

Uniclass follows the international framework set out in ISO Technical Report 14177 Classification of information in the construction industry July 1994. Uniclass offers a systematic scheme to structure and integrate product literature and project information. It incorporates (a) EPIC (Electronic Product Information

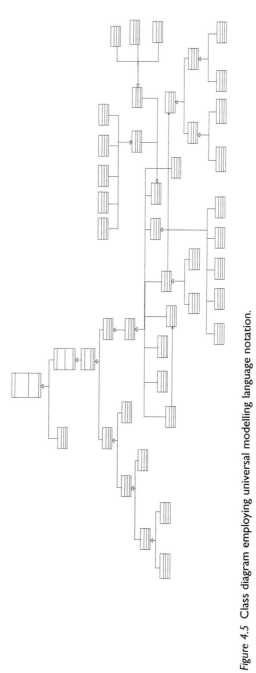

Figure 4.5 Class diagram employing universal modelling language notation.

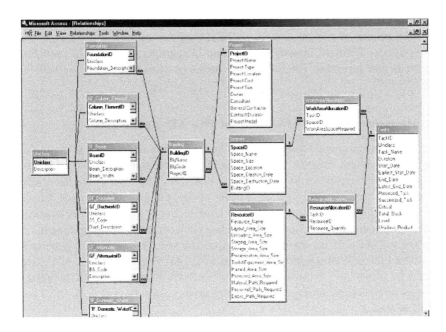

Figure 4.6 Access physical diagram.

Table 4.1 Structure of uniclass facets (CPIC, 1997)

Facets		
A – Form of information	F – Spaces	L – Construction products
B – Subject disciplines	G – Elements for buildings	M – Construction aids
C – Management	H – Elements for civil engineering works	N – Properties and characteristics
D – Facilities	J – Work sections for buildings	P – Materials
E – Construction entities	K – Work sections for civil engineering works	Q – UDC

Co-operation), a new system for structuring product data and product literature and (b) CAWS (Common Arrangement of Work Sections for building works), which became widely used through NBS, SMM7 (Standard Method of Measurement, seventh edition), and NES (National Engineering Specification). It is also intended to supersede CI/SfB, which was last revised in 1976.

Uniclass is structured with a faceted classification system such as CI/SfB rather than a hierarchical classification system such as Masterformat. However, often in classifying items in detail, the hierarchical system is partially used within a facet. The general structure of Uniclass facets is grouped into 15 main subjects, as shown in Table 4.1.

A, B and C facets in Table 4.1 are for general summaries concerning information form or management field. D, E, F, G, H and K facets consist of facilities, spaces, elements and operations for civil and architectural works. L, M, N, P and Q facets are useful to classify information concerning construction products, materials and attributes. The full lists of attributes of each facet are available in CPIC (1997).

Implementation of Uniclass in the 4D database

According to Kang and Paulson (2000) there may be two methods to apply the Uniclass in building projects. One is the partial classification for specific subjects, such as a work section or an element in a WBS, and the other is the integrated information classification through the life cycle of a project. The four facets, including D, F, G and J (from facilities to work sections) can be used for physical objects representing work items in a building project thus facilitating integration of PBS and WBS. If the application is expanded into the information generated from the life cycle of a project, the facets of [L, M, P] and [A, B, C, N, Q] are useful for classifying construction aids or materials. Since the VIRCON project focuses on the planning stage of a building project, only facet D, F, G and J are implemented.

The classification items in each facet can be used with the link to the other facets higher than their levels (see Figure 4.7). For example, the code of [D72111 – G26 – F134:G261:G311 – JG10] means structural steel framing works [JG10] of ground floor column [F134:G261:G311] for superstructure frame [G26] in a school of health building facilities [D72111]. The classification items define work gradually, according to how the items are linked to other facets. Thus, facet J consists of the most detailed operations on which the works are undertaken with resources on construction sites. Finally, D, F, G and J facets can apply to represent the physical objects according to the work being completed through a project. Figure 4.7 shows an example of an integrated PBS and WBS of the School of health project. It is important to note that (1) the codes in the first and second levels, that is [D72111 – G26] represent summary levels, (2) the codes in the third level, that is [F134:G261:G311] – GF column represent product, which are stored inside the Uniclass code library in the database and (3) the codes in the fourth level, that is [JG10] represent process.

AutoCAD database

One of the main objectives of information infrastructure for 4D modelling is to establish a dynamic connection between the database and graphical information in AutoCAD drawings. All related drawing information was obtained from industrial partners including Bond Bryan Architects, SDA Services and Anthony Hunt Associates. This section helps users to populate the graphical components from AutoCAD drawings to the database. The output of this section is also to establish connectivity between AutoCAD and the database.

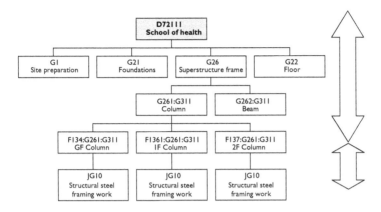

Figure 4.7 An example of integrated PBS and WBS of a building project.

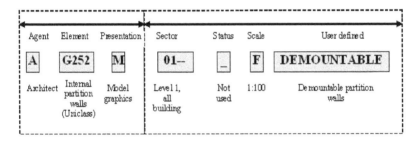

Figure 4.8 Example of layer coding using both mandatory and optional fields.

British standards layering convention

According to Roberts (1998), careful consideration should be made whenever structuring the CAD model in layers. This is a central CAD principle that associates the ability to manipulate the model elements. Moreover, such a principle allows various project members to communicate and exchange CAD files in a more compatible way. Therefore, representation of drawing management through structuring the information in layers in accordance with the current BS 1192-5 was established.

The British Standards committee advises that a project team should agree upon the layering and coding used for CAD files. A format with specific concepts and categories should be used to eliminate the complexity when data are exchanged. It is also possible to use other specific codes only for specific purposes. Coding convention includes a mandatory field and an optional field (see Figure 4.8). The mandatory field is given in an order of alphabetic and/or alphanumeric attributes as explained in BS 1192-5 (see Table 4.2). The optional field is given in the recommended character codes as illustrated in Table 4.2.

Table 4.2 British standard layer code classification (source: BS 1192-5 Guide)

Concept class	Characters	Recommended character codes
Mandatory field		
Agent responsible	1 (alphabetic)	A = architect B = building surveyor C = civil engineer D = drainage, sewage and road engineers E = electrical engineer F = facilities manager G = geographical information system engineer and land surveyor H = heating and ventilating engineer I = interior designer K = client L = landscape architect M = mechanical engineer P = public health engineers Q = quantity surveyor S = structural engineers T = town and country planners W = contractors Y = specialist designers Z = general
Element	4 (alphanumeric)	Use recognised classification system such as Uniclass
Presentation	1 (alphanumeric)	D = dimensioning G = grid H = hatching M = model-related graphics P = page/plot-related graphics T = text - (hyphen) = whole model or drawing definition
Optional field		
Sector	4 (alphanumeric)	-2B2 = level 2, block B, zone 2 --- (four hyphen) = whole project
Status	1 (alphabetic)	N = new work E = existing (to remain) R = remove T = temporary work - (hyphen) = whole project
Scale	1 (alphabetic)	A = 1:1 B = 1:5 C = 1:10 D = 1:20 E = 1:50 F = 1:100 G = 1:200 H = 1:500
User defined	Unlimited (alphanumeric)	According to project specific code. Any number of characters

The detailed process of how to populate the database will be illustrated by the case study in the following section.

Case study – stockport project

Introduction to the selected case

A primary school construction in UK was selected as a case study for the system evaluation in this thesis. The construction contract was based on a guaranteed maximum price (GMP) scheme at £1.2 million. The project involved an approximate 1,600 square metres of one-storey steel-structure building with brickwork facets. The school building was divided into three main zones: the junior wing infant wing and a middle hall that connected the two wings together. Total contract duration including design and construction was 15 months (August 2001– November 2002). It should be noted that the project was in the eighth month (April 2002) at the time of this study. Figure 4.9 illustrates the architectural impression of the project.

Preparation of preliminary data

Converting 2D to 3D CAD model

This first step for preparation of preliminary data involved a process of converting 2D to a 3D CAD model. Primarily, a request of electronic 2D CAD files was issued to the contractor and the designer of the project. Eighteen AutoCAD (.dwg) files

Figure 4.9 The architectural impression of the case study.

covering overall plan, sections, flooring, building services, furniture and land-scape were supplied.

By considering the organisations of 2D CAD layers, it was clearly seen that the layers were not systematically structured. Layer names did not follow any stan-dard layering conventions. It was therefore very difficult to recognise the contents of a particular layer. Furthermore, many layers that contain no object were unnec-essarily created whilst some created layers contain more than one type of object.

To improve the structure of 2D CAD layers, Uniclass code (Crawford *et al.*, 1997) was assigned to each type of object so that each object will be classifiable as a product. In addition, all the CAD files were combined into one file. All the layers were re-organised based on the BS 1192-5, guide for structuring and exchange of CAD data (BS1192-5, 1998), and unnecessary layers were deleted. Table 4.3 illustrates a sample list of the re-organised 2D CAD layers.

Once the 2D CAD layers were properly organised, the next process was to con-vert the 2D files into a 3D CAD model. 2D CAD was basically extruded up from a specified datum. This process of extrusion results in what is sometimes known as 2.5D (McCarthy, 1999), to distinguish it from a true 3D wire frame or solid model. Such models have the considerable advantage that they require much less work in developing the original CAD model, and are appropriate where a 3D model is not justified for other reasons such as analysis of product performance and clashes between products. However, some products including roof trusses and roof covering demand a true 3D modelling for a purpose of proximate reality in the visualisation. Figure 4.10 presents the converted 3D CAD model. The model contained 8,788 objects, which were organised into 45 layers. Total man-hours of 117 was spent for restructuring CAD layers and 3D modelling.

Table 4.3 A sample list of the re-organised 2D CAD layers

New layer name	Content
A-F134_G261_G311-M-010G-_-F-COLUMN	Steel columns
A-F134_G262_G311-M-010G-_-F-BEAM	Steel beams
A-G221_G311-M-0100-_-F-FLOOR	Pre-cast concrete floors
A-F134_G2511_G311-M-010G-_-F-OUTERLEAF	Outer-leaf external walls
A-F134_G2512_G311-M-010G-_-F-INNERLEAF	Inner-leaf external walls
A-F134_G2522_G311-M-010G-_-F-PARTITION	Internal partitioning walls
A-F134_G251_G321-M-010G-_-F-WINDOW	Windows
A-F134_G251_G322-M-010G-_-F-DOOR	Doors
A-G24_G3111-M-010R-_-F-ROOF_TRUSS	Roof trusses
A-G24_G312-M-010R-_-F-ROOF_COVER	Roof cladding
A-G43-M-010G-_-F-KITCHEN	Kitchen furniture
E-F134_G53_G633-M-010G-_-F-CABLE_TRAY	Cable trays
E-F134_G53_G637-M-010G-_-F-DIST_PANEL	Distribution panels
E-F134_G5412-M-010G-_-F-CEILING_LIGHT	Ceiling lights
M-F134_G52_G635-M-010G-_-F-EXTRACT_FAN	Extract fans
M-F134_G52_G66-M-010G-_-F-CONTROLLER	Controllers

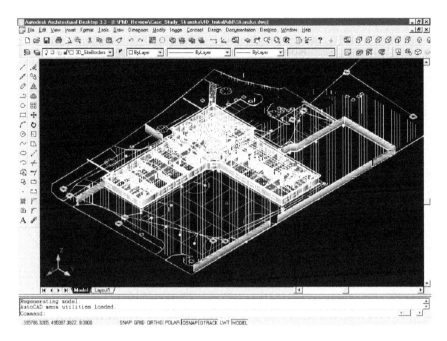

Figure 4.10 The converted 3D CAD model.

Gathering original project schedule

The second step for preparation of preliminary data involved a process of gathering original project schedule. The contractor originally prepared CPM schedule in Power Project software. Since the prototype was developed within the MS Project environment, the schedule data was gathered and manually re-input into the MS Project. All schedule parameters such as activity list, duration, dependency, project start date and calendar were maintained. There were 94 activities in total, 15 of which were pre-construction activities and 79 of which were construction activities. Project start and finish dates were 27 August 2001 and 1 September 2002 respectively. Figure 4.11 illustrated the original schedule in MS Project format.

Gathering original resource allocations

The third step for preparation of preliminary data involved a process of gathering original resource allocations. In the first place, the resource allocations data was not prepared and included in the original schedule. The project planner was asked to identify the resource allocations data for the purpose of this research. Fifteen types of labour resources – general labour, mechanician, electrician, demolition worker, ground worker, steel erector, scaffold erector, cladding worker, bricklayer,

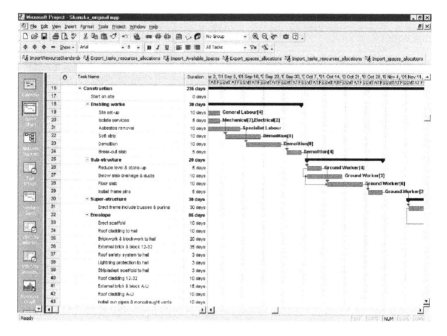

Figure 4.11 Original schedules in MS Project format.

specialist labour, window and door installer, partition and ceiling worker, joiner, decorator and flooring worker – were classified and allocated to each activity in the original schedule. Unit of the allocation was in man/days.

Generating initial 4D model

Developing product-based work breakdown structure

To enable generation of the initial multi-constraint model, product-based work breakdown structure must first be developed. Product-based work breakdown structure is a medium that allows systematic and consistent integration of product and process data. In this case study, it was clearly seen that the breakdown of activities in the original schedule didn't suit this requirement. Many products were not explicitly planned to be constructed. For instance, installation of extract fan, cable tray and distribution panel are hidden in first, second, and final fix M&E activities. This made it difficult to interpret the association between products and processes. Table 4.4 shows a complete list of products that were not explicitly planned to be constructed found in this case study.

In this step, a product-based work breakdown structure was designed by considering the product layers in the 3D CAD model and activity list in the

Table 4.4 List of products that were not explicitly planned to be constructed

No.	Product name	Remarks
1	Temporary perimeter fencing	Products no. 1 and 2 were hidden
2	Site office and accommodation	in site set up activity
3	Column	Products no. 3–5 were hidden in
4	Beam	erect frame activity
5	Trusses and purlins	
6	External doors	Products no. 6 and 7 were hidden
7	Internal doors	in install doors activity
8	External wall outer leaf	Products no. 8 and 9 were hidden
9	External wall inner leaf	in external brick and block
10	Extract fan	activity
11	Controller	Products no. 10–21 were hidden
12	Cable tray	in first, second and final fix M & E
13	Compartment DADO trunk	activities
14	Distribution panel	
15	External lighting	
16	External emergency lighting	
17	External wall lighting	
18	Internal lighting	
19	Internal emergency lighting	
20	Internal wall lighting	
21	Internal ceiling lighting	
22	Drainage manhole	
23	Drainage pipe	Products no. 22 and 23 were hidden in external drainage activity

original schedule. Figure 4.12 illustrates an example of product-based work breakdown structure created in MS Project.

The structure was broken down into four levels as follows: (1) Level 1 represented project title (2) Level 2 represented summary tasks (3) Level 3 represented products and (4) Level 4 represented activities. The complete structure consists of 171 items in total. The number of summary tasks, products and activities were 14, 45 and 111 respectively. It should be noted that a Uniclass code was also assigned to each of the items in the structure.

Grouping CAD products

As a step towards the generation of initial multi-constraint model, this process involved grouping CAD products in relation to the developed product-based work breakdown structure. The detail of product grouping depends on the detail of activities broken down in the schedule (Level 4). For example, three groups of column namely COLUMN1, COLUMN2, and COLUMN3 were generated to represent erection of infant wing column (gridline 12–32), junior wing column

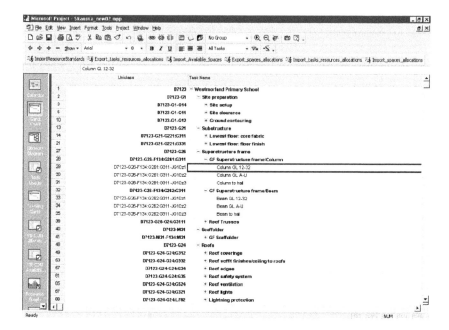

Figure 4.12 A snapshot of product-based work breakdown structure in MS Project.

(gridline A–U) and hall column respectively. To group the product, only one required layer should be turned on so as to avoid difficulty in selecting the members of the product group. By using the 'group' command in AutoCAD, it is relatively simple to create a new product group and select required members of that particular group. Figure 4.13 shows an example of product grouping in the AutoCAD environment.

However, it is normal that some types of activities such as demolition, groundwork, decoration, installation of small fixtures, cleaning and testing and commissioning are not easily represented by CAD objects. Based on the case study, Table 4.5 shows the list of activities that had no associated CAD objects. It is considered that, if necessary, represented symbols may be created to denote locations of these types of activities.

Linking product groups with activities

After all product groups are generated, the next step is to link these CAD groups with activities in MS Project as shown in Figure 4.14.

Figure 4.14 shows an example of linking a group of infant wing trusses (TRUSS1) to an activity of roof trusses GL 12–32 in MS Project. To enable this linking process, a new text field called 'Product Group' was created in MS Project.

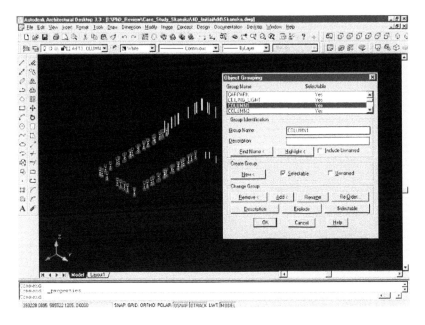

Figure 4.13 An example of grouping of infant wing columns in AutoCAD.

Table 4.5 List of activities that had no associated CAD objects

No.	Activity name	No.	Activity name
1	Site office and accommodation	17	First fix joinery
2	Services isolation	18	Second fix joinery
3	Asbestos removal	19	Tape and joint
4	Demolition	20	Fitted furniture
5	Soft strip	21	Mist coat decorations
6	Breaking out slab	22	Decorations
7	Frame pins installation	23	Underfloor heating
8	Scaffolder	24	Sub-base
9	Roof safety system	25	Paving
10	Install sun pipes and monodraught	26	Site clearance
11	Vents	27	Test and commission
12	Glazed screens	28	Inspection and defect
13	Lightning protection	29	Final clean
14	Second fix partitioning	30	Training
15	Internal doors	31	Handover
16	Suspended ceiling		
	Ceiling tiles		

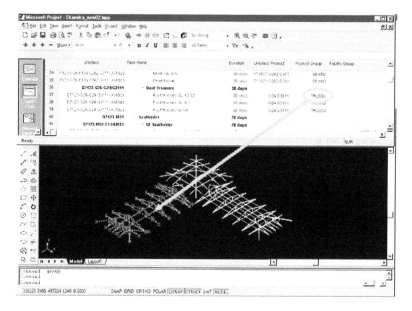

Figure 4.14 An example of linking a CAD group to an activity.

Considering the associated activity, the name of product group was then input into the Product Group field accordingly.

Modifying schedule parameters

Since the level of detail of project schedule was changed in relation to the new developed product-based work breakdown structure, the preliminary data input such as activity duration and resource allocation must be accordingly modified. However, to maintain the sequences of work, peak demand of labour utilisation and total labour cost as inherited in the original schedule, the following rules were applied:

- Same duration and same dependency as appeared in the original summary activity were assigned to all of its new sub-activities so as to preserve the sequences of work in the original schedule.
- The units of resources allocated in the original summary activity were distributed among all of its new sub-activities so as to maintain peak demand of labour utilisation and total labour cost in the original schedule.

An example of modification of schedule parameters based on the above rules is demonstrated in Table 4.6.

Table 4.6 An example of modification of schedule parameters

Activity	Duration	Predecessor	Resource	Cost
Original schedule				
External drainage	35 days	Doors installation	Ground worker [3]	105 units
Modified schedule				
Site drainage				
– Drainage manhole	35 days	Doors installation	Ground worker [1.5]	52.5 units
– Drainage pipe	35 days	Doors installation	Ground worker [1.5]	52.5 units

Table 4.7 Availability and cost of labour resource

No.	Resource name	Availability (max units/day)	Cost (£/hr)
1	General labour	8	8.00
2	Mechanician	6	15.00
3	Electrician	6	15.00
4	Demolition worker	10	10.00
5	Ground worker	10	10.00
6	Steel erector	8	15.00
7	Scaffold erector	4	10.00
8	Cladding worker	8	15.00
9	Bricklayer	14	15.00
10	Specialist labour	6	18.00
11	Window and door installer	6	12.00
12	Partition and ceiling worker	8	12.00
13	Joiner	6	12.00
14	Decorator	6	15.00
15	Flooring worker	4	12.00

Identifying availability and cost of labour resource

Availability and cost of labour resource are two important constraints to be modelled in the initial multi-constraint model. For each type of resource, maximum available units per day were identified as a constraint for resource allocation and levelling problem. Hourly labour costs were also identified. The sum of labour cost based on eight working hours per day will be calculated as the direct cost for time-cost trade-off problem. Table 4.7 presents the identified availability and cost of labour resource to be used in this case study. It is initially assumed that the availability and cost of labour resource are constant throughout the construction period. However, if this is not the case, these data can be easily modified in MS Project. It is important to note that dynamic constraints such as deliveries of material will be incorporated into the model later during the construction stage.

Allocating spaces to activities

For the purpose of space congestion analysis, it was considered adequate to initially divide the case study project into three zones: (1) infant wing – 677 m², (2) junior wing – 643 m² and (3) hall – 424 m². First, a layer called 'Space' was additionally created in the 3D CAD file and three simple rectangles were drawn to represent each space. The areas of these spaces were measured by the CAD property feature. The name and size of each space was then input into the resource table in MS Project. Figure 4.15 presents the classification of spaces as resources in MS Project. Finally, the spaces were allocated to each activity using standard resource allocation feature in MS Project. At this early stage, space requirement of each activity was estimated as this information was not originally available. To accommodate the complexity of space analysis during the construction stage, it should be noted that project planners can either perform detail analysis by classifying more working zones or simply adjust the space requirements to reflex space sharing ability among activities.

Populating 4D database

Once all the data had been prepared, the last step of generating initial multi-constraint model is populating LEWIS database. The data required to be populated in the database include schedule data and resource and space utilisations data.

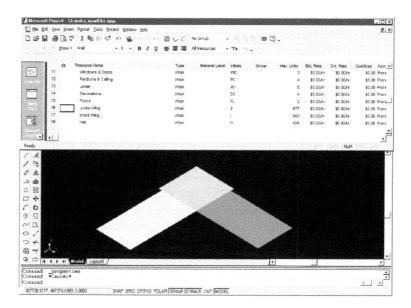

Figure 4.15 Classification of spaces in MS Project.

The population of product data is optional in this case because the real 3D model was not generated in the case study, and hence there was no point in populating the detailed geometrical data in the database. If the model is generated in the real 3D format, the population of product data will enable automated preparation of bill of quantities, purchase orders and progress payments. To populate the schedule data from MS Project into SQL server, an ODBC export feature in MS Project was used and a map between MS Project data and LEWIS database fields were created. Regarding the population of resource and space utilisations data, a VBA macro was written in MS Project to automate this process.

Evaluating the original plan

Examining overall activity sequences

The first step of evaluating the original plan is to examine overall activity sequences with the aid of multi-constraint visualisation prototype system. For this step, ordinary 4D visualisation feature that is the simulation of construction products through time was used. A snapshot of 4D model in week 12 of the original schedule is shown in Figure 4.16.

All the snapshots of potential problems like in Figure 4.17 were pre-prepared and presented to evaluators for confirmation. 64 out of 100 activities (not included activities at summary levels) were confirmed.

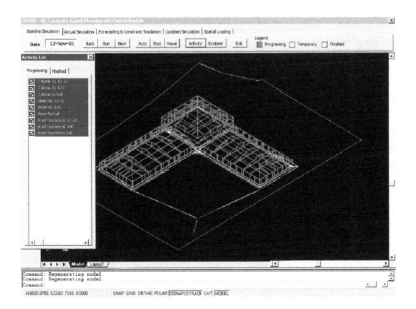

Figure 4.16 A snapshot of 4D model in week 12 of original schedule.

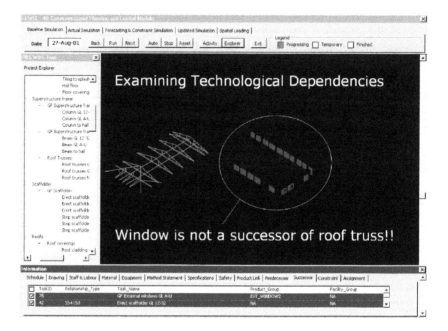

Figure 4.17 An example of illogical link between roof trusses and windows.

Examining site space congestion

In addition to the examination of overall activity sequences and detailed technological dependencies, this step involves an examination of site space congestion
at each period. Space congestion occurs when work crews of different trades
working on concurrent activities have to share a common workspace and therefore interfere with each other. This can decrease their productivity as well as
prevent the execution of one or more affected activities.

Based on the space allocation data identified and populated space utilisation
data shown in Figure 4.14, percentages of space criticality of each zone and in
each period were automatically calculated by the 'SpaceLoading' query in the
LEWIS database (Table 4.8).

Three levels of space criticality, (1) underloading (green shaded) – per cent
space criticality <100, (2) critical (magenta shaded) – per cent space criticality =
100 and (3) overloading (red shaded) – per cent space criticality >100, were then
visualised in the multi-constraint visualisation module. From the case study, a
large number of space congestion circumstances were anticipated. Figure 4.18
shows an example of space congestion between roof cladding and brick/block
external outer-leaf walling activities in the hall zone during 7 January 2002 to 13
January 2002.

Table 4.8 An example of calculation of per cent space criticality

Utilisation ID	Date	Space name	Space size (m²)	Space occupation (m²)	Per cent space criticality
5354	07/01/2002	Junior wing	677	600	88.6262924667652
5702	07/01/2002	Infant wing	643	0	0
6050	07/01/2002	Hall	424	824	194.339622641509

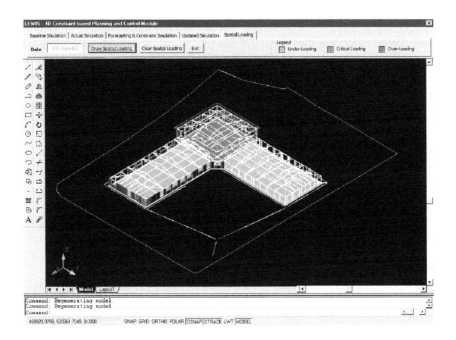

Figure 4.18 An example of visualisation of site space congestion.

Examining project duration, cost and resource utilisations

The last step of evaluating the original plan involves examining whether or not the project duration meets the contractual requirement, whether cost is low enough to yield satisfied profit margin to the contractor and whether utilisation of resource is under-loaded and levelled. For this step, resource availability and cost data provided in Table 4.7 were input into the MS Project original schedule. Given £750,000 material cost and £500/day indirect cost, Table 4.9 summarises project duration, cost and resource utilisation before and after executing resource levelling feature in MS Project.

Table 4.9 Project duration, cost and resource utilisation of the original schedule

Resource levelling	Duration (day)	Labour cost (£)	Material cost (£)	Indirect cost (£500/day)	Total cost (£)	Resource utilisation status
Before	235	253,936	750,000	117,500	1,121,436	Over-loaded and fluctuated
After	425	253,936	750,000	212,500	1,216,436	Under-loaded and leveled

According to Table 4.9, it is obvious that in order to satisfy resource utilisation status, project duration and cost will not be in the acceptable range. To be more specific, the project duration of 425 days will push the project finish date to 2 May 2003 as opposed to the contract finish date on 30 November 2003. In term of cost, total project cost of £1,216,436 (not included penalty cost of delays) will exceeded £1.2 million as agreed in the guaranteed maximum price contract.

Summary

The integrated decision support system for multi-constraint planning and control developed was evaluated using real case study data. The evaluation were categorised into six main phases, which are as follows: (1) preparation of preliminary data input, (2) generating initial multi-constraint model, (3) evaluating the original plan, (4) re-establishing baseline plan, (5) monitoring and forecasting plan and (6) updating plan.

Three dimensions were considered for the system evaluation in each phase. First, the system was verified as to whether it is functioning correctly as intended. Second, the system was validated through diagnosis of its benefits over other available systems. Third, the system was evaluated in terms of its usability and barriers for implementation.

According to the thorough evaluation demonstrated throughout this chapter, it has been verified that programmatic errors in the system are minimised thus allowing generation of accurate output as intended. In addition, it has been validated that the system can greatly support planners to establish and maintain constraint-free optimum plans throughout the course of construction. To be more specific, three major advantages of the developed system over other currently available systems can be summarised as follows:

- Multi-constraint information management feature enables capturing and incorporating constraint information regarding logistics of information and resources into the main contractor's schedule. Once the constraints are analysed, the system manages the workable backlog and facilitates weekly work planning at workface operation level. Furthermore, the system acts as a central project repository where project participants can query up-to-date information and statuses of the project.

- Multi-constraint visualisation feature enhances planner's ability to detect hidden problems in the CPM schedules. Based on the case study, invited evaluators could detect a large number of problems regarding illogical dependencies, site space congestion and non-readiness of information and resources. Furthermore, the system can be utilised to confirm the constraint-free schedule to project participants.
- Multi-constraint optimisation feature enables generation of optimum constraint-free schedules by changing priority and construction method options of each activity. Based on the case study, the system could reduce the project duration from 294 days to 247 days or approximately 16 per cent, while maintaining technological dependencies and constraint-free condition.

Although the benefits of the system have been appreciably realised, a few issues about system implementation were raised by the invited evaluators. The most important one relates to preparation of preliminary data input. It is widely acknowledged that many data such as 3D CAD model and space allocations are not normally available in typical projects. Therefore, considerable efforts must be spent to generate and organise the data in compliance with certain standards such as Uniclass and BS1192-5. It is argued that the use of 2.5D modelling approach in which the 2D drawings are simply extruded up would require much less efforts.

Another major barrier lies in the process of monitoring information and resource readiness due to the difficulty to make all upstream supportive organisations including designers, suppliers and sub-contractors subscribing to the LEWIS system. Fortunately, substantial research projects considering data standardisation have been initiated and could have potential to minimise these barriers for implementation. So far, two different schools in this area have emerged. One supports a view of implementing a universal data standard such as Industry Foundation Class (IFC) while another one supports a view of having mediator to negotiate discrepancy among heterogeneous data sources. However, since the work on IFC has focussed on product models while limited process extensions have been developed, it seems more likely that firms will maintain the use of legacy applications/ databases for process data. As a result, the importance of a research on integrating the LEWIS system with a mediator system (SEEK) has been realised and funded by the US government.

If all the major barriers can be removed in the future, it is envisaged that successful implementation of the developed system will enable generation of reliable plans and constraint-free execution assignments and, in turn, reduce production risks and improve on-site productivity.

Bibliography

Badiru, A.B. (1992) *Expert Systems Applications in Engineering and Manufacturing*, Prentice Hall, New Jersy.

Bailey, R.W., Allan, R.W. and Raiello, P. (1992) Usability testing vs. heuristic evaluation: a head-to-head comparison, Proceedings of the Human Factors Society 36th Annual Meeting, 409–413.

Barry, D.K. (1996) *The Object Database Handbook: How to Select, Implement, and Use Object-Oriented Databases*, John Wiley & Sons, Inc.

Boloix, G. (1997) System evaluation and quality improvement, *Journal of Systems Software*, 36, 297–311.

Borenstein, D. (1998) Towards a practical method to validate decision support system, *Decision Support Systems*, 23, 227–239.

Brown, C.M. (1988) *Human–Computer Interface Design Guidelines*, Ablex Publishing Corporation, Norwood, New Jersey.

BS 1192-5 (1998) Construction drawing practice: guide for structuring and exchange of CAD data, British Standards Institution, ISBN 0 580 29514 1.

Carroll, C., Marsden, P., Soden, P., Naylor, E., New, J. and Dornan, T. (2002) Involving users in the design and usability evaluation of a clinical decision support system, *Computer Methods and Programs in Biomedicine*, 69, 123–135.

Clarke, A., Soufi, B., Vassie, L. and Tyrer, J. (1995) Field evaluation of a prototype laser safety decision support system, *Interacting with Computers*, 7(4), 361–382.

Construction Industry Project Information Committee (CPIC) (1997) Uniclass (unified classification for the construction industry), in Marshall Crawford, John Cann and Ruth O'Leary (eds), Royal Institute of British Architects Publications, London.

Cox, S. and Hamilton, A. (1995) *Architect's Job Book, Sixth Edition*, RIBA Publications.

Davis, E.W. and Patterson, J.H. (1975) A comparison of heuristic and optimum solutions in resource-constrained project scheduling, *Management Science*, 21(8), 944–955.

Finlay, P. (1989) *Introducing Decision Support Systems*, NCC Blackwell, Oxford.

Gass, S.I. (1983) Decision-aiding models: validation, assessment, and related issues for policy analysis, *Operations Research*, 31(4), 603–631.

Harris, R. (1978) *Resource and Arrow Networking Techniques for Construction*, John Wiley & Sons, New York.

Hegazy, T. (1999) Optimization of resource allocation and levelling using genetic algorithms, *Journal of Construction Engineering and Management*, 125(3), 167–175.

ISO 9241-11 (1998) Ergonomic requirements for office work with visual display terminals (VDTs) – Part 11: Guidance on usability, International Organization for Standardization.

Jagdev, H.S., Browne, J. and Jordan, P. (1995) Verification and validation issues in manufacturing models, *Computers in Industry*, 25, 331–353.

Jeffries, R., Miller, J.R., Wharton, C. and Uyeda, K.M. (1991) User interface evaluation in the real world: a comparison of four techniques, Proceedings of ACM CHI'91, New Orleans, LA, 27 April–2 May 1991, 119–124.

Jorgensen, A.H. (1999) Towards an epistemology of usability evaluation methods, Available from: http://cyberg.curtin.edu.au/members/papers/43.shtml (last visited: July 2003).

Kang, L.S. and Paulson, B.C. (2000) Information classification for civil engineering projects by Uniclass, *Journal of Construction Engineering and Management*, 126(2), 158–167.

Karat, C., Campbell, R. and Fiegel, T. (1992) Comparison of empirical testing and walkthrough methods in user interface evaluation, Proceedings of ACM CHI'92, 397–404.

Law, L.-C. and Havannberg, E.T. (2002) Complementarity and convergence of heuristic evaluation and usability test: a case study of UNIVERSAL brokerage platform, Proceedings of NordiCHI 2002, 71–80.

Lewis, C. (1982) Using the 'thinking-aloud' method in cognitive interface design, IBM Research Report RC 9625, IBM T.J. Watson Center, New York, Yorktown Heights.

McCarthy, T. (1999) *AutoCAD Express NT*, Springer, London.

McFarlane, S. (2000) *AutoCAD Database Connectivity (Autodesk's Programmer Series)*, Thomson Learning, Canada.

Miser, H.J. and Quade, E.S. (1988) Validation, in H.J. Miser and E.S. Quade, (eds), *Handbook of System Analysis, Craft Issues and Procedural Choices*, John Wiley & Sons, UK.

Nielsen, J. (1992) Finding usability problems through heuristic evaluation, Proceedings of CHI'92 Conference on Human Factors in Computing Systems, 373–380.

O'Keefe, R.M., Balci, O. and Smith, E.P. (1987) Validating expert system performance, *IEEE Expert*, 2(4), 81–90.

Prochain Solutions, Inc. (2003) Prochain® project scheduling user's guide, Version 6.1.

Rieman, J., Franzke, M. and Redmiles, D. (1995) Usability evaluation with the cognitive walkthrough, Proceedings of CHI'95, ACM, Colorado, USA, 387–388.

Roberts, K. (1998) *Advanced GNVQ, Construction and the Built Environment*, Addison Wesley Longman, England.

Sears, A. (1997) Heuristic walkthroughs: finding the problems without the noise, *International Journal of Human-Computer Interaction*, 9, 213–234.

Silberschatz, A., Korth, H.F. and Sudarshan, S. (1997) *Database System Concepts, Third Edition*, McGraw-Hill, London.

Talbot, F. and Patterson, J. (1979) Optimal methods for scheduling projects under resource constraints, *Project Management Quarterly*, December, 26–33.

Underwood, J., Alshawi, M.A., Aouad, G., Child, T. and Faraj, I.Z. (2000) 'Enhancing building product libraries to enable the dynamic definition of design element specifications'. *Engineering, Construction and Architectural Management*, Blackwell Science Ltd., 7, 4, 373–388.

Wharton, C., Rieman, J., Lewis, C. and Polson, P.G. (1994) The cognitive walkthrough: a practitioner's guide, in J. Nielsen and R. Mack (eds), *Usability Inspection Methods*, Wiley, New York, 105–140.

Chapter 5

Construction scheduling

A Latin American perspective

Leonardo Rischmoller

Introduction

There is consensus among practitioners and academics that Construction Projects involve a high degree of uncertainty that ideally could be reduced if timely, effective and efficient Construction Planning and Scheduling (P&S) is carried out (Figure 5.1). However, despite the unanimous assertion about the importance of P&S; and the agreement on the benefits that good P&S can bring to a construction project and the problems that a bad or non-existent (i.e. improvising) P&S can produce, the compliance of projected construction costs and schedules are an exception rather than a norm in construction projects, showing that the results of P&S are not as effective as it is expected or even desired. Figure 5.1 illustrates how the ability to influence costs in a construction project is bigger while the involvement in planning and scheduling is earlier.

Research carried out by the Production Management Center (GEPUC) in Chile shows that on average, and despite some exceptions, the percentage of Planned Activities Completed (PPC) in Chilean construction projects barely reach 60 per cent (Figure 5.2) and only a few companies, where new management approaches have been applied in the last 5 years (e.g. Last Planner System), have been able to increase plan reliability, remaining however in average below the 70 per cent PPC level. Similar PPC data has been found in countries outside Chile (e.g. Ballard, 2000).

Considerable research effort has been invested during the last few decades in developing or adapting methodologies and theoretical approaches to better carry out construction planning and scheduling (e.g. CPM, PERT, etc. (see Figure 5.3)). The reality has shown that these very important theoretical approaches have not been completely successful when applied to real projects, becoming partially impractical and sometimes used more as a good theoretical framework (a reference) than a solid guide to project execution and later control. Fifty to seventy per cent of compliance with project plans and schedules are used to be considered good results and a 100 per cent compliance is considered an utopia. The special characteristics of the construction industry and other arguments are posed as logic explanations to this results sustaining that nothing can be done for further improvement.

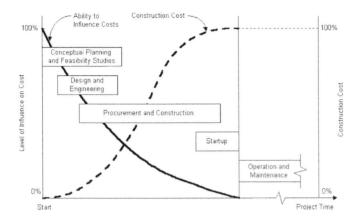

Figure 5.1 Ability to influence cost in construction projects.

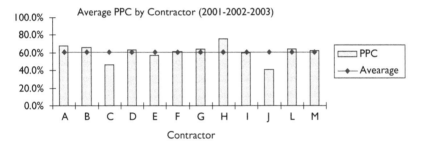

Figure 5.2 Percentage of Planned Activities Completed (PPC) by contractor (Source GEPUC 2003).

As such, more user-friendly and more powerful software tools have been developed that try to overcome the theory – practice dilemma described in the pervious paragraph, promising the ability to, for example, create rapidly and easily, CPM bar charts or PERT diagrams including resources and cost information and providing mechanisms to carry out an optimum construction project control leading to timely decision making and optimum project performance. These tools are mainly developed to automate existing and sometimes new (e.g. last planner system) theoretical approaches. This stance that the tools just merely aid the automation process, thus underestimates the importance of the tools and subordinates its existence to the goodness of the theoretical approaches. At the end a new dilemma, tools-practice, arises resulting in most construction project executors still relying on spreadsheet and paper-based solutions to perform construction P&S rather than taking advantage of available tools specially designed to support construction P&S.

Construction P&S can be then considered as a process that has traditionally been approached within theory–practice and technology–practice dilemmas which has led the industry to conform with PPC data below 70 per cent as very good results. In this scenario, academic groups, software/systems vendors and construction executors seem to be running in parallel rather than converging through a common goal of improving construction P&S. In this chapter, the theoretical background and some practical experience that has emerged from research on how to improve construction P&S using Information Technology is presented. The need and opportunity of a new approach to construction scheduling within a multi-dimensional (nD) context due to new available and coming technology will be discussed.

The traditional construction scheduling process

Gantt bar charts depicting a list of interrelated activities or tasks each with a certain duration calculated using rates or resource assignment are central to traditional construction scheduling processes. The theory behind Gantt bar charts and similar approaches has led us to focus first on the efficiencies of individual activities, using historical productivity rates to support time estimates, floats, resources allocation, costs and so on. The focus is then changed to the relationships between activities in a cumbersome process which, early or late, escapes from our mind capacity to review, understand and use the Gannt bar chart to control and re-schedule our project if necessary during construction execution. This neither induces parts to do what is good for the system as a whole nor direct managers to the point that needs their attention, steering the project leader away and sometimes leading even to failure of projects. Other techniques and approaches to carry out construction scheduling (e.g. line of balance, last planner system, etc.) conceived also under solid theoretical frameworks suffer from the same lack of 100 per cent practical applicability and the emerging levels of uncertainty, complexity and variability inevitably lead to a move away from proactive to reactive behaviours increasing as you come near the jobsite and workface.

Uncertainty at the jobsite is reflected in the form of urgent requirements, non-consistent constructive sequences dependant on immediate available resources (Alarcón and Ashley, 1999), lack of coordination in the supply chain, project scope changes, quality failures and so on. The combined effect between uncertainty and complexity in a project produces variability (Horman, 2000). High levels of variability and uncertainty lead to inconsistent estimations and assumptions, and thus general project performance deterioration. Current practices make use of material inventories, time and cost contingencies, excess of labour and equipment capacity and so on, for example, to deal with variability and uncertainty in intuitive and informal ways (González et al., 2004).

A lot of safety (i.e. floats, contingencies, etc.) is added into construction schedules in order to protect the project from the uncertainty and also from the variability associated with the traditional construction scheduling process. Safety

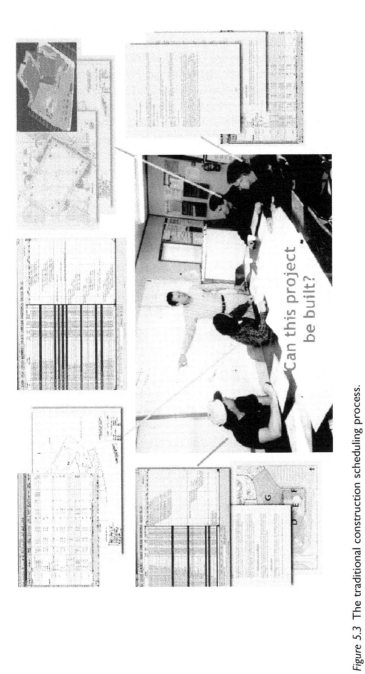

Figure 5.3 The traditional construction scheduling process.

is added to the project as a whole, and safety is also added to each and every step of the project. A limited understanding of safety is embedded in the planning of a project and safety is usually not included rationally in the scheduling process. For estimating the duration of every activity in the schedule, 'realistic estimates' are given according to their worst past experience. Time estimates become then a self-fulfilling prophesy and once they are established, with all their safety, there is little positive incentive, if any, to finish ahead of time, but there are plenty of explanations required when the project is delivered late. There is no point asking how much safety is included in the project because people believe that they give realistic estimations. The problem is in what they call realistic. The higher the uncertainty, the higher the resulting safety. Whenever a step in a project is a collection of several tasks, each done by a different person, the boss of the project asks each person for their own estimates, adds them up and then adds his own safety factor on top. Safety is inserted into the time estimates of almost every step of a project using time estimates based on a pessimistic experience. The larger the number of management levels involved, the higher the total estimation, because each level adds its own safety factor. The estimators also protect their estimations from global cut. When you add it all up, safety must be the majority of the estimated time for a project.

Unlike in production where work centres are protected with inventory, in construction projects, scheduled activities are protected with safety time. In production if there is a stoppage, inventory does not disappear. In projects, time is gone forever.

The core process

A 'core process' means a chain of tasks, usually involving various departments or functions, that deliver value (products, services, support, information) to external customers. Alongside the core processes, each organization has a number of 'support' or 'enabling' processes that provide vital resources or inputs to the value-adding activities. While the idea of a core process may seem pretty straightforward – and it is – it is interesting that this key organizational 'building block' is a relatively recent idea, one of the breakthrough concepts of the Six Sigma system (Pande *et al.*, 2000).

Independent of the tools, techniques and methods to capture construction knowledge and transform it into a construction schedule, a number of questions remain associated to what the construction scheduling core process should be:

• What will at the end need to be constructed?
• How much of this or that will be constructed?
• When will it be constructed?

The mental and practical construction scheduling process is guided by the answers to these questions. But independently of the project type, the answers to these questions seem to be tied to a common approach focused on calculating activities durations, resources, costs and depicting the relationship between these activities through graphical representations (e.g. networks, Gannt charts, etc.) developed by specialized personnel. Currently, construction scheduling is usually a task that involves only specialized estimators and schedulers concentrating on everything related to the construction scheduling process rather than various departments or functions and support or enabling processes working together to deliver value. The resulting schedule is then passed to the contractor who must convert the schedule into practical tasks concentrating on everything, which sometimes is synonymous with not concentrating at all. Then if project leaders use early starts, they will lose focus. If they use late starts, focusing is not possible at all. There are no mechanisms, rules and tools that will enable project leaders to focus.

Existing techniques and supporting tools

Existing scheduling methods and techniques used by the construction industry are based undoubtedly on sound theoretical knowledge which in some cases come from and are also used very successfully in other industries (e.g. aeronautics, manufacturing, etc.). Paper, pencil and electronic calculators have given up their starring roles as supporting tools for these scheduling methods and techniques to computer-based supporting tools, that is, software. For example, specialized software to develop, 'easily' and 'fast', CPM bar charts, PERT networks and Pure Logic Diagrams are commercially available and used with some success in several industries but in construction. These software tools are developed following the scheduling method or technique that they try to automate. There are also some scheduling techniques, that is, Line of Balance, for which software tools are not as widespread but there are several specific efforts around the world trying to automate these construction methods and techniques.

Existing scheduling methods and techniques used by construction are an important part of every project and it would be unfair to sentence them as poor. Most of the built environment that we see around us has been constructed using these methods. However, since at the end a constructed project reflects that the contractor has been effective in achieving a construction project's 'main' goal, this 'success' in achieving our goals, that is, a constructed project, has led us to conform with the low levels of efficiency of the scheduling process accepting them as 'normal' and believing and even feeling that it is an utopia pretending to improve to 100 per cent PPC levels.

The need for change

The construction scheduling process in its traditional form to a great extent is the result of the traditional tools and techniques used for their preparation combined with the related working process. This traditional approach has to some extent led to focus in the efficient preparation of complex bar charts and other forms of representation but not necessarily at its practical use. Dominant ideas associated to existing tools and work processes have been with us for decades providing us with a way of looking at construction projects. Whatever way one tries to look at how construction projects are executed today, is likely to be dominated by the ever present but undefined dominant ideas associated to traditional tools, techniques and work processes that have accompanied us for decades. Unless one can pick out the dominant ideas, one is going to be dominated by them. If one cannot pick out the dominant ideas, then any alternatives one generates are likely to be imprisoned within that vague general idea (de Bono, 1970).

There is nothing wrong in the logic behind traditional construction planning and scheduling. However, there are new efforts tending to build around themselves meanings, contexts and lines of development that represent conscious efforts to pick out the dominant ideas in the construction industry projects processes development, acting as active changing agents trying to push the construction industry to adopt new available technology under a proactive approach instead of waiting to be taken by the storm of new technology (Sawyer, 2004). New available technologies and production principles bring an opportunity to re-think the construction scheduling process and generate alternatives to the 'traditional' approaches. These alternatives consider the multi-dimensional (nD) context of construction projects, moving away from vague general ideas and existing tools and work processes to more explicit and realistic ways to carry out construction planning and scheduling.

Construction scheduling as it is carried out today has not provided a truly proper control mechanism that can help keep project managers focused. And if project leaders are not focused or don't maintain focus, the probability that emergencies will turn the project into a fiasco is high. Everybody knows what a control mechanism is: it measures the progress of the project. The problem is that by the time the progress report indicates something is wrong, it is usually too late. A project report might tell you that 90 per cent of the project is finished in one year and then, the remaining 10 per cent takes another full year (Goldratt, 1997).

The 'traditional solution' for this problem has been the addition of safety to our assumptions (e.g. contingencies, floats, buffers, etc.) in schedules. The safety introduced in schedules (i.e. cost and time) are not the result of bad assumptions but they are the result of our lack of ability to consider the multiple dimensions involved in the construction execution. For example, constructability issues or space–time constraints are difficult to represent into a bar chart. Once safety is introduced into a construction schedule activity, there is no rush to start it until the last minute so that the activity is executed in the assigned period of time and using the assigned resources, including all the safety, which in this way more than used becomes wasted. The dependencies (e.g. predecessors and successors activities) between activities sometimes cause delays to accumulate and advances to be wasted, and again all the safety we put into the schedule vanishes. We need a much better way to manage our projects in which we could visualize what is beyond a bar chart and keep project mangers focused.

An nD construction scheduling process

New information technology tools provide the opportunity to undertake the construction scheduling process in a completely different way with regard to how it is accomplished today. Commercially available visualization tools not only allow a greater understanding of the geometry of the project by looking at time in more realistic 3D Computer Aided Design (CAD) models. When the construction of a 3D CAD model is carried out or guided by those who will be in charge of the construction of the project in real life, the 3D model can be divided in pieces that later on can be associated to construction activities in what is known as 4D modelling. The use of 4D models in an iterative process where at some point construction planners decide on whether a 3D CAD model represents a constructible design or the design need to be refined and/or whether a 4D model represents a construction schedule that is executable or it needs to be refined, has been referred by Rischmoller *et al.* (2000a) as the 4D-PS (4D Planning and Scheduling) work process. The combination of technology, methods and procedures by which to conduct 4D planning pioneered in the first 4D application on a live project in Chile (Rischmoller *et al.*, 2000b) rendered significant and demonstrable savings in man-hours and materials, giving solid credence to the assertion that 4D technology is viable and cost effective. Research efforts around the world has arrived on the same conclusion and 4D modelling is now in the process of crossing over the boundary between academic research and practical implementation by the construction industry (Heesom and Mahdjoubi, 2004).

4D considers the time as the fourth dimension which when added or linked to a 3D CAD model or a 2D digital representation of the project activities (Rischmoller and Valle, 2005) leads to the formation of a 4D model that can be shown in a display device where the spectator can see four dimensions simultaneously. Other dimensions, like for example the cost, the materials quantities take off coming from the 3D model pieces, the temporal–spatial relationships and all the design information useful from the construction planning and scheduling perspective are to some extent implicitly considered when using 4D modelling. However, these and other dimensions could be considered not only implicitly but also explicitly into an nD model.

An nD model is an extension of the building information model by incorporating all the design information required at each stage of the lifecycle of a building facility (Lee *et al.*, 2003). Thus, a building information model (BIM) is a computer model database of building design information, which may also contain information about the building's construction, management, operations and maintenance (Graphisoft, 2003). From this database, different views of the information can be generated automatically, views that correspond to traditional design documents such as plans, sections, elevations and schedules. As the documents are derived from the same database, they are all coordinated and accurate – any design changes made in the model will automatically be reflected in the resulting drawings, ensuring a complete and consistent set of documentation (Graphisoft, 2003).

The nD model concept was developed as one of the main results of the funded 3D to nD modelling project at the University of Salford. The core ideas behind this concept can be traced back to maybe 10–20 years ago (Bertelsen, 1992). However, the advances in research on Information Technology in construction, the availability of new tools and the maturity of the industry to use these tools, have led to conclude that the nD concept could start the process of crossing over the boundary between academic research and practical implementation by the construction industry sooner than expected. Virtual Design and Construction at Stanford University, Scenario Based Project Planning by FIATECH and Building Information Modelling (BIM) by Autodesk are samples of some world class research approaches, strategies and organizations dealing, to some extent, with the same ideas behind the nD concept. In Chile, Rischmoller *et al.* (2000), proposed the Computer Advanced Visualization Tools (CAVT) concept as a new concept that should lead to face options and opportunities offered by IT with a revitalized creativity and in a constructive fashion, when dealing, not only with visualization issues, but mainly with engineering and construction management.

Computer advanced visualization tools

A 1997 study focused on the full range of IT use revealed that design firms tended to equate information technology with CAD (Fallon, 2000). Engineering managers have clearly become more interested in the topic of integrating CAD technology effectively into their work processes and in the specific features required to make CAD effective for everyone in the firm, not just drafters (Fallon, 2000). CAD is however an acronym that in itself is tying us to the past and limiting our ability to develop creative thinking regarding IT developments and their application to engineering and construction processes. The broad spectrum in which the CAD acronym is used today does not match the original concept behind it, limited to receive aid during the design process. A new broader concept that will lead to new ideas and improvements is needed.

2D CAD has almost completely replaced the drawing board, 3D CAD modelling enjoys a well-developed approach in most AEC/EPC leading companies and 3D Visualization has been identified by CAD vendors as the most obvious advanced capability of CAD products that will provide more share of the existing CAD market (Philips, 1999). 4D CAD is a technology starting to cross over the boundary between research and practice. Common to all these approaches is their ability not only to represent visually, the final products aimed by AEC/EPC industry (i.e. 3D model), but also the processes (i.e. 4D model) needed to be carried out to achieve the final products. This, jointly with the need to represent not only the visual animation information, but also underlying information, led to the CAVT concept development.

CAVT is defined as 'the collection of all the necessary tools, which allow for the visual representation of the ends and the means of AEC/EPC needed to accomplish an AEC/EPC design and construction project.' CAVT defined in such a broad sense provides a definition that could evolve over time, since it is not tied to any particular tool. And even CAVT is mainly related with the visual aspect of project representation, it is not limited to such an approach, which constitutes only its ultimate output. The CAVT definition also considers underlying information about facility components and activities that might lead to a 3D rendering, or a 2D plot, or a bill of materials, or a work order report, or a virtual reality environment, each coming from a unique product and process model representation, which can be visualized through a computer-based display device. CAVT understood as a concept aligned with the nD concept demands rethinking construction product and process modelling processes into new approaches that should be not only supported, but intrinsically linked to CAVT.

Product modelling – the digital reality

According to traditional rationalistic philosophy, the difference between 'reality' and our understanding of that reality is not an issue, because it claims that there exists a rather simple mapping between the two. Our ability to intelligently act in the world around us is due to the mental images or representations of the real world that we have in our minds. In fact, studies have shown that we receive approximately 80 per cent of our external information in visual form (Intergraph, 1998).

The level of detail and realistic views that 3D product models can achieve by using CAVT in the construction industry, and that will be achieved in the near future, makes it sometimes difficult to differentiate them from reality. Furthermore, computers allow going beyond the visual aspect of the objects represented and attach to them all kind of attributes and links. In this way, physical properties (i.e. material, mass, density, inertial momentum, etc.) and other characteristics of the 3D product models become an inherent part of them. Environment lights coming from different sources and directions and the response of the objects to them (i.e. shadows, reflections) can complement 3D models representations to make them much more realistic.

Product models are not then just geometric representations of an intended project now, but they indeed 'exist' digitally into a computer, which lead us to name them Digital Realities (DR).

Naming of a project product model as Digital Reality has more than semantic implications. The process of design varies from trying to replicate the future by representing it with the use of computers, to an iterative transformation of a Digital Reality in a new process of refining it. This new approach is developed concurrently in a common, collaborating and multi-disciplinary digital dimension, pursuing an optimum and constructible design. The Digital Reality is in this way 'dynamic', unlike a 3D product model, which is 'static'. Furthermore, future easier and faster CAVT should lead to facilitate the construction of the Digital Reality within the computer, and even outside it with devices like, for example, the workbench response table at Stanford University (Koo and Fischer, 2000).

Future widespread use of CAVT make us envision the result of the design stage as not only geometric information in 3D models, but also complete construction planning and scheduling visualization models (i.e. represented in complete nD models) which may include scope and cost beside time (Staub *et al.*, 1999) and where construction activities, tasks durations, resources and costs could be linked and/or obtained from a project product model (i.e. 3D model and databases). We also envision construction tasks being completely transformed when narrowing the degree of uncertainty existing currently at the job site. CAVT will transform the way we execute the work at the jobsite (i.e. reporting in real time though CAVT and associated hardware mobile devices) so as to have no resemblance to anything we know today in the construction industry. So Digital Reality spoken in two words goes beyond the traditional product modelling definition as oppose to Virtual Reality. If prizes were awarded for best oxymoron, virtual reality would certainly be a winner (Negroponte, 1995). Virtual as opposite to Reality states a big contradiction of both words together. 'Walkthroughs' into a 3D CAD model produces a sense of 'being there', even without using electronic glasses and gloves, typical common devices of virtual reality technology. Digital Reality could represent to the construction industry, the foundation over which completely new paradigms for the design and construction processes can emerge, transforming the way AEC/EPC projects are developed even today with the 'widespread' use of Information Technology.

Process modelling

Designers develop digital realities, and contractors need to construct these digital realities. This can also be done digitally before going to the job site. The construction industry relies on processes of varying complexity to accomplish every task it is related with. These processes are the means that allow the transformation of abstract information into a physical reality, an important goal of a construction project. Simulations have been used widely to represent construction industry processes. In general, simulation refers to the approximation of a system with an abstract model in order to perform studies that will help to predict the behaviour of the actual system. A previous modelling effort is essential to develop any simulation task. Model development efforts must invariably consider the general modelling technologies upon which new models will be based on (Froese *et al.*, 1996).

Within the computer graphics and visualization context, in the last years 3D modelling has reached a high level of development in the AEC/EPC industry, specially in the Plant Design industry where 3D and shaded models have become an inherent part of most of the design process. And currently available CAVT provide the most advanced technologies to visually model the construction process, by allowing the development of 4D models. However, despite its availability, this advanced CAVT feature has not been widely implemented yet in AEC/EPC projects.

A comprehensive study of theoretical and practical approaches of 4D modelling around the world and through several case studies carried out in Chile, lead us to conclude that this is the technology that will trigger the major changes expected from CAVT in the way to an nD Construction Schedule. 4D models reflect the realities of project execution more closely than the approaches used in practice today (Fischer and Alami, 1999). Available 4D modelling tools can be used to support construction planning and scheduling. Field personnel knowledge introduced into 4D models, can be used to generate project schedules with computer software. Furthermore, when a 4D application is carried out and included into work processes specially designed to take advantage of 4D modelling, it can be proved that using 4D is productive and cost effective (e.g. Rischmoller *et al.*, 2001).

Thanks to computer advances, it is expected that CAVT will continuously reduce the construction process modelling effort. CAVT and 4D modeling technology are commercially available today, and yet we are not taking full advantage of them. In the future it is expected that the CAVT and nD approaches will allow the visualization of the mapping scheme developed to support the relationships between the different hierarchical representations of design, cost, control and schedule information represented at different levels of detail (Staub *et al.*, 1999). In this way, AEC/EPC projects, which are highly complex systems with a high degree of connectivity between objects and attributes, will become every time more explicit with CAVT application not limited to the design stage, but also to the construction phase of AEC/EPC projects. The next benefit should extend the CAVT and nD approaches to the whole life cycle of AEC/EPC projects. This will reduce complexity and uncertainty, allowing for increasingly more realistic models of products and processes within the AEC/EPC industry.

The nD construction schedule

The combination of product and process modelling theoretical backgrounds, coupled with the use of available CAVT, can lead to a new construction schedule development process understanding and design based on Information Technology advances. This new approach to construction scheduling can consider now more dimensions than just the time (e.g. quantities take-off, space-time constraints, etc.) with some effort required to overcome the lack of interoperability of currently available CAVT coming from different software vendors. It is expected that in the future the nD Construction Schedule development process will be the result of a single nD modelling effort using interoperable software and specially designed new work processes.

Since 2003, nD modelling is a research topic that has started to gain momentum and interest worldwide. While the 3D to nD research project, at the University of Salford, aims to develop a unique multi-dimensional computer model that will portray and visually project the entire design and construction process, enabling users to 'see' and simulate the whole life of the project, research about CAVT in Chile pursue the same objective, but focused on how to predict and plan the construction process using mainly 3D and 4D commercially available CAVT to develop nD Construction Schedules. The research approach to nD or CAVT used in Chile uses the following commercial tools: Autodesk Architectural Software to 'build' 3D models taking as point of departure 2D drawings usually made in AutoCAD software. Primavera Project Planner software is used to develop the traditional construction schedule, while Intergraph's SmartPlant Review software is used to link the 3D model pieces to the construction schedule activities obtaining 4D models. This research approach cohabits with several IT problems, like for example interoperability, which are not the main part of the research focus. The main objective of the research carried out in Chile is the work processes development needed to take full advantage of available CAVT for nD construction scheduling development and planning and control improvement. Another important objective is the creation of awareness about the future nD tools and work processes that are generated while researching about how to implement and take advantage of available tools.

The research about nD carried out in Chile can the be considered from a 'pull' perspective since it is created into the industry creating awareness that shall contribute to accelerate industry adoption of the future results coming from the 3D to nD research project which at some time will be 'pushed' into the industry. The goal of improving construction management though new approaches to construction scheduling achieved either from 'push', 'pull' or both kind of efforts together shall at the end help to improve the decision-making process and construction performance by enabling true 'what-if' analyses to be performed to demonstrate the real cost in terms of the variables of the design issues.

Conclusion

New IT tools will not only help to automate the things we are doing – continuing doing the same faster – but a big transformation in the way we work is expected in the next years as a result of the application of IT advances to the construction industry. nD Modelling, VDC, BIM and FIATECH are examples of established labels for world class research efforts tending to build around themselves meanings, contexts and lines of development that represent conscious efforts to pick out the dominant ideas in the construction industry projects processes development, focusing with unprecedented opportunity and precision mainly on answering these questions: What will at the end need to be constructed? How much of this or that will be constructed? When will it be constructed? These research efforts act as active changing agents trying to push the construction industry to adopt new available technology under a proactive approach instead of waiting to be taken by the storm of new technology (Sawyer, 2004).

Information Technology for the AEC/EPC industry involves the integration of all product development processes and the management of the information flow between them, irrespective of the data models chosen in the different implementations of one and the same process (Scherer, 1994). Product and process modelling are the main topics related with the former approach, that have gained more attention in the last years within IT research in the AEC/EPC industry.

Although the use of building electronic product and process models is not widespread in the industry at the present time, its need is seen to be self-evident.

Central to product models is the visual representation of the project product, commonly associated to some 3D CAD modelling effort. Different approaches have extended the 'physical' representation of the project product, to include not only geometric information within it, but also the relying data which support the different activities during the several phases involved in an AEC/EPC project. Process modelling has been trying to represent the detailed tasks and transformation of specific entities among simultaneous activities oriented to complete a design into a constructed facility. In the construction industry, product and process modelling efforts have been mainly related with the academic community. Product and Process models coming from the academic world are, however, very different from 3D models being developed within the AEC/EPC industry, especially those dedicated to design. The expectations of real projects are focused on the results and the tools to achieve the results. The tools are mainly the available hardware, software and skills needed to use them. The knowledge about product and process modelling coming from the academic community, has been usually overlooked or developed in a tacit fashion, more guided by the common sense, than for an ordered approach based in techniques, methodologies or prototypes coming from the academic community.

More recent research efforts like CAVT and nD modelling and the availability of more powerful and easy-to-use tools tackles these issues under new approaches that may help to traverse the line from theory to practice faster than expected. New tools require ever-diminishing 'computer' skills from engineers, who don't need to be 'computer operators', but they can focus more on engineering, achieving results faster and with less effort than only a few years ago. CAVT and nD application to the AEC/EPC industry should allow the necessary break between the results and the tools, creating a niche where the accumulated knowledge about product and process modelling could fit, providing great benefits to the AEC/EPC industry, and starting a new phase of research from a practical perspective, instead of the prototype testing approach used commonly to date allowing for the attainment of levels of integration never seen before.

Traditional construction scheduling approaches have proved not to have the capacity to produce and make easily and available on times, the information needed for a latter effective and efficient construction control and execution reflected in a high Percentage of Planned Activities Completed. nD poses barriers and opportunities associated to big transformations which, for example, could lead to relegate traditional construction scheduling approaches like the CPM in which the critical path has become a bottleneck, a constraint of a project. The big change of working with an nD construction schedule can lead to lose less time on the critical path itself, but exploiting this constraint and subordinating everything else to it. Most problems that impact the critical path do not occur on the critical path itself; then subordination is not a nicety, it is a must (Goldratt, 1997). Another big change is that using nD modelling it is possible to alter the order in which the different level of detail schedules are developed for a construction project. Using nD and CAVT, it is possible to develop first the more detailed schedule and then obtaining the master schedule grouping activities in this more detailed schedule. The master schedule can then be adjusted to project requirements and the detailed schedule reviewed in an iterative process. This opposes the way scheduling is carried out today, in which the detailed schedule follows the master schedule development effort and rarely does one go back to review the master schedule deeply, but makes the maximum effort to adjust the detailed schedule to the master schedule. And finally, a very important change is that visualization, planning, analysis and communication capabilities embedded in a new nD construction scheduling development process can improve the traditional scheduling process to the extent that this schedule could finally be used as a real production control mechanism that could lead to tackle problems proactively and re-schedule the project as necessary.

References

Alarcón, L.F. and Ashley, D.B. (1999) Playing Games: Evaluating la Impact of Lean Production Strategies on Project Cost and Schedule. Proceedings of IGLC-7, University of Berkeley, California, USA, 26–28 July.

Alciatore, D., O'Connor, J. and Dharwadkar, P. (1991) A Survey of Graphical Simulation in Construction: Software Usage and Application. *Construction Industry Institute Source Document* 68.

Ballard, G. (2000) The Last Planner System of Production Control. Ph.D. Dissertation, School of Civil Engineering, Faculty of Engineering, The University of Birmingham, Birmingham, UK.

Bertelsen, S. (1992) The Mouse's opinion-on the building design documents. *NIRAS Consulting Engineers and Planners*, NIRAS, Denmark.

CII (1986) Constructability: A Primer. Construction Industry Institute (CII) Publication 3–1, London.

de Bono, E. (1970) *Lateral Thinking – Creativity Step by Step*. Harper & Row, Publishers, New York.

Fallon, K. (2000) IT, AEC and Design Firms. Bentley Wire, www.bentley.com/news/commentary/2000ql/kfallon.htm.

Fischer, M. and Alami, F. (1999) Cost-Loaded production model for planning and control. Proceedings of 8dbmc, Durability of Building Materials & Components 8, NRC Research Press, Vancouver, BC, Canada, Vol. 4, Information Technology in Construction, CIB W78 Workshop, pp. 2813–2824.

Froese, T., YU, K. and Shahid, S. (1996) Project Modeling in Construction Applications. Proceedings of the ASCE Third Congress on Computing in Civil Engineering, Anaheim, CA, USA, pp. 572–578.

Goldratt, E. (1997) *Critical Chain*, The North River Press Publishing Corporation, Great Barrington, Maryland.

González, V., Rischmoller, L. and Alarcón, L.F. (2004) Design of Buffers in Repetitive Projects: Using Production Management Theory and IT Tools. Fourth International Postgraduate Research Conference, University of Salford, Manchester, UK, April 1st–2nd.

Graphisoft (2003) The Graphisoft Virtual Building: Bridging the Building Information Model from Concept into Reality. Graphisoft Whitepaper.

Heesom, D. and Mahdjoubi, L. (2004) Trends of 4D CAD applications for construction planning. *Construction Management & Economics*, Routledge, part of the Taylor & Francis Group, Volume 22, Number 2/February 2004, pp. 171–182.

Horman, M.J. (2000) Process Dynamics: Buffer Management in Building Project Operations. Ph.D. Dissertation, Faculty of ArchWBcture, Building and Planning, The University of Melbourne, Melbourne, Australia.

Intergraph, C.S. (1998) Graphics Supercomputing on Windows NT. *Intergraph Computer Systems*, Huntsville.

Koo, B. and Fischer, M. (2000) Feasability study of 4D CAD in commercial construction. *Journal of Construction Engineering and Management*, 126(4), 251–260.

Lee, A., Marshall-Ponting, A.J., Aouad, G., Wu, S., Koh, I., Fu, C., Cooper, R., Betts, M., Kagioglou, M. and Fischer, M. (2003) *Developing a Vision of nD-Enabled Construction*. Construct IT Report, Salford.

Negroponte, N. (1995) *Being Digital*. Alfred A. Knopf, Inc., New Cork.

Pande, P., Neuman, R.Y. and Cavanagh, R. (2000) The Six Sigma Way-How GE, Motorola, and other Top Companies are Honing their Performance. McGraw-Hill, New York.

Philips, D. (1999) All Eyes in CAD. Computer Graphics World, Vol. 22, No 5, (www.pennwell.shore.net/cgw/coverstory/1999/05_story.html).

Rischmoller, L. and Valle, R. (2005) Using 4D in a new '2D + time' Conceptualization. Proceedings of the CIB W78 22nd Conference on Information Technology in Construction, Edited by: Raimar, J. Scherer, Peter Katranuschkov and Sven-Eric Schapke, Published by the Institute for Construction Informatics, Technische Universitat Dresden Germany, pp. 247–251.

Rischmoller, L., Fischer, M., Fox, R. and Alarcon, L. (2000a) 4D Planning and Scheduling: Grounding Construction IT Research in Industry Practice. Proceedings of CIB W78 Conference on Construction IT, June, Iceland.

Rischmoller, L., Fischer, M., Fox, R.Y. and Alarcon, L.F. (2000b) Impact of Computer Advanced Visualization Tools in the AEC Industry. Proceedings of The CIB-W78, IABSE, EG-SEA-AI International Conference in Construction Information Technology: CIT 2000, Icelandic Building Research Institute, Reykjavik, Iceland, Vol. 2, pp. 753–764.

Rischmoller, L., Fischer, M., Fox, R.Y. and Alarcon, L.F. (2001) 4D Planning and Scheduling (4D-PS): Grounding Construction IT Research in Industry Practice. Proceedings of the Construction Information Technology CIB W78 International Conference: IT in Construction in Africa, CSIR Division of Building and Construction Technology, Mpumalanga, South Africa, pp. 34-1–34-6.

Sawyer, T. (2004) e-construction-INNOVATION-What's Next. McGraw Hill Construction, enr.com, Engineering News-Record (http://www.enr.com/features/technologyEconst/archives/040621n.asp).

Scherer, R.J. (1994) Integrated Product and Process Model for the Design of Reinforced Concrete Structures, in Proceedings of the 1st Workshop of the European Group for Structural Engineering – Applications of Artificial Intelligence, Lausanne (EG-SEA-AI), Federal Institute of Technology (EPFL) Lausanne, 1994.

Staub, S., Fischer, M. and Spradlin, M. (1999) Into the fourth dimension. *Civil Engineering*, 69(5), 44–47.

Chapter 6

nD in risk management

Richard J. Coble

Introduction

Three dimensional drawings have long been used in construction. Even before Computer Aided Design (CAD) was available to the sophisticated construction manager, there were hand-drawn 3D drawings. The use of these 3D drawings was to manage risk, but not in what we today consider the complete gamete of risk management. It was much more a way if representation of mean, method technique, sequence or procedure that would better express cost savings or error elimination potential that would tie to making a job more profitable.

Today construction managers are still tasked with profitability, but also tasked with undertaking this effort with a safety conscious focus. This focus can and has been required to be expressed in hand-written and electronic formats. Just as is the case with complex construction, there is also the need for the same visualization in safety management, which in turn is a subset of the larger issue of risk management.

The use of Computer Aided Design (CAD) in risk management in construction management is a new and not commonly utilized management technique. This is not so much true as to the applicability of 3D visualization, in construction risk management, but more to the onset of risk management in construction management utilizing this tool.

As is the case in general industry and management, construction management lags behind. As it relates to risk management, this was a term and function that was developed by the insurance industry. It was developed for one simple reason, which was increased profitability. In the construction industry, profitability has always been the main goal on a construction project, but it was more a matter of priority. At the turn of the last century, equipment was very crude as compared to today, which meant just dealing with the people and equipment to get the job done was such a variable that it took large portions of the construction managers's time on the project to accomplish that end.

In the early 1970s Boyd Paulson, while at Stanford University, emphasized how computers could be effectively used in construction, but this was more as increased benefit to processing data. In the 1980s computers started to be utilized more in construction companies home offices and people like Dan Halpin, while at Purdue University, developed simulation programs like 'Cyclone'. These programs were solid logic, but lacked the processing capabilities the computers of today now have. These large-in-size computers were not appropriate to utilize in field operations. Even in the early 1990s, laptop computers were being utilized on project sites in very limited applications, even with people like Jimmie Hinze in conjunction with the Centre to Protect Workers Rights (CPWR) bringing some visualization to scheduling for safety. Today, laptop computers are particularly well suited for simulation and planning, both in the office and on project sites, which makes them ideal for construction managers to manage projects in ways that previously were never available. Now the majority of construction managers are utilizing computers on project sites, but they are being used more for scheduling and cost control than for any form of risk management or other use of visualization techniques.

The use of visualization, particularly CAD, in construction risk management is something that can be utilized on a daily basis in field operations to better eliminate and or mitigate, but definitely manage construction risks. Using CAD to plan work, document job history or provide work training is essential for construction managers to maximize their effectiveness. Florida International University has committed an effort to this end in their Center for Construction Risk Management (CCRM), which is showing the potential to provide construction managers with tools that can provide lower cost, safer and more efficient construction, through the knowledge of and use of risk management.

Model development

Modelling risk management issues are appropriate in many applications, but in the area of construction safety, it has particular importance. The following CAD and still photo presentations come from a video clip that was made specifically to illustrate the responsible parties in a construction accident that involved a weld connection failure. This type of issue often involves millions of dollars at risk. To clarify and simplify issues takes away lawyers arguing about issues that can and should have been determined prior to a jury or judge reviewing the issues of dispute. Use of CAD which often is best served in using a time component which is called 4D CAD or other components such as cost often brings clarity to complex issues. These added components are called nD CAD. In the following illustrations, CAD is being used in conjunction with still photographs to show how CAD can be used to fill in where further photographs should have been taken, but were not.

In using CAD for safety illustrations, this process is enhanced by the use of still photographs in combination with the CAD simulation. Still photographs provide authenticity that can be coupled with the CAD to provide for a clear visualization presentation. It can also be tied with time as to when a situation happened or tied to an industry standard as a reference point, which would enhance such an nD presentation in some cases.

Figure 6.1 Accident area.

Figure 6.2 Detail of accident area.

This methodology utilized in Figure 6.1 provides a view of the project site with specific enhanced boxed-in overview of the accident area, which is also the overview of the building. This photographic illustration of the building gives an overview of the problem area and a perspective to the job conditions as they existed at the time of the accident.

The use of 3D CAD can then be used to describe what was left off of, or not a part of, what was documented in the job photographs. This detail of the accident problem is shown in Figure 6.2. This simulation of the job accident area is

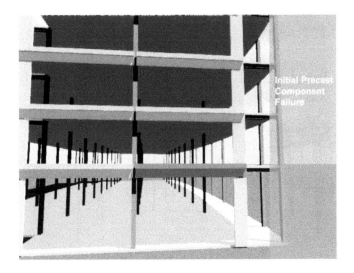

Figure 6.3 Beam failure.

described in the detail provided from such references as shop drawings, actual field measurements and job site interviews where sketches or hand drawings are used to describe a problem.

In some cases, shading can and should be used to delineate the problem or emphasis areas. This is particularly important when these areas are inter-mixed with complicated details. In Figure 6.3, the shading technique brings out the area affected in the failure of the beam that fell because of unsatisfactory welding.

The transition from using CAD to using still photographs is a decision that should provide for flow of sequence. This means that sometimes it is important to use more detailed CAD because the still pictures leave out what would be normal to the telling of a story of what happened. Using still photographs and 3D CAD in the safety perspective of risk management, one needs to be complete in their presentation of the sequence of the activity, regardless of the amount of photographs that are available. In Figure 6.4 it can be seen that the beam was installed and looked to be a satisfactory connection.

Additional jobsite photographs would have been desirable, but often field construction workers do not know and have not been trained in good jobsite documentation as to when to take a photograph. This entire issue of when photographs should be taken for jobsite documentation can be further studied on the Lynx website, at www.lynxpm,com/drcoblearticle.shift; also then refer to the white paper by Richard Coble PhD. In Figure 6.5, a 3D CAD representation was used to show where the failure occurred.

In some cases it is important to emphasize the description of 3D CAD illustration by using text. This is illustrated in Figure 6.6, which explains who was involved in the causation of the accident.

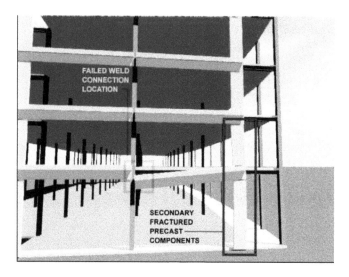

Figure 6.4 Failed connections.

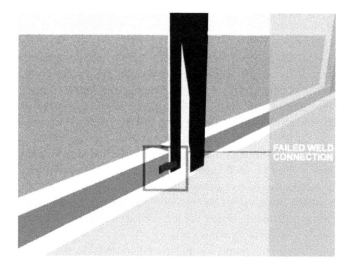

Figure 6.5 Failed weld connection.

In closing a presentation where a combination of still photographs and CAD is used, it is always good to use a still photograph for the finale that shows the project area in a completed form, because the viewing of the project will most likely depict such an image.

Construction safety visualization by the use of CAD has proven itself to be very effective for many purposes, with the primary funding for this work coming

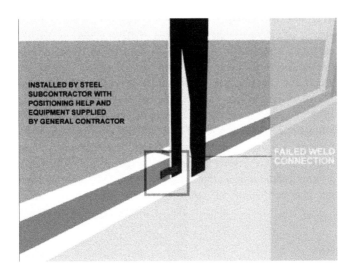

Figure 6.6 Causation of accident.

from litigation support. The opportunities for use of nD CAD in litigation support comes from the basic issues that surround construction contracting and the litigious nature of this business. In the example of the precast beam weld failing, an issue that could be misconstrued and made to be a cloudy issue is made simple by the use of CAD filing in the holes where the incomplete jobsite documentation had failed to document the critical issues.

In Justin Sweet's book on construction law titled *Sweet on Construction*, he states, in a section on the basic rule of risk assumption, '[w]ithout awareness of that residual assumption of risk by the performing party, you cannot understand why contracts are so important, unless there is a contractual protection to the contractor, it is almost certain (not completely certain because some legal doctrines may, on rare occasions, provide protection) that the contractor my bare the risk. It is his business to predict what is likely to happen . . . '. This means that, in many cases although it may be difficult, the door is open for contractors, their attorneys and or insurance company attorneys to pursue or defend litigation in general construction cases and also in safety-related and or personal injury cases that relate to construction accidents. This is where their making a visual representation is vital to their success. Without visualization the judge, jury or arbitrator(s) can make their own mental visualization which may or may not be clear and or correct as to what actually happened when the accident occurred. In the example mentioned using the CAD and still photographs, the simplicity of the presentation takes away all the extemporaneous issues, so that the primary fault is visualized and as described in Figure 6.6 further clarity is brought to bare with respect to the responsible parties as to the causation of the accident, by the written portion being added to the CAD drawing.

There are many and varied ways to provide visualization as it relates to construction safety and risk management, which is further emphasized in the book *The Management of Construction Safety and Health*, edited by Richard Coble, Theo C. Haupt and Jimmie Hinze. This book focuses on issues like interactive computer training techniques and automated technology for construction safety, which are good for the practice of construction safety. The techniques and methodologies described in this book can be coupled with the above referenced CAD and still photograph presentation in very effective ways. Also in this book, there are multiple specific examples of how technology can be used to enhance construction safety visualization, but by using still photographs in combination with 3D, 4D or nD CAD reviewers are provided a foundation of the issues and are guided in forming mental images that provide for the reality of the issue, without all the clutter and confusion that attorneys often interject into the litigation process.

Conclusion

CAD has tremendous risk-management potential in the management of construction projects, but as in most business decisions, it will take a proven successful application to show this value. This will include the industry being convinced that both CAD and risk management are a marriage that are worth the investment and will have mutual synergistic benefits. At this time both risk management and CAD have gained acceptability as individual systems, but very limited acceptance when being considered together.

Risk management in construction, encompasses organizations like the International Risk Management Institute (IRMI) who have taken leadership roles in construction risk management, but more from an insurance perspective, which is not unusual considering the fact that they are primarily funded by the insurance industry. In fact, risk management has a heritage from the insurance industry, wherein it is one of their larger profit centres.

CAD, from the other perspective, is a tool primarily used by architects to visually represent architectural features and or functions. This tool can provide visual presentation that brings clarity and uniform quality to the visualized work product. CAD is so accepted that it is often required to be utilized by those buying architectural design services.

To utilize CAD in scheduling, estimating and potential risk-management avoidance is to take a tool that can be cross utilized in very effective ways. This multi-use of CAD is what has come to be known as 'nD CAD', meaning that CAD can be used for many purposes and applications. In this text the use of CAD will be discussed as it relates to construction risk management. It should be pointed out at this time that proactive construction companies should and are in many cases making it their goal to train every manager to assist every employee in coming to the point that they can identify, anticipate and avoid and or mitigate risks.

Safety is a primary focus area for construction risk managers. In this regard, hazard avoidance is where it starts when it comes to those dealing with hazards. Hazards are risks that need to be identified in the most fundamental ways as it relates to the visual context, if avoidance is going to be accomplished. This is where CAD is a visualization tool that needs to be in the tool box of every proactive construction manager who places risk management as a priority.

Construction risk managers have an opportunity, like no other time in world history, to make a difference in the construction process by the use of nD CAD to enhance their work efforts. It is a time when some risk managers are being hired and others are being developed in companies, because of the realization that risk management can be one of the greatest profit centres within their company.

Acknowledgements

The author acknowledges the support of Robert Blatter, Architect, in his work for RJC Consulting in the area of visualization with special emphasis in 3D and 4D CAD. This is also true as it relates to the work of Dennis Fukai PhD, Architect, for his continued support and brainstorming that has been so valuable to Shiloh Consulting LLC in the development of visualization as a practice area.

Bibliography

Coble, R., Haupt, T. and Hinze, J. (eds) (2000) *The Management of Construction Safety and Health*, Rotterdam, Baelkma.

Construction Industry Institute (1991) *Managing Subcontractor Safety, Publication 13–1* (February).

Dodd, M. and Findlay, J. (2006) *State Guide to Construction Contracts and Claims*, Aspen, New York.

Hinze, J. (1997) *Construction Safety*, Prentice Hall, New Jersey.

Levitt, R. and Samuelson, N. (1992) *Managing Construction Safety*, *Second edition*, John Wiley, New York.

Shiloh Webpage Photographs.

Chapter 7

Construction safety

Khalid Naji

Introduction

The '3D to nD Modeling' project at the University of Salford aims to support integrated design; to enable and equip the design and construction industry with a tool that allows users to create, and share, contemplate and apply knowledge from multiple perspectives of user requirements such as accessibility, maintainability, sustainability, acoustics, crime, energy simulation, scheduling and costing (Lee *et al.*, 2002). This chapter presents the author's initiative to expand such research effort to address safety during construction.

In any business, an injury-free working environment is the most important factor to be considered. This is due to the fact of eliminating financial losses associated with injury claim, the lost time of work, and most importantly the associated schedule delays. It is also recognized that profits in any business are directly related to the degree in which losses are avoided or eliminated (Richard, 1999).

Despite the fact that job site safety and health in construction have improved in recent years, as reflected in the reduction in injury and illness rates since the early 1990s, safety in the contracted environment is under pressure by a variety of industry trends. These trends are the increasing complexity of operating systems, the increased specialization of equipment and the downsizing and outsourcing of work (Richard, 1999). Statistics related to construction site and job safety show that construction is one of the most dangerous industries in the world. Considering the United States as one large example of such industry, the National Institute for Occupational Safety and Health (NIOSH) reports indicate that, on the average, 1079 construction workers were killed on the job each year between 1980 and 1993 (Richard, 1999).

The construction site is one of the primary resources available to the contractor. Therefore, it is an essential task for the contractor to effectively plan the site layout and facilities to produce a working environment that will maximize efficiency and minimize risk (Whyte, 2003).

In the last decade, the need for safety awareness among construction companies has greatly increased (Wilson and Koehn, 2000). This is due to the high cost associated with work-related injuries, compensation insurance and the increased chance of litigation. Every year, a considerable amount of time is lost as a result of work-related health problems and site accidents (Anumba and Bishop, 1997). According to Kartam (1997) the US Bureau of Labor Statistics for the year 1982 shows an average of one death and 167 injuries per $100 million of annual construction spending in the United States. The total cost of these accidents reached $8.9 billion or 6.5 per cent of the $137 billion spent annually on industrial, utility and commercial construction. Similarly, in the United States one out of every six construction workers is expected to be injured. According to Hinze (1997) these staggering numbers translate into more than 2,000 deaths and 200,000 disabling injuries each year in the United States during the late 1980s. There are numerous factors responsible for health problems and construction site accidents. The Occupational Safety and Health Administration (OSHA) examined the causes of construction fatalities that occurred in the United States from 1985 to 1989. The results showed that 33 per cent of the fatalities in construction were caused by falls, 22 per cent struck-by incidents, 18 per cent caught in/between incidents, 17 per cent electrocutions, and 10 per cent caused by other conditions. Various researchers have reported that proper site management is vital for reducing hazards and accidents on construction sites (Anumba and Bishop, 1997). Several causes of construction site accidents and health hazards (such as falls, falling objects, site transportation, site layout and hazardous substances) can be controlled through creating an efficient site layout plan (Hinze, 1997).

Critical safety issues in construction

Every construction project requires mechanical equipment to carry out the equipment-based tasks; however, heavy machinery could introduce a serious source of potential injury during construction due to failure in operation, operator's negligence, overloading and most likely collision with other machinery. Research in the area of construction site safety and management reveals that collision is among the most common causes of accidents during construction (Holt, 2001).

Similarly, early research efforts showed that the following are among the top human-related causes of accidents during construction: (1) operating equipment without authority and/or at unsafe speed, (2) when equipment came into contact with a worker, (3) when equipment came into contact with a building component, (4) when equipment came into contact with temporary structures and (5) when equipment came into contact with another equipment.

In the United Kingdom, early statistics show that during the year 1998–1999 about 13.64 per cent of total construction accidents were due to a person being struck by a moving vehicle within the construction site, 12.12 per cent were due to a person being struck by a moving or falling object and about 3.03 per cent were due to contact with a moving equipment (Holt, 2001).

How would nD modelling improve construction site safety?

As the nD modelling concept expands to integrate many tools, techniques and information bodies related to the building construction industry, nD modelling could play a major role in improving construction safety. This new dimension added to the integrated model would become effective by integrating the proper tools, methodologies and data models for conducting true what-if analysis for improving construction safety.

As construction safety is related to the state of a project under execution, an nD modelling tool used to address safety of the construction site should be realistic, interactive, expandable, scalable and yet very dynamic. In practice, no tool would be sufficient to incorporate these features except a general Virtual Construction Environment (VCE). Additionally, such a VCE should be portable and able to communicate with other n-Dimensions through a common protocol or a language. The nD modelling research team at the University of Salford in the United Kingdom suggests the Industry Foundation Class (IFC) as a generally accepted viewing and modelling language for nD applications (Fu *et al.*, 2005). The IFCs provide a set of rules and protocols that determine how the data representing the building in the model are defined and the agreed specification of classes of components enables the development of a common language for construction. Moving beyond the building model dimensions, for a typical construction site an nD tool should be able to present and process data related to building components, site environment, equipments, labour, temporary structures, facilities and materials since these are the main building blocks of any construction site regardless of its size, type or complexity. The nD tool presented in this chapter is still under development to fully incorporate the IFC 2.x standards for viewing and data control; its development so far will be explained.

System architecture and the required information model

Based on the work undertaken by the University of Salford, the system architecture of the nD modelling prototype tool should include a number of elements as shown in Figure 7.1. These elements are a construction site information model, an IFC model server and a tool that provides safety analysis and decision support.

- Construction site information model: a model that provides information related to the construction site.
- Construction site safety analysis and decision support tool: a tool that provides a sort of safety analysis such as collision and falling analysis for decision support.
- An IFC construction site model: a 3D model representing the construction site based on the IFC classifications and linked to the information model.
- An IFC model server: a server containing the set of protocols required for viewing the IFC model.

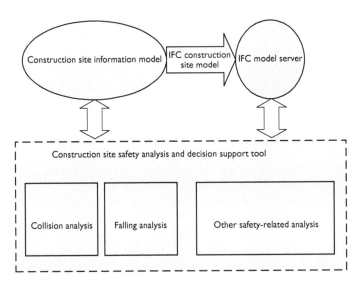

Figure 7.1 System architecture of the nD modelling tool for construction sites.

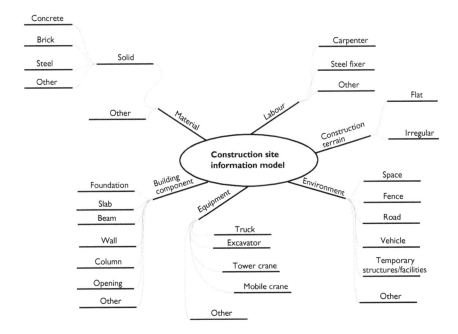

Figure 7.2 Elements of the construction site information model.

The construction site information model should provide information such as building components, the construction site terrain, surrounding environment, material, labour and equipment, as presented in Figure 7.2. Additionally, every element in this information model should also include information related to location, colour, texture and so on.

Current research effort at the University of Qatar

As the term virtual reality (VR) is defined as a technology that enables real-time viewing of, and interaction with, spatial information, similarly, there are other words used in the literature to describe same groups of technologies, and similar concepts including virtual environments (VEs) (Faraj and Alshawi, 1996; Kamat and Matinez, 2001 and visualization, interactive 3D (i3D) (Aouad *et al.*, 1997). Another example is SIMCON which is a 3D VE developed in 1997 (Naji, 2000) as a test pad to have insight of the construction site during the pre-construction phase. The SIMCON application simulates machine-based construction processes, such as material handling, earth removing and crane operations, in real-time object-oriented graphical environment. The main objective of the SIMCON research was to develop a tool by which safety in construction could be enhanced by having a test pad to graphically conduct what-if analysis for different scenarios of building, machine and site configurations, and provide a high level of user interaction (see Figure 7.3). The SIMCON application was further modified, during the last 6 years, to include additional features and to support more elevated level of user-simulation interaction.

After discussions with the University of Salford in 2004, it was decided that the SIMCON application should be migrated towards the nD project based on the following:

- Phase 1: updating of the current version towards more industry standards by including a VRML model viewer and a Windows-XP-based interface.
- Phase 2: developing of the safety what-if analysis framework.
- Phase 3: updating the current version with a tool for conducting safety what-if analysis.
- Phase 4: updating towards the nD environment by including the IFC 2.X protocols and an IFC model viewer.

SIMCON+ system architecture in depth

The current version of the SIMCON+ application was updated to become a stand-alone application, and the ability to store the state of the visual simulation in a data file where information could be retrieved later for further analysis.

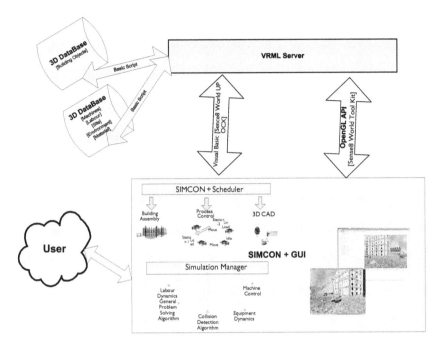

Figure 7.3 System architecture of the SIMCON+ application.

Original development of the SIMCON application was done using the Sense8-WorldUP (WorldUP is a registered trademark of Sense8 Corp) virtual reality authoring tool. The latest version SIMCON+ was developed based on the WorldUP ActiveX (OCX) control and Microsoft Visual Basic. Figure 7.4 presents the stand-alone SIMCON+ application. More than 36 application modules written in Visual Basic and Basic Script were embedded into the SIM-CON+ application to control the overall user-simulation interface on the Window-XP operating system. For sample code of the embedded SIMCON+ modules the reader is encouraged to refer to Appendix-A following this chapter.

The two key components of SIMCON+ application are the internal *Scheduler* and *Simulation Manager.* These two components manage and control the resulted 3D visual simulation. The internal *Scheduler* of the SIMCON+ application is the main element by which the user is engaged into an interactive 3D graphical simulation once the application is launched. The *Scheduler* consists of a 3D CAD Database, a process control module, and a building assembly module. These modules within the SIMCON+ application are the heart by which many different scenarios of construction site configurations could be tested and

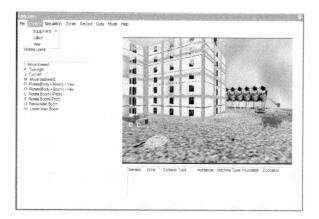

Figure 7.4 SIMCON+ graphical user interface on Windows XP.

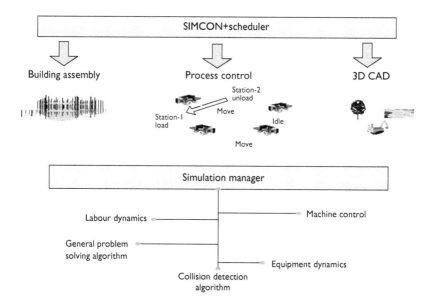

Figure 7.5 SIMCON+ internal components.

analysed with respect to the available site space. Additionally, the user has the ability to interactively interact with machines to analyse their impact on the overall safety of the construction site in a visual interactive format. Figure 7.5 presents components of the SIMCON+ *Scheduler* and *Simulation Manager* modules.

3D CAD and building assembly

In the SIMCON+ application the 3D CAD Database contains all 3D CAD object definitions, such as building components, machine prototypes, terrain and so on, and these 3D CAD objects are retrieved into the simulation based on object-oriented nature. This means that these objects are re-usable and feature inheritance within the simulation. Once the simulation is started the building assembly module is fired and all 3D visual objects are placed within the environment based on their 3D CAD definition or user input. Figure 7.6 presents the mechanism of building/objects assembly in SIMCON+.

Process control module (PCM)

When the SIMCON+ application starts, the *Scheduler*'s PCM retrieves information regarding machine type (truck, excavator, crane or mobile crane), stations within the construction site and the type of operation needed to be performed. This information is stored as a script file loaded into the application upon user selection. The process control module PCM manages the sequence of executing these scripting files during the simulation. Additionally, simulation of machine processes using the SIMCON application could be done through what is called a non-autonomous mode. In this mode of operation user-simulation

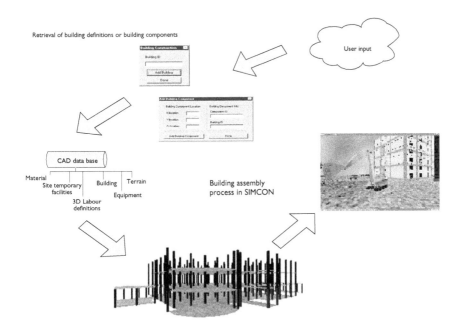

Figure 7.6 Building/objects assembly in SIMCON+.

interaction is done based on user inputs to the simulation through a number of 2D and 3D input devices.

SIMCON+ simulation manager (SM)

The Simulation manager (SM) in SIMCON+ manages all of the real-time simulation tasks necessary to for the Interactive Visual Simulation. Within the SM both machine and labour dynamics are achieved through General Problem Solving (GPbS) and collision-detection algorithms. The GPbS algorithm consists of five different actions to be done by each object in the following sequence: (1) find target and move towards it, (2) if obstacle exists on the path move opposite from the target direction, (3) rotate towards a free-obstacle orientation, (4) move forward and (5) find the target and move towards it. Figure 7.7 shows this embedded GPbS algorithm.

Additionally, machine and worker objects are checked against collusion every cycle once the SIMCON+ application is started. This is done through a general collusion-detection algorithm applied to all machine, worker and building objects. Once the SIMCON+ application is started the user has the ability to record from and play back motion paths for any machine and/or labour objects within the simulation, and collision checks are fired every simulation cycle.

Construction site safety: what-if analysis framework

As mentioned, there are very critical issues regarding safety in construction. The focus here is on safety with respect to equipment-based construction processes. In practice, during construction equipment-based processes are among the most dangerous factors that lower the safety level in any building project. This is due to the fact that equipment-based processes usually interact with building components, workers, materials and workers. In this regard, a safety what-if analysis tool should provide information related to falling and collision incidences on the construction site and their location, type and cause. Figure 7.8 presents the system components of such tool.

Simulation scenarios and what-if analysis

This section presents a framework for conducting what-if analysis based on the SIMCON+ data output. This framework is illustrated in Figure 7.9. The methodology of this framework is concentrated on the fact that the SIMCON+ application is able to retrieve 3D CAD objects representing the actual construction site in terms of building structures and actual site dimensions. Typically, these 3D CAD definitions are regular scaled-down outcomes from the design process

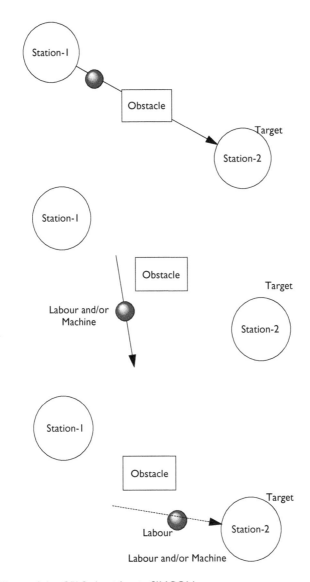

Figure 7.7 Steps of the GPbS algorithm in SIMCON.

during the pre-construction phase. As shown in Figure 7.9, once the simulation starts the user has to define one scenario, and activate one machine object at a time. The user then should naturally engage himself into the simulation by activating a number of machine and labour-based animations. The SIMCON+ application at any time during the simulation will record data with respect to

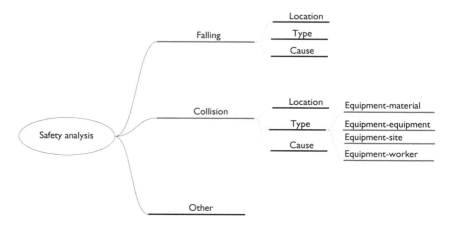

Figure 7.8 Components of a safety analysis tool.

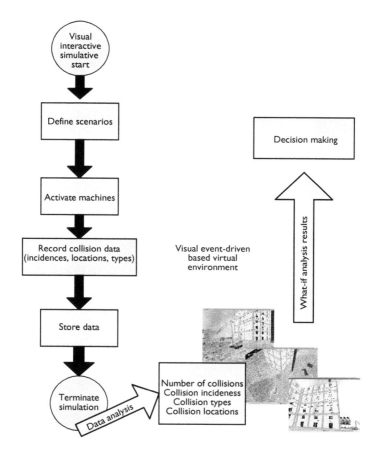

Figure 7.9 Framework for conducting what-if analysis based on SIMCON+.

collision incidences that took place by the driving machine, locations of these collision incidences and the type of each collision incidence.

The recent SIMCON+ application added a new feature for storing and retrieving data representing different simulation scenarios. This feature stores simulation data with respect to the number of collision incidences that occurred during the simulation. Additionally, this feature stores the exact location of these collision incidences on the construction site terrain, and the construction site objects that caused these incidences.

Moreover, SIMCON+ stores a new variable representing the construction site zone based on user definition during the simulation. Figure 7.10 presents a sample output Welcome to SIMCON+Colloision data file.

A demonstration case study

The case study in this section represents a small size turn-key residential building project in the city of Doha, capital of the State of Qatar. The total value of the project is approximately $2.82 million (QR 8,330,000) based on a lump-sum contract type. The project is located in Bin-Mahmoud area with a total built up area of 7,200 sq m distributed on five identical floors and one basement. The total project duration as agreed in the contract was 13 months. The main contractor of the project is a local Grade (B) contracting company with an approximate total value of annual projects equal to $5.5 million. This project was investigated with respect to construction site safety using the SIMCON+ application during summer and fall 2005. The main focus was to study the available construction site space with respect to the pre-planned operations before construction. Figure 7.11 presents a 2D CAD representing the construction site space with respect to the surrounding environment, and a plan view representing the final building object as considered in the SIMCON+ application (Naji, 2005).

This case study was tested in SIMCON+ considering one building location scenario due to the city of Doha building regulations, and few other scenarios were considered based on different site configurations with respect to machines, location of temporary site structures and zones for workers-based operations. Figure 7.12 presents a sketch of a typical scenario considered using the SIMCON+ application.

As mentioned earlier, a number of scenarios were tested for this case study using the SIMCON+ application. Figure 7.13 presents one scenario output data that resulted from the SIMCON+ application. These results show the total number of collision incidences recorded during the visual simulation run, and in this particular scenario they were equal to 22 incidences that resulted from three different collision types. Similarly, the results show the average for each collision incidence type with respect to the total number. The analysis shows that the highest number of collision incidences where based on an impact of a working machine with a temporary site object on the construction site terrain. Additionally, the analysis shows the exact special location of each collision incidence within the

(a)

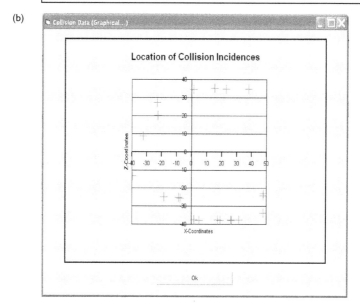

Welcome to SIMCON+, Interactive Construction
Site Visualizer & Simulator
Developed by; Dr. Khalid K Naji
Spring 2005
....

Scenario Zone Collision Type incidence Machine   X-Location Z-Location

Scenario-A South west machine-machine   1 truck -2904.401 787.1602
Scenario-A South west machine-material   2 truck -2732.063 3369.133
Scenario-A South west machine-site object 3 truck -4546.041 552.4695
Scenario-A South west machine-machine   4 truck -3281.041 550.1841
Scenario-A South east machine-machine   5 truck -3257.013 -371.2122
Scenario-A South west machine-worker   6 truck -2655.316 2903.299
Scenario-A South west machine-traffic   7 truck -2005.963 4271.428
Scenario-A South east machine-machine   8 truck -4574.847 -1228.796

(b)

Collision Data (Graphical...)

Location of Collision Incidences

X-Coordinates

Ok

Figure 7.10 (a) Sample of SIMCON+ collision data file text output. (b) Sample of SIMCON+ collision data file graphical output.

virtual construction site space created on the SIMCON+ application. These locations represent exact points on the construction site terrain since the original 3D CAD file representing the case study was imported into SIMCON+. The output column representing the truck machine type is simply indicating the virtual machine used by the user during the simulation cycle in a non-autonomous mode (fully interactive) in order to drive through the virtual constructed site.

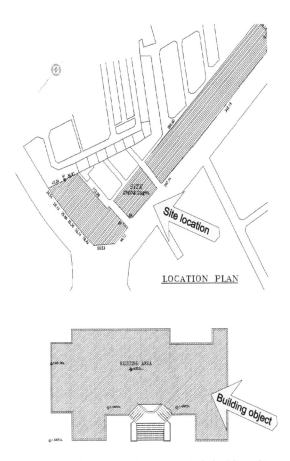

Figure 7.11 Construction site location and the case study building object.

Moreover, the SIMCON+ output produces a 2D graphical representation of the collision incidence locations within the virtual site as shown in Figure 7.13. Finally, this analysis suggests that a decision rate that equals to 63 per cent should be given to consider the safety of machine-based operations with respect to the temporary site objects virtually simulated within the simulation. Similarly, a decision rate of 31.8 per cent should be given to consider the safety of machine-based operations with respect to other machines.

Research status

Despite the fact that 4D models are time consuming to create, manipulate and visualize, it is the only possible way to address the construction knowledge for

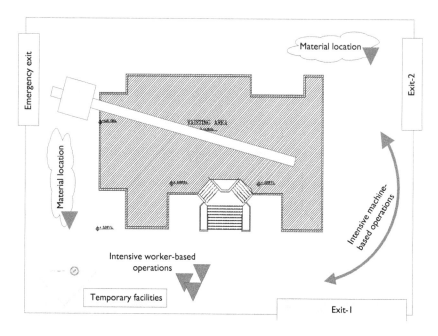

Figure 7.12 Case study typical scenario as tested in SIMCON+.

the past two decades. Therefore, it was decided that the SIMCON+ project at the University of Qatar should go through a transition period from fully dependable on VRML and OpenGL towards the IFC 2.x standards. This transition period is the current state of the SIMCON+ project where it is currently being updated to include a VRML–IFC converter and an IFC model viewer. As few publications are released out on the techniques required for building an IFC-based model viewer (Fu *et al.*, 2005), and IFC-based single database (Tanyer and Aouad, 2005), it is expected that by the end of summer 2006 the SIMCON+ will utilize both features towards being an nD-enabled modelling tool for improving safety of construction job sites.

Future directions

Although the SIMCON+ application was built on top of a very intuitive OpenGL and VRML-based authoring systems like the Sense8 WorldUP and World ToolKit, there are more VR authoring systems that have been released out in the market during the past 3 years. A very powerful VR authoring system, and yet one which requires much less API (Application Programming Interface) programming is the EON Professional Studio 5.x developed by the Swedish EON Reality Inc., which feature's a very powerful CAD tool, an event-based programming interface, advanced shading algorithms and advanced visual simulation techniques such as particle/human dynamics and fire creation. It is expected that the first migrated version of the SIMCON+ application could be presented during late 2006.

(a)
```
Welcome to SIMCON+, Interactive Construction
Site Visualizer & Simulator
Developed by; Dr. Khalid K Naji
Spring 2005
.............................................
Collision Incidences Data
.............................................
Total collision incidences                  = 22
Total collision due to machine-machine      = 7
Total collision due to machine-worker       = 0
Total collision due to machine-material     = 1
Total collision due to machine-building     = 0
Total collision due to machine-site object  = 14
Total collision due to machine-traffic      = 0
Average Collision Incidences Data by Type
.............................................
Average collision due to machine-machine = 0.3181818
Average collision due to machine-worker = 0
Average collision due to machine-material = 4.545455E-02
Average collision due to machine-building = 0
Average collision due to machine-site object = 0.6363636
Average collision due to machine-traffic    = 0
Senario    Zone Collision_Type Incidence Machine_Type X_Loc. Z_Loc.
Scenario-A West side  machine-machine      1 truck    -2252.206 2039.103
Scenario-A West side  machine-site object  2 truck     147.5826 3461.46
Scenario-A West side  machine-machine      3 truck    -3939.83 -1312.644
Scenario-A East side  machine-machine      4 truck    -1854.226 -2491.14
Scenario-A East side  machine-site object  5 truck    2655.586 -3771.665
Scenario-A East side  machine-site object  6 truck    1910.277 -3793.611
Scenario-A East side  machine-machine      7 truck    -757.6066 -2557.025
Scenario-A East side  machine-machine      8 truck    -856.5674 -2509.952
Scenario-A East side  machine-site object  9 truck     156.902 -3768.048
Scenario-A East side  machine-site object 10 truck    2638.12 -3759.712
Scenario-A East side  machine-site object 11 truck    3163.762 -3768.588
Scenario-A East side  machine-site object 12 truck    1754.696 -3763.658
Scenario-A East side  machine-site object 13 truck     508.2935 -3781.838
Scenario-A North side machine-site object 14 truck    4767.698 -2418.307
Scenario-A North side machine-site object 15 truck    4798.963 -2302.45
Scenario-A North side machine-site object 16 truck    4789.249 -3367.854
Scenario-A West side  machine-site object 17 truck    2296.533 3483.143
Scenario-A West side  machine-site object 18 truck    3790.958 3487.943
Scenario-A West side  machine-site object 19 truck    1517.989 3527.857
Scenario-A West side  machine-material    20 truck    -2288.88 2742.03
Scenario-A West side  machine-machine     21 truck    -3225.151 871.5676
Scenario-A West side  machine-machine     22 truck    -3225.151 871.5676
```

Figure 7.13 (a) Case study data output file.

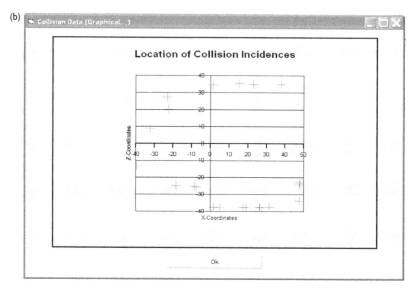

Figure 7.13 Continued. (b) Case study graphical output data.

Appendix-A

Sub terrain_following ()

    Dim m as Truck
    Dim debug as Boolean
    debug=False
    Dim distance(4) as Single
    Dim tx(4) as Vect3d
    Dim startnode as Node
    Dim p2 as Vect3d
    Dim poly(4) as Long
    Dim foundgeom(4) as Boolean
    Dim geom as Geometry
    Dim height as Single
    Dim tdim as Vect3d
    Dim tmid as Vect3d
    Dim tpos as Vect3d
    Dim sum as Single
    Dim num as Single
    Dim pitch as Single
    Dim roll as Single
    Dim q as Orientation

```
Dim ori as Orientation
Dim dir as Vect3d
Dim head as Single

Set m=GetTruck(which_machine)

' Get equipment's orientation
m.GetRotation ori
OrientToDir ori,dir
dir.y=0
DirToOrient dir,ori
Vect3dNorm dir
head=ATN(dir.x/dir.z)
If dir.x<0 and dir.z>0 Then
      head=head*180/pi
ElseIf dir.x<0 and dir.z<0 Then
      head=head*180/pi
      head=head+180
ElseIf dir.x<0 and dir.z>0 Then
      head=head*180/pi
      head=head+180
ElseIf dir.x>0 and dir.z<0 Then
      head=head*180/pi
      head=head+180
Else
      Exit Sub
End If

' Get Truck's position and geometry information
m.GetTranslation tpos
m.GetMidpoint tmid
m.GetDimensions tdim

' Save equipment's corner points in xz plane
tx(1).x=tmid.x+(tdim.x*0.5)
tx(1).y=tmid.y
tx(1).z=tmid.z+(tdim.z*0.5)
tx(2).x=tmid.x+(tdim.x*0.5)
tx(2).y=tmid.y
tx(2).z=tmid.z-(tdim.z*0.5)
tx(3).x=tmid.x-(tdim.x*0.5)
tx(3).y=tmid.y
tx(3).z=tmid.z+(tdim.z*0.5)
tx(4).x=tmid.x-(tdim.x*0.5)
```

```
tx(4).y=tmid.y
tx(4).z=tmid.z-(tdim.z*0.5)

Set startnode=GetNode('terrain_land-1')
p2.X=0.0
      p2.Y=1.0
      p2.Z=0.0
      num=0.0

      For i=1 to 4
            tx(i).y=tx(i).y-m.Radius
            poly(i)=RayIntersect(startnode,tx(i),p2,geom,height)
            If geom is not nothing Then
                  distance(i)=height/1
                  sum=sum+distance(i)
                  foundgeom(i)=True
                  num=num+1
            End If
      Next i

      If num>0.0 Then height=sum/num

      If (height>0) Then
                  m.Translate 0,height-m.radius-(tdim.y/3),0
      End If

      If (foundgeom(1) and foundgeom(2) and foundgeom(3) ) Then
                  pitch=ATN( (distance(2)-distance(1) )/tdim.z)
                  roll=ATN( (distance(1)-distance(3) )/tdim.x)
                  OrientSet q,pitch*180.0/pi,head,roll*180.0/pi
      End If
      m.SetRotation q
End Sub
```

Bibliography

Abdelhamid, T. and Everett, J. (2000) Identifying root causes of construction accidents, *Journal of Construction Engineering and Management, ASCE*, 126, 52–60.

Anumba, C. and Bishop, G. (1997) Importance of safety considerations in site layout and organization, *Canadian Journal of Civil Engineering*, 24(2), 229–236.

Aouad, G., Child, T., Marir, F. and Brandon, P. (1997) Developing a virtual reality interface for an integrated project database environment, Proceedings of the IEEE International Conference on Information Visualization (IV'97), London.

Faraj, I. and Alshawi, M. (1996) Virtual reality in an integrated construction environment, The ITCSED'96 Conference, Glasgow, pp. 14–16.

Fu, C., Aouad, G., Lee, A., Mashall-Ponting, A. and Wu, S. (2006) IFC model viewer to support nD model application, *Journal of Automation in Construction*, Accepted for publication, In Press.

Hinze, J.W. (1997) *Construction Safety*, Prentice-Hall, London.

Holt, A. (2001) *Principles of Construction Safety, First Edition*, USA: Blackwell Science.

Kamat, V. and Matinez, J. (2001) Comparison of Simulation-Driven Construction Operations Visulization and 4D CAD, Proceedings of the 2002 Winter Simulation Conference, pp. 1765–1770.

Kaminetzky, D. (1991) *Design and Construction Failures*, McGraw-Hill Inc, NewYork.

Kartam, N. (1997) Integrating safety and health performance into construction CPM. *Journal of Construction Engineering and Management*, 123(2), 121–126.

Koo, B. and Fischer, M. (2000) Feasibility study of 4D CAD in commercial construction. *Journal of Construction Engineering and Management, ASCE*, 126(4), 55–62.

Lee, A., Betts, M., Aouad, G., Cooper, R., Wu, S. and Underwood, J. (2002) *Developing a Vision for an nD Modeling Tool, CIB w78 Conference Proceedings*, Aarhus School of Architecture, Denmark.

Naji, K. (2000) Simulating equipment-based construction operations using virtual environments, Seventeenth Annual International Symposium on Automation and Robotics in Construction, Taipei, Taiwan.

Naji, K. (2005) Applications of Virtual Construction Environments in Improving Construction Site Safety, (electronic proceedings) of the 10th International Conference on Civil, Structural and Environmental Engineering Computing, Rome, Italy.

Occupational Safety and Health Administration OSHA (1990, 1998), *Analysis of Construction Fatalities*, U.S. Department of Labor, Washington, DC.

Richard, D. (1999) *Construction Site Safety: A Guide for Managing Contractors*, Lewis Publishers, New Jersey.

Rischmoller, L. and Matamala, R. (2003) Reflections about nD Modeling and Computer Advanced Visualization Tools (CAVT), *Construct I.T. Report/Develop a Vision of nD-Enabled Construction*, University of Salford, Salford.

Tanyer, A.M. and Aouad, G. (2005) Moving beyond the fourth dimension with an IFC-based single project database, *Journal of Automation in Construction*, 14, 15–32.

Waly, A. and Thabet, W. (2002) A virtual construction environment for preconstruction planning, *Automation in Construction*, 12, 139–154.

Whyte, J. (2003) Industrial applications of virtual reality in architecture and construction, *IT in Construction*, 8, 43–50.

Wilson, J.M. and Koehn, E. (2000) Safety management: problems encountered and recommended solutions, *Journal of Construction Engineering and Management, ASCE*, 126, 77–79.

Chapter 8

Automated code checking and accessibility

Robin Drogemuller and Lan Ding

Introduction

Ever since the introduction of architectural CAD (Computer Aided Design) systems, building designers have dreamt that the onerous task of checking building designs for compliance with the relevant building codes and standards could be automated. The reasons for this are obvious – the Building Code of Australia (BCA), as an example, contains over 4,000 clauses filling several thousand pages of text. These clauses reference over 1,000 separate standards, with options available in the standards that need to be explicitly selected within the contract documentation. Failure to comply with provisions in the building code or the referenced standards can mean expensive rework on site or even after completion of the project. This may have significant financial impacts on all parties involved in the building project.

The ability to automate code checking would also reduce problems with human interpretation of codes, where different people understand the same text in different ways. Reading the history of modern architecture leads to the conclusion that innovative and unusual buildings are often given a more rigorous analysis than more 'normal' buildings. Automating code checking could thus reduce the likelihood of errors in projects and improve the interpretation and application of codes.

The interest in automated code checking has led to a relatively long history of research in this area. Work at the NBC in Canada (National Building Codes) and at NIST in the United States (National Institute of Standards and Technology) started in the 1980s, as did research at CSIRO in Australia. There have been a number of implementations of knowledge-based systems arising from these research projects, some of which, such as BCAider, were commercialised and used in live projects. This first generation of implementations required the user to enter the characteristics of the project since they did not support automatic extraction of information from CAD data. The major reason for this was the lack of semantics (computer interpretable content) in the CAD systems available at that time. The introduction of object-oriented CAD systems since the 1990s has changed this situation where it is now feasible to build automated code checking systems. Three of these are described in this chapter.

Building codes and standards

Internationally, building codes and standards can be divided into two groups, prescriptive and performance-based codes. A prescriptive code states the requirements explicitly. A designer must comply with the provisions as given. Checking against a prescriptive code is relatively easy. A design either complies with one of the provisions explicitly stated within the code or it does not. In the latter case, the design will need to be modified. The rapid change in the available technologies within the building industry has lead to a realisation that prescriptive codes potentially restrict innovation and the uptake of new technologies. The problem with prescriptive codes is captured in the title 'Does an Ice Hotel Require a Sprinkler System?' (Rauch, 2005). The limitations of prescriptive codes have led to a movement internationally to performance-based codes.

While there is no one structure that matches all performance-based codes, the structure of the Building Code of Australia provides a useful example (Figure 8.1). At the top level are the objectives, specifying the goals of the code, then the functional statements, setting out how this should be achieved and then the performance requirements, stating how the functional requirements will be measured. Typically, the number of provisions increases as the hierarchy deepens. All of these are specified in qualitative terms. Since qualitative descriptions are open to interpretation, and hence dispute between designers and certifiers, deemed-to-satisfy solutions are also provided that give quantitative requirements to meet the requirements of the code. Where a designer wants to use a different method than that allowed in the deemed-to-satisfy provisions, the designer may propose an alternative solution. This is the major difference between performance-based codes and prescriptive codes. In a prescriptive code there is little or no flexibility in what may or may not be constructed.

Alternative Solutions, under the BCA can be certified in various ways (Figure 8.2). Comparable systems exist in other performance-based regulations. Figure 8.3 gives extracts from the BCA showing the hierarchy of requirements for disabled access from objectives down to deemed-to-satisfy. In this case, AS1428.1 is accepted as one possible solution to the stated objective. Of course, other solutions that meet equivalent requirements are equally acceptable.

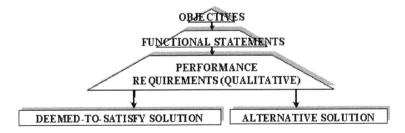

Figure 8.1 Formal structure of the Building Code of Australia.

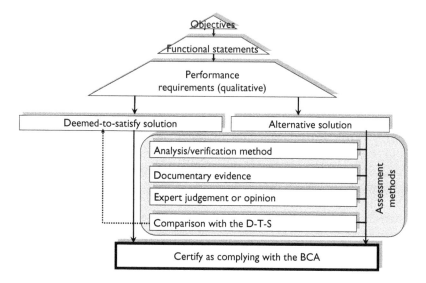

Figure 8.2 Alternative solutions under the BCA.

The type of building code has a significant effect on how an automated code checking system is implemented. Automated checking is simpler under a prescriptive system since the requirements are explicit and the design either meets these requirements or it does not. Under a performance-based system, there can never be a definitive answer since it is only possible to check against deemed-to-satisfy provisions and a designer may be proposing an alternative solution. This means that there must be a higher degree of user interaction in a system that checks against a performance-based code so that the designer can identify which methods of compliance have been chosen. In a fully automated system, meta-data would need to be stored to indicate which parts of a code or standard are met by a particular deemed-to-satisfy.

User requirements

There are three groups of people whose needs must be recognised when developing a code checking system. Designers will use code checking systems to test whether a design meets the requirements. Ideally, the system will also assist in checking partial designs to avoid backtracking when a detailed design is found not to comply.

Building compliance checkers need to validate completed designs when the design is submitted for building approval. In jurisdictions where the building compliance checking system has been de-regulated, such as some states of

Objective:
DO1
The Objective of this Section is to –

(a) provide, as far as is reasonable, people with safe, equitable and dignified access to –

   (i) a building; and
   (ii) the services and facilities within a building; and

(b) ...

Functional statement:
DF1
A building is to provide, as far as is reasonable –

(a) safe; and
(b) equitable and dignified, access for people to the services and facilities within.

Performance requirements:
...
DP7
Accessways must be provided, as far as is reasonable, to and within a building which –

(a) have features to enable people with disabilities to safely, equitably and with dignity –

   (i) approach the building from the road boundary and from any carparking spaces associated with the building; and
   (ii) access work and public spaces, accommodation and facilities for personal hygiene; and

(b) ...

Deemed-to-satisfy provisions:
D3.2
General building access requirements

(a) Buildings must be accessible as required Table D3.2.
(b) Parts of buildings required to be accessible must comply with this Part and AS 1428.1.
(c) ...

Figure 8.3 Hierarchy of requirements within BCA.

Australia, building compliance checkers often provide advice to design teams, in which case their needs are identical to designers.

During development of Design Check (described later), both designers and compliance checkers indicated that they wanted to be able to analyse a design in various ways, through identifying which objects did or did not comply with specific clause(s) and also by selecting types of objects and checking which clauses complied/did not comply. Both groups also preferred that the entire code should be stored and all requirements met in one operation. They did not want to remember that some clauses were covered by an automated system while they would have to manually remember to check other clauses.

Compliance checkers will often audit designs more thoroughly than designers. Compliance checkers indicated that they also wanted to be able to examine individual objects or types of objects and to check which rules had or had not been run against the objects.

The final group of users is the people writing the rules which encode the provisions of codes and standards, whether these have legal status or are informal. They need systems and user interfaces that make the writing of rules simpler and assist in checking that everything is correct.

Automated checking systems

Three code checking systems have been implemented against the IFC (Industry Foundation Classes) model (IAI, 2004):

- Solibri Model Checker
- ePlanCheck
- Design Check

Solibri Model Checker (www.solibri.com) was the first of the IFC-based systems to be released for industry use. It consists of a constraint set manager module that allows users to build 'rule sets' (Figure 8.4), the model checking/reporting engine itself (Figure 8.4) and a viewer that allows users to see which building component is causing a problem (Figure 8.5).

Solibri Model Checker was tested as a method of storing the requirements of AS1428.1, 'Design for Access and Mobility – General Requirements'. The software was easy to develop rules in, easy to use and produced excellent reports with the ability to embed 'snap shots' from the viewer of identified problems. The drawbacks were that there was no method for a user to access attributes and values that were not already exposed through the Solibri rule interface and it was not possible to write algorithms, for example to define an unobstructed path.

e-PlanCheck is the most widely known and publicised example of code checking since it was sponsored by the Singapore government (Khemlani, 2005). It covers the architectural and building services aspects of the Singapore code (Solihin, 2004b). It is based on the Express Data Model system (Solihin, 2004a).

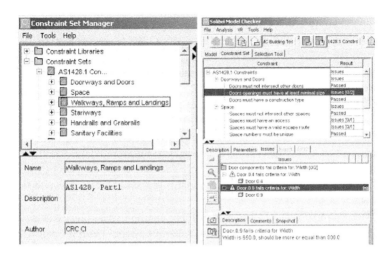

Figure 8.4 Solibri constraint set manager and reporting engine.

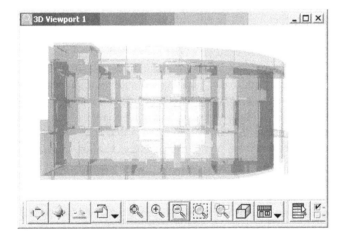

Figure 8.5 Solibri viewer with identified problem.

In 2004, 92 per cent of the clauses in the architectural requirements were met and 77 per cent of the building services requirements. Most of these were removed due to the impracticality of checking rather than an inability to do so (Solihin, 2004b). While ePlanCheck provides 24 hours-a-day by 7-day-per-week access, it does not allow designers to define their own rules. It is consequently limited to checking against the provided requirements.

Checking for compliance

The Cooperative Research Centre for Construction Innovation (www.construction-innovation.info) launched the first stage of a two-stage project to develop code checking using AS1428.1, 'Design for Access and Mobility – General Requirements' as the requirements in 2001 and completed the second stage in 2005. The system uses the Express Data Model System like the Singaporean ePlanCheck system. The rules for checking compliance are encoded using Express-X, a constraint language defined as part of the ISO10303 (STEP) suite of standards.

Implementing code checking requires a combination of capable building modelling software and the rules to check against the stored data. The IFC model is commonly used as the means of encoding data due to the support by the major architectural CAD vendors and its compatibility with the STEP (ISO10303) standards.

Checking against some clauses is trivial. For example, AS1428.1 Clause 10.2.5 states that 'the toilet seat shall be of the full-round type, i.e. not open fronted.' While CAD systems do not normally support this level of detail, an extra property stating that the toilet seat was 'closed' was exported for toilet seats that had to be compliant with this clause.

Some clauses are unlikely to ever be fully supported from CAD. For example, clause 10.5.4 (c) states that 'floor tracks for sliding screens shall be flush across the shower recess opening'. It is difficult to imagine someone going to the trouble of building a building model to such detail.

Clause 5.1.2 states 'there shall be a continuous path of travel to and within any building to provide access to all required facilities.' Resolution of this clause requires an algorithmic approach. The method used to implement this rule was a recursive function that started at a complying entrance and then iterated through the spaces within the building model to find other spaces which were accessible (had a complying door or opening) from a space that is known to be accessible. Ramps and lifts were treated as complying openings to allow changes in level and storey to be accommodated (Figure 8.6).

The most difficult type of checking requires geometric reasoning. Figure 8.7 (from AS1428) shows a series of doors along a corridor with constraining dimensions. This type of situation is identified by finding spaces, locating the doors into the space, finding the direction of swing and then overlaying the 2D shapes to calculate the minimum distance between the door swings.

Implementing a code checking system

As mentioned earlier, some additional information must be added to the CAD model to allow full checking. These are passed through the IFC export interface as extension properties. These are not part of the official IFC model but can be handled as user-defined extensions. The IFC model allows the definition of 'proxy' objects. These are objects which are not part of the official IFC model but are understood in the context of a one-to-one information exchange. Some CAD

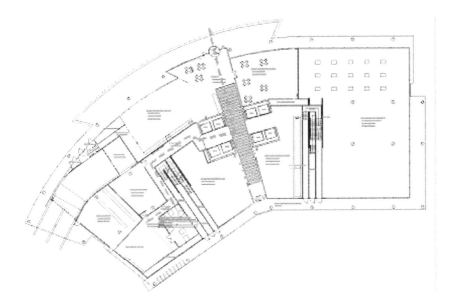

Figure 8.6 Checking continuous routes.

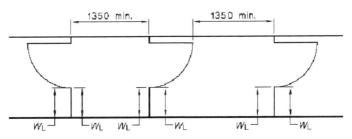

Figure 8.7 Door tolerances from AS1428.1.

systems also support library objects as proxies since they cannot determine their type explicitly.

A variation to the IFC model was defined that had additional objects that were necessary to cover the required scope. It also added the attributes that were referenced through the rules to the express definition of the object rather than to attached property sets to make the writing of rules more intuitive. This is called the 'internal model'. Processing a file consists of the following steps:

- export the model from CAD as an IFC file;
- read the model into EDM;

- convert all proxy objects of interest into objects of known types;
- map (convert) the IFC data to the internal model;
- run the rules as selected by the user and produce the reports.

Using design check

Using design check is a five stage process:

- select the rule set to be checked against;
- select the building model to check;
- select the rules or objects to check;
- run the check;
- examine and respond to the report.

The user can then make changes to the model if any are necessary and run Design Check again.

Figure 8.8 shows the code selection dialogue. Each code has a unique name under which the relevant rules are stored. The dialogue shows the two codes that are currently stored, AS1428 part 1, the current Australian standard for access for disabled access, and BCAD3, the draft provisions that are currently out for public comment as the next version of the disabled access provisions.

The user then selects the project to be checked from the project database using a standard file open dialogue. Checking can be performed based on either rules or on the relevant object types (Figures 8.9–8.12). When checking against rules the user can check against the entire code by selecting all clauses, or can select individual clauses or sub-clauses. The sub-clauses are accessed by drilling down through the tree view (Figure 8.9). The user can select as many rules as desired in whatever combinations are appropriate. This ability to choose relevant clauses supports the checking of designs at various stages of development. If, for example, the user has just finished laying out the spaces and doorways, access to all appropriate spaces can be checked by selecting the appropriate clause, without being overwhelmed in the reports by information about missing detail that has not been considered yet.

Once the 'Check Project' button is hit the rule checking is run and the results displayed (Figure 8.10). There are five possible results for each selected clause.

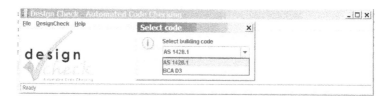

Figure 8.8 Selecting the desired code for checking.

Figure 8.9 Drilling down into clause hierarchy and selecting clauses.

'Passed' means that the clause is satisfied. 'Pass given' indicates that additional information has been added to the model to explain why a particular clause has not been met. This may be because a particular requirement has been met through an alternative method, as allowed under performance-based codes. 'Specification needed' indicates that information is required that is not supported by current architectural CAD systems. It is expected that this information would be covered in the project specification in some manner. References to a specification clause or other documents can be added and explained later. This facility allows the full 'design' to be checked even if a requirement is not supported by current technology. 'Model information missing' indicates that objects that are expected to be in the model are not there. The user can either add the information to the next iteration of the model or can add a reference to the project specification or other information (i.e. product manufacturer's brochure). This provides flexibility in the location of information. Within Design Check there is no requirement that all required information be entered in a particular manner. The final possible result, 'some issues' indicates that all of the required information is in the model, but does not comply with the clause requirements.

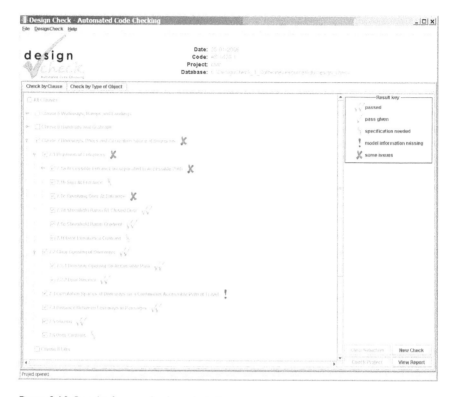

Figure 8.10 Results from a check against clauses.

Designers will normally need more information about the results which can be accessed from the 'View Report' button. A report is generated (Figure 8.11) which allows the user to see in more detail each issue. Filters at the top of the report allow the user to view parts of the report. 'Non-compliance' shows all of the problems that must be fixed. 'Compliance' shows all objects that comply with the requirements. This is provided for auditing purposes. Code compliance checkers required the capability to check whether a particular clause had fired against an object and passed. This allows checking of the accuracy of the model by the user. The final filter is 'Specification needed' which allows the user to quickly identify where additional information is required and then add it.

Three tab panes are displayed below the results. 'Model information missing' allows the addition of comments to cover expected information that is not in the model, 'Specification needed' allows the addition of references to documents outside of the CAD model and 'Checker designer comments' allows other people to add comments. All of the information entered in these tab panes is added to the model and can be viewed by anyone accessing the model thereafter.

Figure 8.11 Online report.

This information can also be retained if the model is 'round tripped' through an architectural CAD system.

Checking a particular object type(s) follows a similar path to checking against clauses. Selecting the 'Check by type of object' tab brings up a listing of all of the objects of concern for this code (Figure 8.12). The objects to be checked are then selected from the list, with the ability to drill down into the object sub-type hierarchy through the tree view. The door objects are selected which is complementary to the query chosen previously (Figure 8.9).

The results are shown in the same way as for clauses (Figure 8.13) and the report is similar (Figure 8.14). Note that the same object, RevolvingDoor_01, is found in both analyses as would be expected. The number of results typically increases when checking by a single object type against checking by a single clause that has that object type in scope. Checking by clause returns the results of the application of a single clause against all of the objects, with the domain of application of a clause normally being restricted by the intent of the clause. Checking by object type applies all clauses against that particular object type.

The standard reports are presented in a fairly dense format set up for interactive use to make them easier to use when adding comments and data to the model. These are implemented as a Java user interface. 'Printer friendly' reports provide

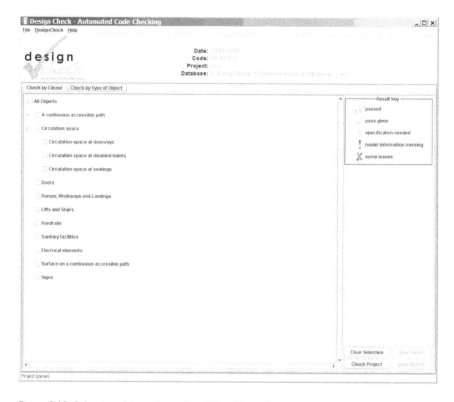

Figure 8.12 Selecting objects from the object hierarchy.

a format that is preferable for printing or for online examination and use html so that they can be displayed in any standard web browser (Figure 8.15). Both reports are generated from the same results data that is stored within the model so they will always be consistent – a single source for two different presentation methods.

Further development for design check

There are two major areas where design check needs improvement:

- For users, integration with a viewer application, as Solibri Model Checker provides, is necessary. This is underway.
- The second area is to support the writing of rules. As mentioned previously, writing sets of rules that are accurate and consistent is difficult. The current use of Express X within a text editor is not optimal. Better support for rule writing is necessary before the 'normal user' can be expected to write complex rules and sets of rules.

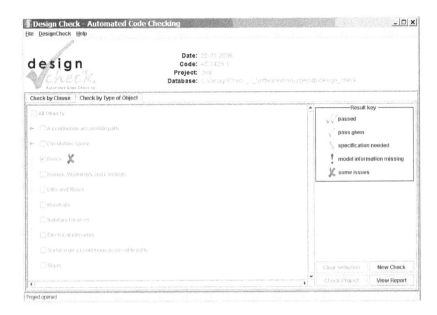

Figure 8.13 Results from checking doors.

Figure 8.14 Online report for check on doors.

Figure 8.15 Printable report.

Testing of Design Check on live projects has also indicated that the amount of information generated in a scan for all but small projects is very large and difficult to handle. Systems such as Design Check will need to identify new ways of handling large amounts of data and presenting this information in understandable and usable ways.

Conclusion

The release of CAD software that supports the IFC standard is giving a new lease of life to the concept of automated code checking. The three systems described earlier all take different approaches and have their own strengths and weaknesses. As the use of building information models increases in industry, the number and range of code checking systems is likely to increase. In the long term this will lead to efficiencies in the design process and reduction in the cost of errors during design, construction and use of buildings. The full implications of these changes will take years to emerge. However, it is totally reliant on the concept of the nD model to make such a system viable.

Acknowledgements

The authors acknowledge the input from Moshe Gilovitz (Building Commission) and David Marchant (Woods Bagot Architects) for their valuable advice and coordination of input from others, and Cheryl and McNamara, Kevin McDonald, Dr John Mashford and Fanny Boulaire of CSIRO and Professor John Gero, Dr Mike Rosenman, Julie Jupp, Wei Peng, JiSoo Yoon and Nicholas Preema of University of Sydney for significant contributions to the Design Check system.

References

IAI (2004) 'IFC 2x Edition 2 Addendum 1', www.iai-international.org/Model/R2x2_add1/index.html

Khemlani, L. (2005) CORENET e-PlanCheck: Singapore's Automated Code Checking System, http://www.aecbytes.com/buildingthefuture/2005/

Rauch, K. (2005) 'Does an Ice Hotel Require a Sprinkler System? Building Code Interpretations in the Provinces and Territories', Building Envelope Forum 3rd Ed, http://www.buildingenvelopeforum.com/sprinklersystems.htm

Solihin, W. (2004a) novaCITYNETS Seals Partnership with Norwegian Company, http://www.novacitynets.com/news_2004oct1.htm

Solihin, W. (2004b) IAI VIU Workshop, 'Lessons learned from experience of code-checking implementation in Singapore', http://www.iai-international.org/NewsEvents/IAI_VIU_Workshop_Oct_2004/Lessons%20learned%20from%20experience%20of%20code_checking.pdf

Chapter 9

Acoustics in the built environment

Yiu Wai Lam

Introduction

Acoustics is the science of sound, and sound is of course an important element of everyday life. There are however two opposite facets of sound. On the good side, sound is an important instrument for communication, via speech, and is a medium to convey enjoyment, through music and other desirable soundscape. Since the passage of sound is significantly affected by the surrounding environment, good acoustic design of a built environment is essential to avoid degrading the functioning and the enjoyment of sound.

On the other hand, unwanted sound, which is generally referred to as noise, is a nuisance that substantially lowers the quality of life, interferes with the enjoyment of normal activities and at high levels can damage health and even create safety hazards. Unlike optical vision which can be shut off by closing one's eyes or by turning away from the offending scene, noise cannot be as easily avoided. One hears noise from all directions and there is no mechanism to close the ears! Some forms of external means are often required to mitigate noise nuisances. Indeed noise from an environmental stance is of increasing concern. The control of noise insulation in residential buildings, the prevention of excessive noise exposure in sensitive areas such as schools and workplaces are important topics in environmental noise impact assessments and are regulated by law. Controlling such noise is therefore an important aspect of modern building design.

This chapter aims to provide a brief introduction to various aspects of acoustics that should be taken into account in the design of a built environment – from the basic science to legislative requirements. It then illustrates some of the basic modelling methods that can be used to estimate the acoustic performance of building elements and building spaces. Some of the concepts of more advance modelling techniques will also be introduced.

Basic acoustics

Speed of sound

The quantity that we perceived as sound is caused by minute oscillatory changes in the ambient static atmospheric pressure. The term sound pressure refers to the small difference between the instantaneous pressure and the static pressure. Typical conversation produces a sound pressure of about 0.04 Pascal (Pa), while the atmospheric pressure is about 101 kPa. It is a dynamic wave quantity, as air particles oscillate about their equilibrium position. When the particles move towards each other they create an over-pressure and a positive sound pressure, and when they move apart a rarefaction and hence a negative sound pressure occurs. Note that although the sound pressure is negative, the total pressure (atmospheric plus sound) is still positive since the sound pressure is several orders of magnitude smaller than the atmospheric pressure. Because the air particles move in an oscillatory motion, there is no net movement of the air particles. The speed of sound therefore only refers to the speed the sound wave or energy that is transmitted. Because the sound wave manifests itself as the oscillation of the particles in the air medium, the speed of sound depends on the property of the medium, which is in turn affected by the atmospheric pressure and temperature. For air, the speed of sound, denoted by the symbol c, in meter per second (m/s) is given by

$$c = 331.5\sqrt{1 + \frac{T}{273}}$$

Sound pressure level

We hear sound when our ear drum is set to oscillate by the sound pressure. Both the amplitude and the rate of this oscillation greatly affect our perception of the sound. Because the instantaneous sound pressure oscillates between positive and negative values and changes considerably over time, it is more convenient in practice to characterise the amplitude of sound by the root mean square (rms) of the sound pressure, averaged or weighted over a suitable time period. Descriptions of some standard time averaging and weighting functions can be found in national and international standards for sound measuring equipments, for example the British Standard (BS EN 61672-1, 2003). It suffices to say that the time averaging or weighting is chosen to best represent the characteristics of the sound wave relevant to human hearing and is dependent on the type of sound.

The human ear responds to energy input proportional to the square of the sound pressure. The dynamic range it can detect is amazingly large. A young, healthy human can hear sound that is as weak as 20×10^{-6} Pa, or 20 μPa, while the sound pressure that the ear can tolerate before experiencing pain is about 60 Pa, which is more than a million times higher than the weakest detectable sound. To cope with such a large dynamic range, a logarithmic scale, the decibel or dB scale, is used to describe the sound pressure amplitude. The dB scale is a measure of the logarithmic ratio between the rms sound pressure p, and a reference sound pressure value p_o, which is chosen to be the hearing threshold, that is 20 μPa. The resulting quantity is termed the sound pressure level, commonly represented by the symbol L_p.

$$L_p = 10\log_{10}\left(\frac{p}{p_0}\right) dB$$

Thus the hearing threshold is at 0 dB and the pain threshold is at about 130 dB. A doubling of the sound pressure amplitude corresponds to a 6 dB increase in L_p. The SPL equation is also frequently written in the form of

$$L_p = 10\log_{10}\left(\frac{p^2}{p_0^2}\right) dB$$

Since sound energy is proportional to the square of the sound pressure, this latter equation relates the dB scale explicitly to the energy of the sound pressure, and the factor 10 explains the deci prefix of the decibel. Moreover, it can be seen that a doubling of sound energy only gives a 3 dB increase in the sound pressure level L_p. The human ear however does not respond linearly to sound pressure increases. Roughly speaking, a 10 dB increase in L_p is required before the sound subjectively appears to be twice as loud.

In the assessment of noise that may vary with time, it is often necessary to calculate the equivalent continuous sound pressure level, $L_{eq,T}$ over a time period T that runs from t_1 to t_2:

$$L_{eq,T} = 10_{10}\left[\frac{1}{t_2 - t_1}\int_{t_1}^{t_2}\frac{p^2(t)}{p_0^2}dt\right]$$

Frequency

The frequency of the sound refers to the rate at which the air particle or sound pressure oscillates, and is measured in Hertz or Hz. One Hz is one cycle per second. A related quantity that is also often used in acoustics is the wavelength λ.

$$\lambda = \frac{c}{f}$$

Wavelength is the distance travelled by one period or one cycle of the wave. It is a useful quantity since how a sound wave reacts with an object depends on the size of the object relative to the wavelength. When the size is small relative to the wavelength, it has very little effect. This explains why it is more difficult to absorb or shield sound at lower frequencies where the wavelength is large. A larger and more costly (in terms of space and construction requirements) object is required to do the job at lower frequencies.

A young and healthy human ear can detect sound with frequencies as low as 20 Hz and as high as 20 kHz. The ear is however not equally sensitive at all frequencies. It is particularly sensitive to sound of frequency between 1,000 and 5,000 Hz,

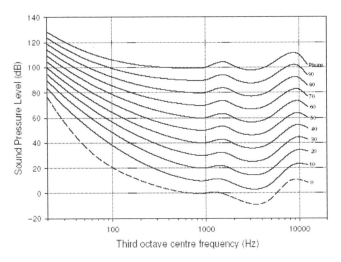

Figure 9.1 Equal loudness curves.

which is not surprising since this is the main frequency range of human speech. The sensitivity drops off sharply at frequencies below 250 Hz and at frequencies higher than 5 kHz. The sensitivity is also dependent on the sound pressure level. Figure 9.1 shows the equal loudness curves at different loudness levels, which is measured with the unit of *phons* (ISO 266, 1997). An equal loudness curve gives the sound pressure level at each frequency that sounds equally loud subjectively to a normal healthy human being, and the *phon* level is defined by the sound pressure level of the curve at the reference frequency of 1 kHz. One can see that the perceived loudness, in another word the hearing sensitivity, greatly depends on the frequency and also on the sound level. At the hearing threshold, which corresponds to the 0 *phon* curve, the sensitivity at 20 Hz is more than 70 dB (more than 3,000 times in terms of sound pressure) worse than at 1 kHz. On the other hand, at very high levels, say for example the 120 *phon* curve, the difference in sensitivity at these two frequencies is less than 20 dB. It is therefore important that we take both the sound level and the frequency of the sound into account when we assess the human perception of sound. The sound pressure level as a function of frequency is referred to as the frequency spectrum of the sound.

Octave and 1/3 octave frequency bands

Most noise events are made up of components with many different frequencies. In the simplest case the overall sound pressure level of the noise is measured by passing the noise through a linear bandpass filter of frequency range 20 Hz–20 kHz.

This overall level is however rarely used because of its poor correlation with subjective response. The frequency components of the noise may be separated into discrete narrow frequency bands using techniques such as Fast Fourier Transform, for example, to enable analysis of tonal noise contributions. However, in the majority of cases the noise will be filtered into groups of frequencies or frequency bands to be examined together. The two most commonly used groupings of frequency bands are the *octave* and 1/3 *(onethird) octave* series of frequency bands. The definitions of these frequency bands are contained in international standards (ISO 266, 1997; BS EN 61672-1, 2003) and are incorporated in most noise-measuring and analysing equipments. The centre frequencies of these frequency bands are established by successive intervals above and below the reference frequency of 1,000 Hz. In the series of octave frequency bands, the centre frequency of a band is double that of the immediate lower band. The lower and upper frequency limits of an octave band are given respectively by dividing and by multiplying the centre frequency of the band by a factor of $\sqrt{2}$. The series of 1/3 octave bands is similarly defined, but with the factors 2 and $\sqrt{2}$ replaced by $2^{1/3}$ and $2^{1/6}$ respectively. Table 9.1 shows the commonly used centre frequencies of the octave and 1/3 octave bands.

One of the reasons for the popularity of octave and 1/3 octave bands is that the human ear can be considered to act as a set of overlapping constant percentage bandpass filters, and the 1/3 octave bands correspond approximately to the frequencies and the bandwidths of most of these filters. The octave bands are generally used to reduce the amount of data to be handled in cases where the frequency content of the sound does not vary sharply.

Frequency weightings

Because of the differences in the sensitivity of human hearing to sound of different frequencies, it is essential to take this into account when the impact of sound is assessed. A few standard weighting systems have been defined in international standards (BS EN 61672-1, 2003) to adjust the sound pressure levels to approximate the ear response, so that the resulting level can be related to subjective perception. These weighting systems are incorporated as an electrical filtering network into calibrated noise-measuring equipment, like a sound level meter, so that the weighted sound pressure levels are usually measured directly. The most

Table 9.1 Centre frequencies of octave and 1/3 octave frequency bands

Octave bands	Centre frequencies (Hz)								
	31.5	63	125	250	500	1,000	2,000	4,000	8,000
	25	50	100	200	400	800	1,600	3,150	6,300
1/3	31.5	63	125	250	500	1,000	2,000	4,000	8,000
Octave bands	40	80	160	315	630	1,250	2,500	5,000	10,000

commonly used weighting in noise analysis is the A-weighting, and the resulting sound level measured using this weighting system is called the *A-weighted* sound pressure level, and is denoted by a subscript 'A' in the symbol, for example L_{pA}. The *A-weighting* is a frequency response curve defined to follow approximately the inverse of the equal loudness curve (Figure 9.1) with a reference level of 40 dB at 1 kHz. The overall *A-weighted* sound pressure level, from 20 Hz to 20 kHz, has been shown to correlate very well with subjective responses and is widely used as a single-figure descriptor of noise.

The *A-weighting's* choice of an equal loudness curve with a 1 kHz reference level of 40 dB is biased towards lower levels of noise. This is appropriate for most noise annoyance assessments where the noise levels are expected to be low to moderate. However, since hearing sensitivity, and hence the shape of the equal loudness curve, is dependent on the absolute sound level, the *A-weighting* may not be appropriate for cases involving high noise levels. Consequently other weighting systems have been defined on equal loudness curves with higher reference levels. For an example the *C-weighting* is defined on the equal loudness curve with a 1 kHz reference level of 100 dB. Figure 9.2 shows the relative frequency response of the *A* and *C-weighting*. It can be seen clearly that the *A-weighting* applies a much more severe penalty on noise at low frequencies, due to the insensitivity of human hearing to low-level, low-frequency noise.

There are other weighting systems which are defined for specific applications, such as the SI-weighting for the assessment of speech interference. Among all the noise weighting systems, the *A-weighting* is by far the most widely used for general purpose noise assessments and is incorporated in almost all noise-measuring equipment. Most noise limits set in environmental planning are specified in *A-weighted* sound pressure levels.

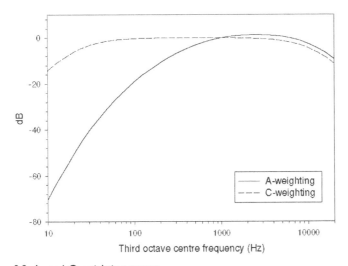

Figure 9.2 A- and C-weighting curves.

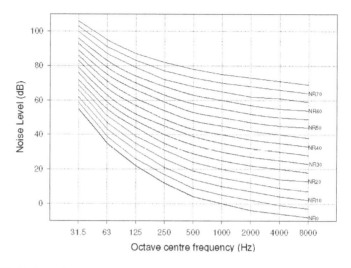

Figure 9.3 Noise rating curves.

The *A-weighting* and *C-weighting* are applied to represent human hearing sensitivity to frequency. There are also other weightings that are used to represent the frequency characteristics of noise sources, which are useful to rank acoustic performance of building elements against noise. For example the standard traffic noise spectrum (BS EN ISO 717-1, 1997; BS EN ISO 1793-3, 1998) is used in the assessment of noise insulation performance of building elements of residential buildings. Another way of using frequency weighting is to see if the actual frequency spectrum of the noise falls below a set of standard rating curves. A popular set of curves, used originally for rating environmental noise but is now frequently used for rating background noise in buildings, is the Noise Rating Curves (BS 8233, 1999), shown in Figure 9.3. The Noise Rating number is given by the level at 1 kHz. For example the NR30 curve has a sound pressure level of 30 dB at 1 kHz, and to achieve NR30, the frequency spectrum of the actual noise in the building must fall below the NR30 curve.

Acoustic requirements for the built environment

A good quality of life demands suitable protection against noise nuisance. Hence, most of the legal requirements on sound concern noise. On the other hand, requirements on critical listening spaces, such as auditoria, are generally specified according to individual designs, although codes of practice do exist for more common cases such as schools and offices. This section gives a brief introduction to some of the requirements that have the widest impact on common building design. Although the requirements quoted are mainly from UK sources, most other countries should have similar standards.

Planning and noise

Perhaps the first piece of regulation that one would face in the building of residential properties will be the Planning Policy Guidance 24, PPG24 (1994), that one will need to consider in the planning stage. The PPG24 (1994) gives guidance to the local authorities in England on the use of their planning power to accept or reject planning applications based on the impact of existing noise on the planned developments. It introduces four Noise Exposure Categories (NECs) for residential developments. The NECs are based on day-time (07.00 to 23.00) and night-time (23.00 to 07.00) noise exposure levels from different types of traffic and mixed noise sources. There is also a special proviso that limits occurrence of noise events with maximum levels equal to or exceeding 82 dB A-weighted during night time. The advice for each Noise Exposure Category is as follows:

- **NEC A**: Noise need not be considered as a determining factor in granting planning permission, although the noise level at the high end of the category should not be regarded as a desirable level.
- **NEC B**: Noise should be taken into account when determining planning applications and, where appropriate, conditions imposed to ensure an adequate level of protection against noise.
- **NEC C**: Planning permission should not normally be granted. Where it is considered that permission should be given, for example because there are no alternative quieter sites available, conditions should be imposed to ensure a commensurate level of protection against noise.
- **NEC D**: Planning permission should normally be refused.

The range of noise levels corresponding to each of the Noise Exposure Category is given separately for different types of noise sources to account for the dependence of noise nuisance on source characteristics. These are reproduced in Table 9.2. However, no specific advice is given in the Planning Policy Guidance 24 for industrial noises alone as there is insufficient information to allow detailed guidance to be given. In such cases specific advice has to be sought from other procedures. Only when the industrial noise is part of a mixture of noise and is not the dominant component can it be considered under this Guidance.

Although it is tempting to use the NEC procedure to assess the impact of introducing new noise sources, such as a new road, into an existing residential area, the Planning Policy Guidance 24 specifically advises against such applications. One of the reasons is that the options for noise control measures that are required under NEC B and C will be more limited under these circumstances. In such cases individual environmental assessment will normally be required and the impact of noise will be considered in that context.

Building regulations

The Approved Document E of the Building Regulations for England and Wales (2003) provides practical guidance with respect to the requirements of building

Table 9.2 The Planning Policy Guidance 24, PPG24 (1994), noise levels ($L_{Aeq,T}$ dB) corresponding to the Noise Exposure Categories. These are free field levels, with corrections for possible ground reflection, and not facade levels

Noise source	Noise exposure category			
	A	B	C	D
Road traffic				
07.00–23.00	<55	55–63	63–72	>72
23.00–07.00	<45	45–57	57–66	>66
Rail traffic				
07.00–23.00	<55	55–66	66–74	>74
23.00–07.00	<45	45–59	59–66	>66
Air traffic				
07.00–23.00	<57	57–66	66–72	>72
23.00–07.00	<48	48–57	57–66	>66
Mixed traffic and industrial sources				
07.00–23.00	<55	55–63	63–72	>72
23.00–07.00	<45	45–57	57–66	>66

elements to provide adequate resistance to the passage of sound. It applies to newly built buildings or buildings that have a 'material change of use' for residential purposes. This is done in the context of addressing the welfare and convenience of building users. Document E details four requirements:

- **E1** – Protection against sound from other parts of the building and adjoining buildings.
- **E2** – Protection against sound within dwelling houses, flats and rooms for residential purposes.
- **E3** – Reverberation in the common internal parts of buildings containing flats or rooms for residential purposes.
- **E4** – Acoustic conditions in schools.

The Requirement E4 is a new addition in 2003 to bring schools under the legal framework of the Building Regulations Approved Document E.

Requirement E1

To satisfy Requirement E1, the relevant building elements need to achieve the sound insulation values given in Table 9.3. Pre-completion testing using field tests of noise insulation performance of building elements on-site is necessary to demonstrate compliance. This is to ensure that all the flanking transmission (transmission through paths other than the direct path through the building

Table 9.3 Performance standards required under E1 for separating walls, separating floors, and stairs that have a separating function

	Airborne sound insulation sound insulation $D_{nT,w+}C_{tr}$ dB (Minimum values)	Impact sound insulation $L'_{nT,w}$ (Maximum values)
Purpose built dwelling-houses and flats		
Walls	45	
Floors and stairs	45	62
Dwelling-houses and flats formed by material change of use		
Walls	43	
Floors and stairs	43	64
Purpose built rooms for residential purposes		
Walls	43	
Floors and stairs	45	62
Rooms for residential purposes formed by material change of use		
Walls	43	
Floors and stairs	43	64

element in question) and construction effects are taken into account in the evaluation of performance. Laboratory testing on building elements samples can be used to evaluate the likely performance of a building element but it is not possible to use laboratory test results to demonstrate compliance even if the laboratory include flanking paths. The emphasis on pre-completion field tests has a significant implication on cost. There are suggestions from the construction sector that perhaps some sufficiently robust constructions can be prior approved to reduce the requirement on pre-completion testing. However, at the time of writing this has not yet been accepted.

In the table, the quantitiy $D_{nT,w}$ is the weighted standardised level difference. It is a single figure number obtained by comparing the standardised level difference D_{nT} values measured in 1/3 octave frequency bands to a reference curve using the procedure described in ISO 717-1 (1997). The standardised level difference D_{nT} is determined from the measured difference in sound levels, D, between two adjoining rooms corrected for the reverberation time in the receiving room (BS EN ISO 140-4, 1998):

$$D_{nT} = D + 10 \log_{10} \frac{T}{T_0} dB$$

where T is the reverberation time of the receiving room and T_0 is the reference reverberation time which is defined as 0.5 s for dwellings. Note that the values are given in dB scale. In linear terms, a 40 dB reduction corresponds to a 10^{-4} reduction of the energy level. So a lot of the sound energy has to be isolated in order to achieve the required specifications.

The term C_{tr} is a spectrum adaptation term using a typical A-weighted urban traffic noise spectrum. The addition of this term means that the requirement assumes predominately traffic noise sources.

The quantity $L'_{nT,w}$ is the Weighted standardised impact sound pressure level. It is a single-figure number obtained by comparing the standardised impact sound pressure level, L'_{nT} values measured in 1/3 octave frequency bands to a reference curve using the procedure described in ISO 717-2 (1997). The standardised impact sound pressure level L'_{nT} is determined from the measured impact sound levels, L_i, corrected for the reverberation time in the receiving room (BS EN ISO 140-7, 1998):

$$L'_{nT} = L_i - 10 \log_{10} \frac{T}{T_0} \text{dB}$$

Requirement E2

To satisfy the Requirement E2, the relevant building elements need to achieve the sound insulation values given in Table 9.4. In this case laboratory testing is to be used to demonstrate compliance.

In the table, the quantity R_w is the weighted sound reduction index. It is a single-figure number obtained by comparing the sound reduction index, R, whose values are measured in 1/3 octave frequency bands, to a reference curve using the procedure described in ISO 717-1 (1997). The sound reduction index, R, is measured using the procedure described in ISO 143-3 (1995).

Requirement E3

The purpose of Requirement E3 is to prevent more reverberation around the common parts than is reasonable (Table 9.5). This is controlled by ensuring that

Table 9.4 Performance standards (laboratory values) required under E2 for new internal walls and floors within: dwelling-houses, flats and rooms for residential purposes, whether purpose built or formed by material change of use

	Airborne sound insulation R_w dB (Minimum values)
Walls	40
Floors	40

Table 9.5 Absorption required to satisfy Requirement E3 for reverberation in the common internal parts of buildings containing flats or rooms for residential

	Method A Area covered by rated absorber (Minimum values)	Method B Absorption area per cubic metre volumn (Minimum values)
Entrance halls	Entire floor are by Class C absorber	0.2 m²
Corridors or hallways	Entire floor are by Class C absorber	0.25 m²
Stairwalls or a stair enclosure	Entire combined areaª by Clas D or 50% by Class C absorber or better	Not applicable

Note
a Combined area of the stair treads, the upper surface of the intermediate landings, the upper surface of the landings (excluding ground floor) and the ceiling area on the top floor.

a sufficient amount of sound absorption is in place in the common parts of the building. The Requirement is given in terms of the minimum amount of absorption required in the area. Two alternative methods are described to satisfy the requirement:

* Method A – Cover a specified area with an absorber of an appropriate class that has been rated according to BS EN ISO 11654:1997 Acoustics Sound absorbers for use in buildings Rating of sound absorption (BS EN ISO 11654, 1997).
* Method B – Determine the minimum amount of absorptive material using a calculation procedure in octave bands.

Except for stairwells and stair enclosures in which Method B is not applicable, Method B generally allows more flexibility in the choice and application of absorbers. The absorption area is defined as the hypothetical area of a totally absorbing surface, which if it were the only absorbing element in the space would give the same reverberation time as the space under consideration. A larger absorption area will produce a shorter reverberation time in a given space. The reverberation time is defined as the time in seconds for the sound pressure level to decay by 60 dB after the source of the sound is switched off. It is a common measure of how reverberant a space is. A longer reverberation time means that the sound energy will remain longer in the space and is therefore potentially more likely to interfere with other activities. The total absorption area, A_T, is calculated by the sum of the absorption area of all the surfaces in the space.

$$A_T = S_1\alpha_1 + S_2\alpha_2 + \ldots + S_n\alpha_n$$

where n is the total number of surfaces in the space, each with a physical area S in m^2 and an absorption coefficient α. The absorption coefficient represents the fraction of sound energy absorbed by the surface when sound hits the surface.

Requirement E4

With the introduction of Requirement E4, the Building Regulations extends its scope to cover schools as well as residential buildings. The Requirement aims to ensure that each room or other space in a school building shall have the acoustic condition and the insulation against disturbance by noise appropriate to its intended use. Therefore the performance criteria given under the Requirement cover not only noise insulation, but also internal ambient noise levels and reverberation time. However Document E itself does not give explicit guidance for this Requirement. Instead, it states that the normal way of satisfying Requirement E4 will be to meet the values for sound insulation, reverberation time and indoor ambient noise which are given in Section 1 of Building Bulletin 93 'The Acoustic Design of Schools' (2003). We will therefore look at the specific details in the following section on Building Bulletin 93.

The acoustic design of schools

The Building Bulletin 93 (2003) aims to provide a regulatory framework for the acoustic design of schools in support of the Building Regulations in the United Kingdom. It gives advices and recommendations for planning and design of schools in the context of acoustic conditions suitable for school activities. It is well recognised that teaching and learning activities requires suitable acoustics and are adversely affected by excessive noise. Teachers will need to be able to communicate clearly with students through speech and students will need to be able to concentrate on their intended activities without undue disturbance form noise. However, a school is a dynamic environment with a great variety of activities many of which produce high levels of noise themselves. Hence it is essential to be able to reach a right balance of acoustics in this complex environment. The Bulletin aims to achieve this by providing guidance on performance standards on four acoustic items: the internal ambient noise level, sound insulation, reverberation time and speech intelligibility.

Internal Ambient Noise Level (IANL)

The IANL is the highest equivalent continuous A-weighted sound pressure level, $L_{Aeq,T}$ over a 30-minute time period, $T = 30$ min, that is likely to occur during normal teaching hours. This is a measure of the background noise level created by external noise sources (e.g. traffic noise) and building services (e.g. mechanical ventilations). Since the noise coming in to a room depends on the opening of windows, the ventilators or windows should be assumed to be open as required to provide adequate ventilation. On the other hand, if the room is mechanically ventilated, the ventilation plant should be assumed to be running at its maximum operating duty. The IANL however excludes noise from teaching activities and from machines (e.g. machine tools, computers, projectors, etc.) used in the space for the teaching activities. Noise from these are expected to be dealt with in the consideration of sound insulation and the design and specification of the machines. Very high noise levels from machines are also under the control of health and safety legislations.

Interestingly, although rain noise is known to create problems sometimes in schools with lightweight roofs and roof lights, the current Bulletin exclude rain noise in the consideration of IANL. It does indicate its intention to introduce a performance standard for rain noise in a future edition of BB93. This situation is caused by the limited data available on the subject and the lack of a standard for the measurement of insulation against rain noise. A draft international standard, ISO/CD140-18 (2004) is being prepared on rain noise measurement methods. It is reasonable to expect that BB93 (2003) will be revised to include rain noise performance requirements once the standard is accepted and manufacturers are able to provide product data from laboratory measurements.

Because of the large variety of spaces and activities in school, there are a large number of values given by the Bulletin for the limit of IANL for each type of space to cater for its intended use. Details are not reproduced here but can be found in Table 9.1 of the Bulletin (BB93, 2003). As examples, the IANL limits range from 30 dB for large lecture rooms (more than 50 people) and classroom for use by hearing impaired students, 35 dB for small lecture rooms and study rooms, to a maximum of 50 dB for swimming pools and kitchens. These school spaces are also rated (using Low, Medium, High, etc.) individually as noise-producing rooms according to their activity noise, and as noise-receiving rooms according to the noise tolerance of their intended use. These ratings are then used to specify the sound insulation required between rooms.

Sound insulation

For air-borne sound insulation, the required minimum values between rooms are given in terms of the weighted standardised level difference, $D_{nT(Tmf,max),w}$ between two rooms (Table 9.6). This weighted difference is similar to the $D_{nT,w}$ used in Approved Document E of the Building Regulations except that the reference reverberation is not 0.5 s but the upper limit, $T_{mf,max}$, of the mid-frequency reverberation time, T_{mf}, of the specific receiving room. Values of T_{mf} are given in Table 9.5 of Bulletin as performance requirements for all types of teaching and study spaces in finished but unoccupied and unfurnished conditions.

In line with Approved Document E of the Building Regulations, the air-borne sound insulation requirements for the walls and doors connecting to common circulation spaces are given in terms of the laboratory tested weighted sound reduction index, R_w. For ventilators, because of the small size of the openings,

Table 9.6 Minimum weighted standardised level difference, $D_{nT(Tmf,max),w}$ between two rooms as specified in Building Bulletin 93 (2003)

Noise tolerance in receiving room	Activity noise in source room			
	Low	Average	High	Very high
High	30	35	45	55
Medium	35	40	50	55
Low	40	45	55	55
Very low	45	50	55	60

Table 9.7 Performance standards for airborne sound insulation between circulation spaces and other spaces used by students as specified in Building Bulletin 93

Type of space used by students	Minimum R_w (dB)		Minimum $D_{n,e,w}-10\,log_{10}N$ (dB)
	Wall including any glazing	Doorset	
All spaces except music rooms	40	30	39
Music rooms	45	35	45

the requirement is given in terms of the weighted element-normalised level difference, $D_{n,e,w}$, $-10\,log_{10}\,N$, where N is the total number of ventilators with $D_{n,e,w}$. $D_{n,e,w}$ is a single-figure number obtained by comparing the element-normalised level difference, $D_{n,e}$, values measured in 1/3 octave frequency bands to a reference curve using the procedure described in ISO 717-1 (1997) (Table 9.7). The standardised level difference $D_{n,e}$ is determined from the measured difference in sound levels, D, between two adjoining rooms corrected for the absorption area A in square metres in the receiving room, and normalised to a standard element size of $A_0 = 10\,m^2$ (BS EN ISO 140–4, 1998):

$$D_{n,e} = D + 10\,log_{10}\frac{A_0}{A}dB$$

The impact sound insulation of floors is given in terms of $L'_{n(Tmf,max),w,}$ which is similar to the $L'_{nT,w}$ used in Approved Document E of the Building Regulations except that the reference reverberation is not 0.5 s but the upper limit, $T_{mf,max}$, of the mid-frequency reverberation time of the specific receiving room. Again the required performance standards are given for each type of room to cater for their intended use. As examples, this ranges from 55 dB for lecture rooms and music rooms to 65 dB in circulation spaces and swimming pools.

Reverberation time

There are two aspects of reverberation that affects school activities. A larger amount of reverberation, indicated by a long reverberation time, will interfere with communication between teachers and students. On the other hand, a reasonably long reverberation time is desirable to give a good ambience to musical performances. It is therefore necessary to have an appropriate design of reverberation time for different types of spaces. Table 9.5 of the Bulletin gives the upper limit requirements for the mid-frequency reverberation time (the arithmetic average of the reverberation times in the 500 Hz, 1 kHz and 2 kHz octave bands) in all types of teaching and study spaces in finished but unoccupied and unfurnished conditions. These ranges from as low as 0.4 s for classroom designed for use by hearing impaired students to 1.5 s for music performance rooms and 2 s for swimming pools. A typical value is about 0.8 s for small lecture rooms and study rooms.

For corridors, entrance halls and stairwells the requirements described in the Approved Document E of the Building Regulations are taken.

Speech intelligibility

In open plan spaces, the requirements on the reverberation time, IANL, and sound insulation alone are not sufficient to guarantee adequate speech intelligibility. This is because of the often complex acoustic behaviour of these spaces and the noise created by individual groups of people working independently. A direct measure of speech intelligibility is required. The measure used is the speech transmission index (STI). The STI is based on an evaluation of the degradation of the modulation in a sound signal in its transmission from the source (speaker) to the receiver (listener). The sound signal is chosen to be representative of the frequency content and modulation frequency of human speech. Thus more degradation of the modulation indicates worse intelligibility. The STI is calculated such that it is within a value of 0–1. A value of 0 means totally degraded signal and therefore totally unintelligible. A value of 1 means no degradation and therefore perfect intelligibility. A value higher than 0.6 is generally accepted as good to excellent intelligibility and is therefore usually chosen as the performance criterion, as is the case in BB93.

Because of the complexity of the activities that may occur in an open plan teaching and study space, prediction of the measure from the activity plan is preferred.

Requirements in other building spaces

The PPG24, the Approved Document E of the Building Regulations, and the Building Bulletin 93 (2003) gives comprehensive guidance to the acoustic requirements in residential buildings and schools. For other types of built environments, the requirements are usually related to health and safety, environmental impact issues and specific needs in critical listening spaces.

In industrial spaces, the overriding concern is the protection of workers against hearing loss. Hearing loss comes naturally as a result of age but can accelerate due to prolonged exposure to even medium level of noise (about 80 dB and above). Many of us have experience of temporary loss of hearing after hearing a short burst of loud sound. Fortunately if we are not exposed to the loud sound regularly the loss is temporary and the hearing will recover in a short time. However the hearing loss due to prolonged exposure to noise is permanent and cannot be recovered. Since even musical performances can easily exceed 80 dB, it is not uncommon for people to be exposed to noise that can create hearing loss in the long run. Since the workplace is a place that workers have to stay in for long periods of time, it is essential that the noise exposure in the workplace be controlled to reduce the risk of hearing loss. For example in the European Union the Physical Agents (Noise) Directive 2003/10/EC on the minimum health and safety requirements regarding the exposure of workers to the risks arising from physical agents (noise) provides limits and action levels for noise exposure in the workplace. The exposure is measured in terms of equivalent continuous A-weighted sound pressure level over a working day (8 hours) or week (in justi-fiable circumstances where the noise level varies markedly from one working day to the next). The Directive limits the noise exposure to 87 dB(A) or a peak sound pressure of 200 Pa, whichever one is breached first, after taking account of the attenuation provided by hearing protection worn by the worker. The Directive also requires actions to be taken at two noise exposure levels. The lower action level is 80 dB(A) or a peak sound pressure of 112 Pa. At this level hearing pro-tection, information and training will have to be made available to the worker. Audiometric testing should also be made available where there is a risk to health. The upper action level is 85 dB(A) or a peak sound pressure of 140 Pa. At this level the use of hearing protectors is mandatory, and there is a requirement to establish and implement a programme of technical and/or organisational mea-sures intended to reduce exposure to noise, and to provide markings, delimiting and restriction of access to areas of the high noise level. The worker will also have a right to hearing checks by a doctor. In many industrial spaces the action levels and even the limit may be easily reached due to the large amount of machines used. Although noise control measures can be implemented after the discovery of the problem, it is generally more cost effective for new buildings to work with the building designers to reduce the build up of noise in the building, for example by proper use of sound absorbers.

Another concern for industrial buildings is the breakout of the noise into the environment. Nowadays almost all industrial developments will require an environmental impact assessment which includes noise in the application for planning permission. The noise control aspect usually comes as a requirement for the noise from the planned buildings not to exceed certain limits at its boundary perimeter and at nearby noise-sensitive spots. The limits are generally based on the consideration that the addition noise should not raise the existing background noise level in the area. This will require the developer to implement careful noise control design, usually in the form of choosing lower noise machineries, placement and shielding of noise sources and proper sound insulation to prevent noise from getting out of the buildings.

In non-industrial offices, the concern is less of hearing loss but more on productivity and privacy. Productivity will be lower if the office worker cannot concentrate on working due to constant disturbance of noise. Constant annoyance due to noise can potentially create health problems as well. There is an even higher requirement for lower background noise in meeting and conference rooms. On the other hand a certain amount of background noise can provide privacy to workers in open plan offices by masking private conversations. It is therefore a balance to provide a suitable soundscape for an office. There are several guidelines published by professional bodies such as the United Kingdom's Building Research Establishment (BREEAM 98) and the British Council for Offices (BCO Guide 2000) that provides advice on the acoustic conditions. Table 9.8 gives some typical values.

In built spaces that require critical listening, the requirements generally depend on whether the listening is for speech or music. For speech the prime requirement is intelligibility. There are surprisingly large amounts of built environments that fall within this category. The obvious ones are large lecture rooms and conference halls, but airports, train stations, football grounds and even shopping malls require reasonable speech intelligibility for their public announcement systems. In large gathering spaces there is in fact a legal requirement to provide intelligible announcements for warnings and information in case of emergency. The requirement generally comes in the form of a minimum speech transmission index (STI). Table 9.9 gives some indication to the relationship between STI and intelligibility. In more critical spaces such as a lecture room, a minimum value of 0.6 is usual. For less critical space a minimum value of 0.5 may be sufficient.

Although STI has been shown to have a very good correlation with speech intelligibility, it may fail in extreme cases such as when there is a strong echo and large fluctuation of background noise levels. In such cases expert advice will be needed.

The other type of critical listening spaces are those that are designed predominately for music, such as concert halls or auditoria. In these spaces the acoustic

Table 9.8 Suitable noise climates in offices

General offices	L_{Aeq} (dB)	Noise rating	
Open plan offices	45–50	NR40/NR45	Moderately noisy
Cellular offices	45–45	NR35/NR40	Moderately quiet
Meeting/executive rooms	30–40	NR25/NR35	Quiet

Table 9.9 Speech transmission index (STI) and approximate speech intelligibility rating

STI	Intelligibility
0.75–1.0	Excellent
0.6–0.75	Good
0.45–0.6	Fair
0.3–0.45	Poor
0.1–0.3	Bad

Table 9.10 Recommended mid-frequency (500 to 1 kHz) reverberation time of critical listening built spaces

	Reverberation time (s)
Concert halls	1.8–2.2
Opera houses	1.0–1.5
Speech rooms (<1,000 m³)	0.5–0.8
Speech rooms (<1,000 m³)	0.7–1.5
Churches	2.0–5.0

quality in enhancing the music sensation is more important than intelligibility. Subtle characters such as reverberance, clarity, spaciousness, envelopment, intimacy, balance and loudness are all important for the sensation. There are many measures that can be applied to judge the quality but they are too complex to be described in a short chapter. Interested readers can find more information in books such as Kuttruff (1991); Barron (1993) and Beranek (1996). Nevertheless a simple criterion that fits most of the good quality auditoria is the reverberation time. Table 9.10 shows the range of values that seems to provide good acoustic conditions for music-orientated spaces. Note that the recommended values are dependent on the size of the space which is not shown in the table.

Modelling the acoustic properties of building elements

We have seen from earlier sections that most of the acoustic requirements are placed on the sound insulation and sound absorption properties of the building elements. In a design point of view, it will be very useful to be able to predict these acoustic properties from some basic construction details as part of a modelling process in the design stage. The prediction of field measured quantities, such as the $D_{nT,w}$ and so on, is a complex matter because these quantities depend on the inter-locking relationship between different parts of the building, and one will need to take account of all the transmission paths through the building structure. However some of the laboratory-measured quantities, notably the sound reduction index R, may be modelled relatively easily to give indicative values for the evaluation of the likely performance of the building element in the field. Here we will look at some simple formulae that can be used for this purpose. However one should be reminded that the approximations involved are significant and the predicted results are for indicative use only and should not be used for demonstration of compliance.

Air-borne sound insulation

In most cases the most important elements of a building that insulate it from external noises or prevent internal noise from breaking out are the walls and roof. When sound hit's a wall some of the sound energy will be reflected back, some will be absorbed internally by the wall and the rest will be transmitted to the opposite side of the wall. There are a few possible paths for the transmitted sound. In a well-constructed wall, the dominant path is usually the direct transmission through the wall – the incident sound sets the wall to vibrate which in turn radiates sound to the other side. If the wall is not mechanically isolated from the rest of the building structure, then significant flanking, structural borne transmission can occur – sound energy that enters into the building structure from the source room can propagate as structural vibration to different parts of the building structure and re-emerge at far away parts of the building. Structural-borne transmission is usually confined to the low-frequency region where structural vibration can be significant. If the wall is poorly constructed and there are large gaps in or around the wall, then sound can pass directly through the gaps to the other side, resulting in significant direct air-borne noise leakage. Because of the usually small size of these gaps, such noise leakage is mostly confined to the high frequency range where the gaps are large compared to the wavelength of the sound. Of these transmission paths, the direct path will be dealt with in this section.

In standard laboratory tests (BS EN ISO 140–3, 1995), the flanking transmission and leakages are minimised so that the measurement is of the direct transmission through the wall sample itself. The performance is then described by the Sound Reduction Index, R. The index is a measure of the sound power transmitted to the receiver room relative to the sound power hitting the wall in the source room.

$$R = -10 \log_{10}\left(\frac{\text{sound power transmitted}}{\text{incident sound power}}\right)$$

The index is in dB scale and is usually given for each 1/3 octave frequency band with centre frequencies from 100 to 5,000 Hz. Additional information may be provided in the lower frequency bands 50 Hz, 63 Hz and 80 Hz in cases where low frequency noise is important. This is increasingly the case in modern society as the content of low-frequency noise increases and at the same time we are getting better at insulating against the higher frequency noise. However, measurements using the standard method and the application of the results at such low frequencies are problematic due to the dependence of the sound transmission on local conditions, such as how the sample is fixed and the configuration of the test rooms. Great care and expert advice are needed in such cases.

The higher the sound reduction index, the more sound insulation the wall provides. Generally speaking the index varies significantly with frequency and is generally poor at low frequencies and good at high frequencies. It is therefore more difficult to insulate low-frequency noise than high frequency noise. This is of particular concern in cases of neighbourhood noise where the noise can be of predominantly low-frequency content (e.g. bass music). For convenience the 1/3 octave band values can be combined into a single-figure number, the weighted sound reduction index R_w by shifting the R frequency curve against a reference curve, as described in ISO 717-1: 1997. The reference curve incorporates some degree of subjective perception and typifies the noise insulation characteristics of common wall elements.

Mass law and single-leaf walls

For a single-leaf wall construction that is made up of isotropic materials, such as a sheet of metal or a plaster board, the air-borne sound insulation performance (described by R) is determined by the panel's impedance seen by the incident sound wave. The higher the impedance the more difficult it is for the sound to be transmitted through and hence the better the sound insulation performance. Figure 9.4 shows the expected sound reduction index of an isotropic panel. At frequencies below the fundamental resonance of the entire panel, the panel

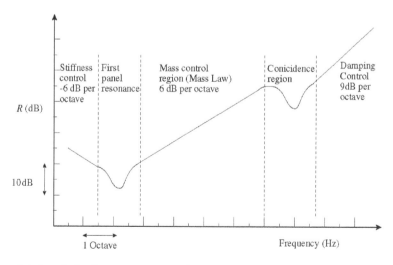

Figure 9.4 An idealised representation of typical sound insulation performance of a uniform single-leaf wall.

impedance is controlled by the bending stiffness of the entire panel and the fixing conditions at the panel edges. The stiffness-controlled impedance, and hence R, is higher at lower frequencies. For most common building elements this region occurs at low frequencies below 100 Hz and changes significantly with local conditions. In practice it is safer to assume the R stays flat rather than increases with decreasing frequencies. At the fundamental resonance the panel impedance falls off to a low value and is controlled by mechanical damping. Around this region the panel offers very little resistance to sound transmission and R is small. Because it is controlled by damping, panels with little internal loss, such as a metal sheet will have a more noticeable drop in R than one with more inherent loss, such as a plaster board or a brick wall. At frequencies higher than the fundamental resonance there is first a region where the impedance is controlled by the mass area density (mass per unit area), m in kg/m^2, of the panel. The mass-controlled impedance increases with frequency, and the sound reduction index can be predicted fairly well by the following formula, with f being the frequency in Hz:

$$R = 20 \log_{10}(m\, f) - 48 \text{ dB}$$

This equation indicates that the sound reduction index increases by 6 dB per doubling of the mass density, and by 6 dB per doubling (an octave increase) of frequency. This behaviour is termed the 'mass law' behaviour of homogeneous, isotropic materials. Significantly, this region usually covers a large part of the frequency range where sound insulation is important. It is therefore common to use the equation as a rule-of-thumb evaluation of the likely air-borne sound insulation performance of a wall panel. Real-life single-leaf building elements do tend to follow a similar trend as predicted by the equation but exact numbers maybe different. For example, some brick walls may have a 5 dB instead of a 6 dB increase per doubling of frequency.

Since the mass area density of a uniform panel is given by the product of its mass density, ρ in kg/m^3, and its thickness, h in m, one may think that better sound insulation can be achieved by simply increasing the thickness of a panel. Unfortunately this is not always the case. In practice increasing the thickness h can lead to some undesirable side effects on the sound reduction index due to coincidence, and careful considerations are needed. The safest way is to increase the density ρ to obtain higher values of R.

In the higher-frequency range, the sound insulation is adversely affected by the coincidence phenomenon. This is caused by the matching of the trace acoustic wavelength in air with the structural bending wavelength of the panel. The velocity of sound in air is constant while the velocity of the bending wave increases with frequency. At low frequencies the bending wavelength is shorter than the acoustic wavelength and the two do not match. In a simple term this means that the sound cannot excite the panel efficiently and vice versa. Since sound transmission is through sound exciting the panel to vibrate on the source side, and in turn the panel excites the air particle to radiate sound to the receiving side, sound transmission is weak if the acoustic and vibration wavelengths do not match. As frequency increases, there comes a point when the bending wavelength equals the acoustic wavelength. Sound waves incident at grazing angle will have a trace wavelength that matches exactly that of the bending wave. Because of this matching the panel offers very little resistance to the sound transmission and the SRI drops considerably. The frequency at which this first happens is termed the critical or coincidence frequency, f_c, and for a uniform panel can be calculated from the following, where E and ν are respectively the Young's modulus and Poisson's ratio of the panel material.

$$f_c = \frac{c^2}{2\pi} \sqrt{\frac{12(1 - \nu^2)\rho}{E h^2}}$$

At higher frequency the acoustic wavelength is shorter than the bending wavelength. There will always be an incident angle at which the trace of the wavelength along the length of the panel will match the bending wavelength and give rise to coincidence. However at these high frequencies the coincidence effect progressively diminishes and the sound reduction rises up at a rate that is dependent on a combination of panel damping, mass density and stiffness.

Among the different frequency regions that characterise the SRI of anisotropic panel, the Mass Law region occupies the most important frequency range in noise control. A good sound insulation design will aim to have the Mass Law region covering the entire frequency range of interest. The lower frequency regions, namely the stiffness-controlled and fundamental resonance regions, are usually below the frequency range of interest for most wall constructions. They are also very dependent on edge-fixing conditions and therefore can vary widely between different samples of the same wall material.

Since the coincidence effect can drastically reduce the SRI in the coincidence region, it is important in noise-control design to make sure that the coincidence frequency is beyond the frequency range of interest. This is doubly important since coincidence sometimes occurs at mid- to high frequencies where human hearing is also most sensitive, between 1,000 and 5,000 Hz. For a fixed material, it is important that the thickness h is not large enough to cause the coincidence frequency to drop within the frequency range of interest. This is the reason why the scope of using h to increase the area mass density and hence the sound reduction in the Mass Law region is very limited. For a fixed thickness h, a larger ρ/E ratio is desirable to push up the coincidence frequency. A large ρ will also improve the sound reduction in the Mass Law region. Hence heavy materials with low stiffness are generally best for noise insulation. Some of the best noise insulation materials are lead cladding and heavy brick walls, which have very high mass densities. Unfortunately as the mass increases the building cost also increases, and a compromise is normally required.

Another complication of the matter is that Figure 9.4 is only applicable to uniform, isotropic panels. Some single-leaf constructions, notably lightweight metal cladding panels, are made to have a much higher structural strength in one orientation compared to the other and are therefore orthotropic. The downside is that the stronger structural strength creates an additional coincidence frequency at a much lower frequency. The sound insulation performance in almost the entire audio frequency range is then affected by the undesirable coincidence effect, and the Mass Law region is almost entirely wiped out. As an example Figure 9.5 shows the measured sound reduction of a single-leaf metal cladding. It can be seen that at frequencies above 500 Hz the sound reduction does not rise at the Mass Law predicted rate of 6 dB/octave. Since the sound insulation performance of a single leaf lightweight metal cladding panel is poor, it is seldom used when high-quality sound insulation is required. The sound insulation performance can be greatly improved by employing double-leaf systems which have two profiled metal sheets separated by an air gap cavity filled with mineral fibre insulation. The added advantage of a double-leaf construction is that it will also improve the thermal performance and is therefore desirable in practice.

Double-leaf walls

The sound insulation of a single-leaf partition is mainly determined by the mass of the partition (Mass Law) in the majority of the frequency range of interest, and limited by finite panel resonance at low frequencies and coincidence at high frequencies. Therefore, a very high SRI is difficult to achieve. A double-wall construction benefits from the isolation created by the presence of the air cavity. Sound has now to go through one panel into the cavity and then through another panel to get transmitted. Therefore if the sound is attenuated by 90 per cent

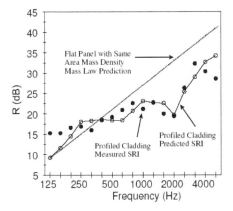

Figure 9.5 Sound reduction performance of an orthotropic, single-leaf metal cladding sheet.

(10 dB reduction in energy) each time it passes through one of panels, then the energy transmitted will only be 10% × 10% = 1%, or a 20 dB reduction. However this simple multiplication of the fraction of energy transmitted (or addition of the sound reduction in dB) is only valid when the two constituent panels are totally isolated from each other. In practice the air cavity serves as a flexible link that couples the panels together to form a mass-spring-mass type system. Other structural fixings that are necessary to fix the wall in place will also bridge the isolation and allow more sound energy to go through. At very low frequencies the coupling by the air cavity is very strong and the whole system works as a single panel with the combined mass. As frequency increases the air cavity becomes more flexible. The system then resonates at the mass-spring-mass resonance frequency f_0. Further up the frequency, the mass-spring-mass system becomes increasingly isolated. The sound reduction rises at a rapid rate until it reaches a limiting frequency f_L above which the air cavity cannot be consider as a spring any more and the sound reduction becomes essentially the addition of the two panels' individual sound reduction in dB. However this ideal behaviour assumes that there is sufficient sound absorption in the cavity to prevent cavity resonances. Such resonances can amplify the sound in the cavity and therefore negate the sound reduction. It is therefore important in practice to have some form of mineral fibre in the cavity which serves as a good sound absorber as well as a good thermal insulator. Assuming that the air cavity has sufficient absorption to prevent resonances in the air cavity, and that the noise bridging effect of mechanical fixings between the panels can be ignored, then the ideal double-leaf sound reduction can be calculated by the following set of equations depending on the frequency region,

$$R = R_{m_1 + m_2} \qquad\qquad\qquad\qquad f < f_0$$

$$R = R_{m_1} + R_{m_2} + 20 \log_{10}(fd) - 29 \qquad f_0 < f < f_L$$

$$R = R_{m_1} + R_{m_2} + 6 \qquad\qquad\qquad f_L < f$$

where m is the mass area density with the subscripts 1 and 2 indicating the two leafs of the construction, d is the depth of the air cavity and f_0 is the mass-spring-mass resonance frequency given by

$$f_0 = \frac{1}{2\pi}\sqrt{\frac{1.8\rho c^2(m_1 + m_2)}{d\, m_1 m_2}}$$

and f_L is the wave field limiting frequency,

$$f_L = \frac{c}{2\pi d}$$

In the first frequency region ($f < f_0$), the stiffness of the air cavity is high and the two leafs behave as one single mass. In the second frequency region ($f_0 < f < f_L$), after passing the mass-air-mass resonance, the stiffness of the air cavity becomes less and less important and the two leafs are increasingly isolated, hence a sharp rise in the sound reduction index. In the third frequency region ($f_L < f$), the cavity acts like a normal sound field and the two leafs work independently to provide high sound reduction. However the performance of a double-leaf wall in practice is limited by the sound bridging created by fixings between the two leafs and other flanking transmission routes and the actual performance is likely to be significantly lower. A better estimate can be obtained by the graph and equations shown in Figure 9.6 which are derived from observations on laboratory-measured data on a variety of double-leaf constructions (Bies and Hansen, 1996).

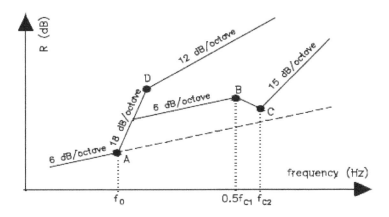

$$\text{point } A \quad R_A - 20\log_{10}(m_1 + m_2) + 20\log_{10}f_o - 48$$

$$\text{point } B \quad R_{B0} - R_A + 20\log_{10}(f_{cl}/f_o) - 6$$
$$\text{line--line } R_{B1} - 20\log_{10}m_1 + 10\log_{10}b + 30\log_{10}f_{c2} + 20\log_{10}\left(1 + \frac{m_2\sqrt{f_{cl}}}{m_1\sqrt{f_{c2}}}\right) - 77$$
$$\text{line--point } R_{B2} - 20\log_{10}m_1e + 40\log_{10}f_{c2} - 99$$
$$\text{point--line } R_{B3} - 20\log_{10}m_2e + 40\log_{10}f_{cl} - 99$$
$$\text{point--point } R_{B4} - R_{B2} + 2$$

$$\text{point } C \quad R_C - R_B + 6 + 10\log_{10}\eta_2 \qquad f_{c2} + f_{cl}$$
$$\phantom{\text{point } C \quad} R_C - R_B + 6 + 10\log_{10}\eta_2 + 5\log_{10}\eta_1 \qquad f_{c2} - f_{cl}$$

$$\text{point } D \quad R_D - R_{m1} + R_{m2} + 6 \qquad \text{at } f_L - 55/d$$

Figure 9.6 Simple calculation of the sound reduction performance of double-leaf partition taking into account sound bridging. m and η are respectively the mass area density and damping coefficient of the two leafs indicated by the subscripts 1 and 2. Curve AD represents perfectly isolated leafs such that point D defines the limiting frequency f_L. Curve ABC includes bridging. If no absorptive materials are placed in the cavity point B is determined at R_{B0}. Otherwise the closest description for panel–panel support is selected, where b represents the spacing between line supports and e is the average distance between point supports on a rectangular grid. Point B is then the largest value of R_{B0}, R_{B1}, R_{B2}, R_{B3} and R_{B4}.

Modelling accuracy and examples

The methods described in the previous sections provide some simple means to estimate the air-borne sound insulation properties of common building elements. They are simple enough for practical use without relying on expensive modelling software. However they are derived from very idealised situations and there are considerable simplifications in the process that could lead to large errors in the estimation, and should therefore be used with extreme caution. At best they give the indicative performance of the building element but in occasions they could fail totally because of unseen complexity in the structure – either because the method cannot account for the complexity or the user fails to specify the complexity properly. It is therefore important to see how they perform in some practical examples.

Single glazing windows

A plain, single-glazing window pane is a uniform structure that fits in well with the assumption of an isotropic plate described earlier. Figure 9.7 shows the laboratory-measured sound reduction of a 6 mm laminated windowpane. As can be seen the sound insulation behaviour of the windowpane does follow closely, the generic pattern described in Figure 9.4. At the very low end of the frequency range the sound reduction is very small. The behaviour there is dependent on how the sample is mounted (which affects the effective stiffness of the sample) as well as the specific conditions of the laboratory. It is not well related to the behaviour in situ and should not be relied on too much. As frequency increases, the sound reduction follows the characteristic mass law behaviour for a large part of the frequency range. Also shown in the figure is the data calculated by means of the mass law equation. The measured data follows the trend of the mass law prediction quite closely until the coincidence effect appears at about 2 kHz. It should however be noted that although the mass law predicted the trend, the absolute values of the sound reduction is over-predicted by a few dB. In the mass law equation, the sound is assumed to be incident onto the pane at all angles. However in reality the contribution of sound at angles near parallel to the pane surface (near grazing angles) could change significantly with sample size, mounting conditions and laboratory settings. Hence it is not uncommon to see a deviation of this magnitude in the absolute values of the sound reduction. As always one should treat the prediction only as an indication of the maximum realisable performance. At coincidence the sound insulation performance drops considerably as expected. The mass law equation unfortunately cannot predict this drop in performance. It is possible to modify the simple mass law equation to allow for the coincidence effect. An example of such a modified prediction is also shown in Figure 9.7 which shows better agreement with the measured data. However, such modifications require knowledge of the fraction of energy dissipated, represented by the damping factor η, which is difficult to obtain without measurements on the specific sample. For simple structures though, a typical values of $\eta = 0.1$ may be assumed.

Double glazing windows

A double glazing window is typically formed by two glass panes separated by an air cavity. Since the sound insulation behaviour of a single glazing window pane is well described by the model in earlier sections, one may think that a double glazing window can be described equally well by the ideal double-leaf model that is based on a combination of the performances of its constituent components. Unfortunately this is far from the case. Figure 9.8 shows the laboratory measured sound reduction of a double glazing window constructed from two 6 mm glass panes separated by a 12 mm air cavity (6–12–6 configuration). The measured sound insulation performance is nowhere near the ideal performance. In fact the double glazing window behaves as if it is a single pane with the combined total mass density of the two constituent panes. The expected additional insulation provided by the air cavity does not materialise. This is because, in this particular case, there is no sound absorber inside the air cavity. The air inside the cavity is excited by the sound passing through and resonates at certain frequencies called the cavity's natural or resonant frequencies. The sound will be amplified when the cavity resonates and negates any sound reduction increases due to the cavity. These natural frequencies depend on the geometry of the cavity but generally occur throughout the mid- to high frequency range and increase in number as

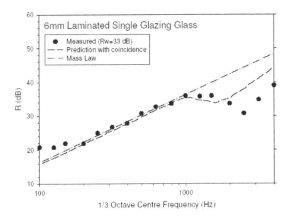

Figure 9.7 Model prediction compared with laboratory-measured sound reduction index of a 6 mm single glazing window.

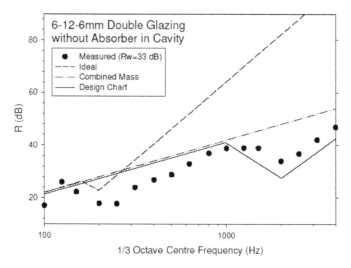

Figure 9.8 Model prediction compared with laboratory-measured sound reduction index of a 6–12–6 mm double glazing window

frequency goes up. Most of these resonances can be damped out effectively if sufficient absorption is placed in the cavity. However without absorption they will have a considerable detrimental effect. To demonstrate this, the design chart of Figure 9.6 is applied assuming no absorption in the cavity. It can be seen that the prediction now matched more closely with the measured data, although the latter is still noticeably worse than predicted at the lower frequencies. This illustrates the need for careful selection of models to apply to particular scenarios. A design will not work according to expectation if certain critical conditions are not met.

Masonry

Another common type of building element is masonry walls. Heavy brick walls are popular for outer building shells to provide high sound insulation. Because of their heavy mass, the sound reduction is high even at low frequencies, and coincidence is not much of a problem in masonry walls due to their inhomogeneous construction and heavy mass. They are created by cementing blocks and therefore do not conform with the assumption of a homogeneous panel. Nonetheless their behaviour generally follows the mass law fairly well. Figure 9.9 shows the laboratory-measured sound reduction index on a 215 mm lightweight concrete blockwork wall and compares them with the prediction using the mass law. It can be seen that the trend follows the mass law although the actual values are lower.

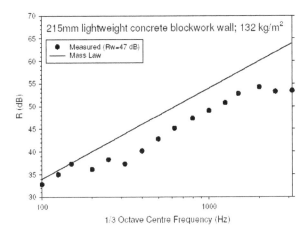

Figure 9.9 Model prediction compared with laboratory-measured sound reduction index of a 215 mm lightweight concrete blockwork wall.

For the purpose of better thermal insulation, most outer walls of a building shell will be constructed as double-leaf walls with thermal insulation in the cavity. One may expect that this would also produce better sound insulation as indicated by the theory of sound insulation of double-leaf walls as described earlier. Unfortunately this is not always the case. Although the sound insulation may have damped out some of the cavity resonances that lower the sound reduction, as seen in Figure 9.8 for a double glazing window, a double-leaf masonry wall still does not achieve the ideal double-leaf sound reduction. This can be caused by a variety of reasons such as the bridging of the sound insulation by the wall ties that connect the two leafs together. Figure 9.10 shows the laboratory-measured sound reduction of a masonry wall constructed with two 100 mm concrete blocks separated by a 50 mm cavity, with 13 mm plaster on each side. Again the sound insulation performance is far from the ideal prediction, and is in fact closer to the simple mass law behaviours with the mass given by the combined total mass of the construction. Notice though that the sound reduction value is actually quite high, reaching about 40 dB even at the lower end of the frequency range. The weighted sound reduction index R_w is 53 dB which is quite sufficient for most sound insulation requirements in building regulations. Hence the inability to achieve the ideal double-leaf sound insulation performance is not too much of a concern for masonry walls.

Metal cladding

Profiled metal cladding is a proper choice for many industrial and commercial constructions, even for schools. It is light weight and the profiling gives it strength in the desired direction. Unfortunately, as noted earlier, both the light

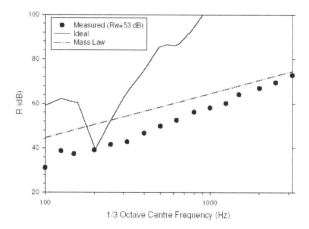

Figure 9.10 Model prediction compared with laboratory-measured sound reduction index of a masonry wall constructed with two 100 mm concrete blocks separated by 50 mm cavity, with 13 mm plaster on each side.

weight and the orthotropic nature of a cladding sheet give rise to rather poor sound insulation performance. Figure 9.11 shows the laboratory-measured sound reduction of a typical profiled metal cladding sheet. The performance is noticeably worse than the mass law prediction and there are significant drops in the sound reduction at several frequencies. The weighted sound reduction index, at 22 dB, is very poor compared with, say the 47 dB of the single concrete wall of Figure 9.9.

Fortunately, single-leaf metal cladding is seldom used in practice for the reason of poor performance in both sound and thermal insulation. Metal cladding walls are generally constructed in a double-leaf fashion, with the inner panel being another profiled metal sheet or even plasterboard. The double-leaf construction improves the sound insulation considerably. Figure 9.12 shows the laboratory-measured performance of a standard double-leaf metal cladding construction. The weighted sound reduction index R_w is now 37 dB, which is still low but a considerable improvement from the 22 dB of a single metal cladding sheet. In fact the R_w is now better than those of the single and double glazing windows shown in Figures 9.7 and 9.8. In here the sound insulation performance of the metal cladding construction approaches the ideal performance predicted by the double-leaf model up to about 1 kHz in Figure 9.12.

At higher frequencies, the ideal performance is compromised by the sound transmission through the fixings between the two leafs and the inherent sound reduction drops in the constituent metal cladding sheets that are created by the profiling. In practice because the R_w of a typical double-leaf cladding construction

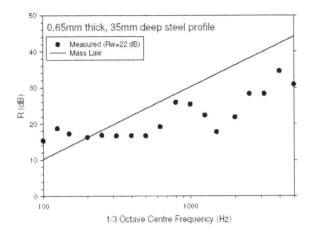

Figure 9.11 Model prediction compared with laboratory-measured sound reduction index of a typical single-leaf profiled metal cladding sheet.

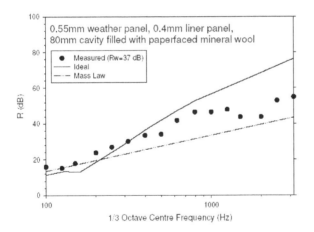

Figure 9.12 Model prediction compared with laboratory-measured sound reduction index of a typical double-leaf profiled metal cladding sheet.

is often close to or just below the value required by building regulations or planning restrictions, it is necessary to tune the construction design to give better performance. The simple ideal double-leaf prediction is still not sufficiently accurate for this purpose. Better prediction will need to be achieved through a more sophisticated model (Lam and Windle, 1996).

Modelling field sound level difference and impact sound level

The preceding sections described some simple models that can be used to provide an indication of the sound reduction performance of common building elements. However they only predict the air-borne sound reduction index in laboratory conditions. That is, they do not take into account possible flanking and other structural-borne transmissions that occur in a completed building. As a result they cannot be used to calculate the weighted standardised level difference, $D_{nT,w}$, in the field that is required in building regulations. The prediction of $D_{nT,w}$ will require the consideration of the construction design of the entire building in order to identify all the significant sound transmission paths and local conditions (such as reverberation) that affect the sound level. It is therefore a difficult task that cannot be taken easily and should only be attempted by specialists with sufficient experience. There is an international standard BS EN 12354-1 (2000) available that describes a practical method for calculating air-borne sound transmission between adjacent rooms of masonry construction that takes into account flanking transmission. The procedure is based on the method of statistical energy analysis (SEA) but simplified to the restricted set of specific conditions of the standard. In simple terms SEA is based on the balancing of energy input, output and dissipation within and between elements (sub-systems) that are connected to each other. As the name implied, the energy is taken as a statistical value, or the mean value over the sub-system and over frequency. It therefore requires both the mean response and the mean properties of the sub-system to be representative and meaningful, which is often the case for most acoustic sub-systems at high frequencies. At low frequencies, individual characteristics such as those created by modal resonances are too separated for such statistical means to be meaningful. Hence SEA is predominantly a mid-to high frequency method and is dependent on how well a system is modelled by the sub-systems. It is nonetheless a very powerful method provided one can fully characterise the energy transfer and dissipation relationships within and between elements. Some of these relationships can be estimated from theory but for complex systems most will need to be measured. Hence in practice the application of SEA, and hence BS EN 12354-1 is a semi-empirical approach that requires considerable experience and data on the sub-systems. Currently SEA is the only realistic approach for calculating in situ sound level differences that take into account flanking transmission but should not be used without expert advice.

Since impact sound transmission inevitably involved structural-borne transmission that takes similar paths as flanking transmission, its calculation can also be done by SEA. BS EN 12354-2 (2000) describes such a procedure based on SEA. However the same caution described in the previous section on the application of SEA also applies here.

Modelling absorption coefficient

A comprehensive set of data on sound absorption coefficient of common building materials can be found in the Building Bulletin 93 (2003). Except for purposely designed absorbers, the absorption tends to be small at low frequencies and increase with frequencies. For example the absorption coefficient of a thick carpet is only about 0.15 (15 per cent energy absorption) in the 125 Hz 1/3 octave frequency band and rises up to 0.75 in the 2000 Hz band. For most hard surfaces the absorption coefficient is small (<0.1) and typical values can be taken without problems. High absorption typically comes from soft, fibrous or porous materials such as carpet, curtain seating and people (human with clothes on is rather absorptive!). In here it should be pointed out that the absorption coefficients quoted are normally those defined according to the procedures of the international standard ISO 354 (2003). The coefficient is measured in a reverberant sound field (a diffuse sound field) meaning that the sound energy is coming to the surface of the absorbent sample equally from all directions. Since the absorption of sound by an absorber generally depends on the angle of energy arrival (the incident angle), the coefficient is therefore an average over all arrival angles. Hence it is sometimes referred to as the diffuse field absorption coefficient.

The exact values of the absorption coefficient of absorbers will depend on the size and shape of the absorber and for some will depend on individual proprietary designs. For a simple, flat layer of fibrous material laid on top of a hard surface, the absorption coefficient can be calculated with reasonable accuracy from the flow resistivity, σ of the material. The flow resistivity (Ns/m^4) is the pressure drop (N) per unit metre (M) per unit air flow velocity (M/S) when air flows through the material. Typically σ is between 1,000 and 50,000 for common fibrous materials. First we calculate the characteristic impedance, Z_{ch}, of the material, normalised by that of air, at a frequency f from the following equations:

$$Z_{ch} = R_{ch} + jX_{ch}$$

$$R = 1 + 9.08 \left(\frac{\sigma}{1000f} \right)^{0.75}$$

$$R = -11.9 \left(\frac{\sigma}{1000f} \right)^{0.73}$$

The impedance is a measure of how hard it is for a sound pressure to move the surface. Since energy dissipation depends on movement, the impedance is therefore a determining factor for absorption. Generally the larger the impedance the smaller the absorption. Note that the impedance is defined as a complex number with a real part R_{ch} and an imaginary part X_{ch} (indicated by the preceding symbol j). The real part corresponds to the resistance, much like the resistance of a viscous door damper to the movement, and is responsible for the energy dissipation. The imaginary part corresponds to the reactive component, much like the reaction of a spring to a force – energy is stored up in the spring but released undissipated when the spring is released. Most common materials have both of these components.

When the material is laid on top of a hard surface, which is commonly the case when fibrous materials are used, the impedance at the surface of the material is modified by the standing wave created by the sound travelling through the material and then reflected back by the backing hard surface. For a layer of material with a thickness d, the impedance at the surface becomes

$$Z = Z_{ch} \coth(\gamma_1 d)$$

where γ_1 is the sound propagation constant of the material:

$$\gamma_1 = \eta_1 + j\beta_1$$

$$\eta_1 = \frac{2\pi f}{c} 10.3 \left(\frac{\sigma}{1000f} \right)^{0.59}$$

$$\beta_1 = \frac{2\pi f}{c} \left[1 + 10.8 + \left(\frac{\sigma}{1000f} \right)^{0.7} \right]$$

For a sound wave coming in at an angle in the same direction as the normal of the surface, the (normal incidence) absorption coefficient is given by

$$\alpha_1 = \frac{4R}{((1 + R)^2 + X^2)}$$

Figure 9.13 shows the predicted and measured normal incidence absorption coefficient on a 40 mm thick fibreglass laid on top of a piece of hard wood. The prediction matched the measured value fairly well and shows the characteristic rise of the absorption with frequency. This is however only the absorption coefficient for one particular wave – one that comes in at normal incidence. It is therefore not the same as the *diffuse field* absorption coefficient that we need. We can convert the coefficient to that for a sound field similar to a diffuse field, by averaging the absorption over all incident angles, under the locally reacting assumption (i.e. each small area of the surface only reacts locally to the sound pressure on that area, and is not affected by what happens elsewhere). This results in the following:

$$\alpha = \frac{8}{A^2} \cos \beta \times \left[A + \frac{\cos 2\beta}{\sin \beta} \tan^{-1} \left(\frac{A \sin \beta}{1 + A \cos \beta} \right) \right.$$

$$\left. - \cos \beta \ln (1 + 2A \cos \beta + A^2) \right]$$

where $A = (R^2 + X^2)^{1/2}$ is the magnitude of the impedance Z, and $\beta = \tan^{-1}(X/R)$ is the phase angle of the impedance. However even with this conversion the absorption coefficient calculated in this way is not quite the same as that measured in

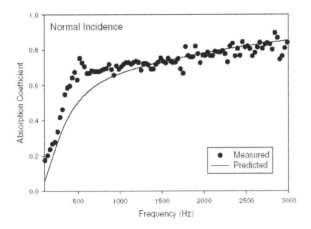

Figure 9.13 Normal incidence absorption coefficient of a 40 mm thick, rigidly backed fibrous material with flow resistivity of 80,000 Ns/m^4.

the reverberant field. In this conversion the surface is assumed to be flat and of infinite extent. In measurements the sample is of finite size. The edges of the sample will also cause a change in the sound field near the edges and result in a more complicated field than that assumed in the conversion. For example, in practice it is not unusual for a highly absorptive sample to have an absorption coefficient larger than 1 measured in the reverberant field method.

Since the sound absorption of most common materials is small at low frequencies and rises up with frequencies, it is difficult to obtain large absorption at low frequencies with common elements. Specially designed absorbers are often required. A special form of absorber is to create a structure that resonates strongly at a certain frequency. When it resonates, it offers very little resistance to the incoming sound and so almost all sound energy will be sucked into the structure. Provided that there is sufficient internal absorption in the structure, the sound energy will be absorbed inside and not get back out. This is called a resonant absorber. It is particularly useful for low-frequency absorption because the frequency at which it resonates can be adjusted by the geometry of the structure rather than relying on the material properties. A simple example is an air cavity of a certain depth d. The cavity will resonate at the frequency where the wavelength λ is 4 times the depth of the cavity. Therefore a deeper cavity will resonate at a lower frequency. Energy dissipation can be provided by a thin layer of resistive (e.g. fibrous) materials placed inside the structure. Figure 9.14 shows an example of a resonant absorber in the form of a 10 cm deep cavity with a resistive wire-mesh placed at the top. The absorption peaks created by the resonances are clearly visible. There are more than one absorption peaks in this example

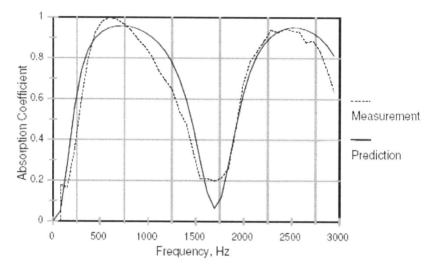

Figure 9.14 Normal incidence absorption coefficient of a 10 cm deep cavity covered with a resistive layer.

because the cavity resonates at harmonics of its first resonant frequency. The prediction is done by the method described by Wu *et al.* (2000).

More optimised absorbers can be achieved through proprietary designs. In practice all purposely built sound absorbers should come with absorption data measured according to one of the international standards and there should be no need to rely on predicted values. The descriptions earlier are to give a basic understanding of the parameters that control sound absorption so that one may make a better choice of absorbers to suit a particular situation.

Modelling the internal sound field

Although most of the building regulations put requirements on the acoustic properties of the building elements, some also put requirements on features of the internal sound field. For example the Building Bulletin 93 (2003) specifies limits in the reverberation times of different school spaces. Moreover guidelines on the design of office spaces and critical listening spaces often put specifications on the sound field itself. It is therefore necessary to know how we can model the internal sound field of these spaces.

Diffuse sound field model

The simplest of the sound field model, which nonetheless works fairly well in most non-critical cases, is the diffuse field model. This is a statistical model that considers only the sound energy and not the wave nature of the sound. In this model the sound energy is assumed to be statistically uniformly distributed within the space. There is an equal amount of energy incident on each unit area of the internal wall surfaces and hence the energy absorbed is given directly by the amount of absorption area or absorption coefficient. The sound energy in the room can then be calculated by balancing the energy input from sound sources, if any, against the absorption on the walls, and the energy increase in the room. Solving this problem leads to two useful expressions; one for the reverberation time, RT, and one for the mean square sound pressure, p^2, from a constant source.

$$RT = \frac{0.161V}{A}$$

$$p^2 = \frac{4W_o\rho c}{A}$$

where A is the total absorption area of the room which can be calculated by simply summing the individual absorption areas of each surface, W_o is the constant total sound power from all the sound sources, V is the volume of the room and ρ is the density of air which is typically $1.21\,kg/m^3$. The above equation for RT is often referred to as Sabine's reverberation time formula in honour of Wallace Clement Sabine who was a pioneer in architectural acoustics in the late nineteenth and early twentieth centuries.

Sound absorption by the passage of the sound through the air medium can be incorporated into the RT equation:

$$RT = \frac{0.161V}{A + 4mV}$$

where m is the air attenuation factor which can be calculated at different temperatures and humidity using the international standard ISO 9613 (1993). In practice the air absorption term can be safely ignored in most cases except for large rooms (e.g. concert halls) or at very high frequencies (above 2 kHz).

The diffuse field model is however difficult to achieve in reality. In actual rooms the absorption is generally not uniformly distributed and often concentrated on the floor area (carpets, seating and audience). In such cases the energy is not quite uniformly distributed and an alternative formula, called Eyring's reverberation time formula, tends to give better results in larger rooms:

$$RT = \frac{0.161V}{-S\ln(1 - \bar{\alpha})}$$

where S is the total surface area in the room and $\bar{\alpha}$ is the area weighted average of the absorption coefficient over all the internal surfaces in the room. The diffuse field model provides an extremely simple means to estimate the reverberation time and sound level inside a building, requiring only simple knowledge of the volume and absorption area of the space. Together with the models on the acoustic properties of the building elements, the building designer will be able to calculate most of the requirements specified in building regulations and guidelines. Even the speech transmission index (STI), required by the Building Bulletin 93 (2003), can be estimated from the RT. Unfortunately, it turns out that although the diffuse field model is adequate for common, non-critical spaces, it is in the critical spaces in which it fails that the designer is in most need of an accurate model to work out a proper design. It is in cases where the sound is not uniformly distributed, such as in a large concert hall or a large open-plan office, or when the subtlety of the sound field at particular locations is of interest and statistical averages cannot be taken that we face most of our room acoustic problems. A more sophisticated model is needed for these applications.

Geometrical room acoustics computer modelling

At present the state-of-the-art method of modelling complex room acoustics is based on geometrical room acoustics. As the name implied, the propagation of the sound waves is approximated by geometrical energy rays. Figure 9.15 illustrates the approximation. When a sound wave is radiated out from a simple point source, the wave expanded out spherically in all directions. The interaction of this wave with the wall surface is theoretically rather complex. As an approximation, one can take the analogue with light and represent the radiation as composed of a large number of sound rays coming out from the source. The diagram in Figure 9.15 shows a few of these rays. Just like in optics when a light ray hits

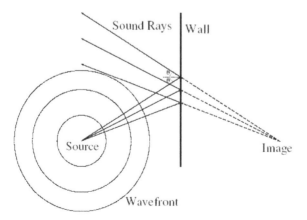

Figure 9.15 Geometrical ray representation of sound wave.

a mirror, the sound rays are assumed to be reflected specularly when they are reflected off a wall surface – the angle of reflection is equal to the angle of incidence. An alternative interpretation is that the wall surface acts as a mirror that creates a mirror image from which the reflected rays are originated. The calculation of the sound propagation is therefore a simple matter of tracing the rays around the room, which can be handled easily by a computer. An example of the tracing of a ray over several reflections is shown in Figure 9.16.

Unfortunately there is the matter of wavelength that makes sound not behaving exactly like light. The wavelength of sound is in the order of 1 m at 300 Hz which is not small compared with the dimension of most rooms. When the wavelength is comparable to the size of the reflecting object, the reflection is not specular, and diffraction and scattering of the sound wave occur. This is compounded by the fact that the impedance of the wall surface is generally complex and further invalidates the assumption of specular reflection. In reality the sound reflected from a wall surface will be spread out in a much wider angle, or in another word the mirror image will be blurred. Therefore, in order for the geometrical models to work, the calculation should be done over many reflections such that the error associated with a single reflection can be smoothed out. Fortunately this is usually the case in room acoustics when the room is reverberant. The smoothing out of errors is also better when only sound energy rather than sound pressure is considered, and the calculation is over a wide frequency band such as an octave rather than a 1/3 octave band. Furthermore additional calculations can be added to allow for the scattering of sound energy into non-specular directions (Lam, 1996) based on the definition of a new scattering coefficient of the surface (ISO 17497-1, 2004). This is sometime referred to as diffuse reflections to approximate the scattering of rays. Once all the rays are traced, or all the images are found, all the sound energies arriving at the receiver are collected to form the time history of the energy arrivals at the receiver – the energy impulse response of the room. Detailed room acoustics parameters can then be calculated from these impulse responses.

This modelling is particularly useful for critical listening spaces where the acoustic quality depends on more than just the reverberation time and the sound

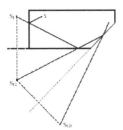

Figure 9.16 Tracing of sound rays and generation of sound source images in a room.

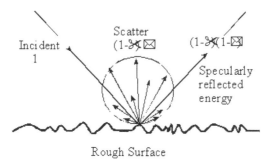

Figure 9.17 Illustration of sound scattering from a surface into non-specular direction and the definition of the scattering coefficient in relation to the absorption coefficient.

level (see Figure 9.17). The distribution of the energy arrivals, especially within about 50 to 100 ms after the arrival of the direct sound, has been shown to have very strong influences on the perception of acoustic quality for music and speech. A number of room acoustics parameters have been developed to quantify these influences and for music a list is given in the Annex of the international standard ISO 3382 (2000). For speech, the speech transmission index (STI), which has a strong correlation with speech intelligibility, can also be calculated from the impulse response. In fact the calculated impulse response can be used further to create auralisation – a physical rendering of the sound field using computer simulation that includes the sound source, the room acoustics and the influence of the human head and ear. Auraliation is a form of virtual reality presentation that allows the designer to hear how the space sounds like before it is built, and is a valuable tool for communicating the acoustic quality directly to the designer and especially to people who do not have specialist knowledge in acoustics. Hence geometrical room acoustics computer modelling has become a very useful computer-aided tool for building designers concerning themselves with acoustics. Nowadays almost all architectural acoustics consultants use computer modelling in one form or the other.

With the widespread use of computer modelling in room acoustics design comes the question on the accuracy and reliability of the models. In an attempt to answer this important question, a series of international round robin tests were undertaken. The first round robin tests started in 1994 with 16 participants from 7 countries. A real auditorium is used as the object of the modelling and the prediction results from all the participants on a number of room acoustics parameters were compared with data measured in the real hall (Vorländer, 1995). Rather disappointingly, the result showed strong discrepancies between participants. There are many reasons for the discrepancies. A common cause of the problem, which still exists today, is the different ways that were used to represent the rather complex real geometry in a computer model. Because the objective is to represent the nature of the sound reflection using an approximated geometrical acoustic model, precise modelling of small geometrical details is not adequate and in fact will almost certainly lead to errors. The basic assumption of geometrical reflection is that the wavelength is small compared with the size of surface. The assumption of geometrical reflection is not going to hold on a small surface or object. It is therefore necessary to simplify the geometry to suit the need of individual computer models. Experience and data accuracy on the geometry both play important roles in determining the accuracy of the results. Furthermore, at the time of the first round robin test the concept of diffuse reflections, or the approximation to account for the effect of scattering from surfaces, was still not well established and there were large uncertainties in how each model implemented or even not implemented diffuse reflections. There were also no reliable data on the scattering coefficients for the surfaces in the auditorium at that time. All in all the first round robin tests served the important purpose of highlighting the problems but was not sufficient to give a good picture on the capability of the computer modelling methods.

Consequently in 1996–1998, a second round robin test was organised (Bork, 2000). This time there were 16 participants from 9 different countries and 13 different computer modelling programs were used. The object of the modelling is the Elmia concert hall in Sweden. The approximate length, width and height of the hall are respectively 40 m, 30 m and 16 m. This time data on both the absorption coefficients and scattering (diffusivity) coefficients were supplied to the participants to avoid uncertainties. With better experience of using the computer models and more established understanding and modelling of diffuse reflections, the result of this second round robin comparison was much better than that of the first round robin. Figure 9.18 shows the comparisons on the parameters reverberation time and sound level (normalised to a reference value according to BS EN ISO 3382, 2000). The graphs in the figure show variations with different source and receiver combinations. There are still noticeable discrepancies between the models, as expected, especially at low frequencies when the wavelength is large compared to the hall size. There are also occasional odd behaviours in some of the models. Nevertheless it is promising to see that the modelling results overall tend to group around the measured values. In room acoustics it has been established that an average listener is unlikely to hear a difference in the acoustics when the difference in the reverberation time is smaller than about 10 per cent of the reverberation time, or typically a 0.1–0.2 s difference. From the graphs in the figure, it is obvious that the discrepancies among a number of the models still exceed this limit. There are also comparisons on other room acoustics parameters (established for assessment of different acoustic quality for critical listening) that can be found on the round robin web site (Bork, 2000), and some of these comparisons show larger discrepancies than that found in Figure 9.18. It is obvious that the models are not perfect and there is still room for improvement. This is not surprising since the assumption of geometrical ray propagation is a rather big

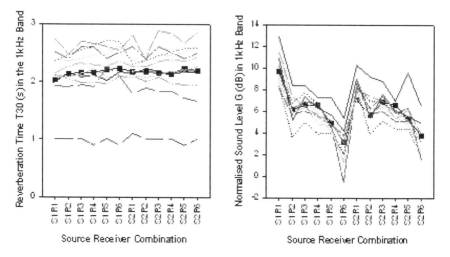

Figure 9.18 Reverberation time and sound level results from the second international round robin test of room acoustics computer models. The black squares are measured data. The other lines represent the modelled results from individual participated models.

assumption. Generally though, the test results suggest that the computer models can give a fairly reasonable impression of the what the sound field is like and will therefore be able to help design decisions.

One of the reasons for the discrepancies in the second round robin test was thought to be caused by the complex geometry of the Elmia hall. As mentioned before, computer models need to simplify the geometry to adequately represent the nature of the reflections to suit the individual modelling approximations. Hence a complex geometry gives rise to the possibility of variations in the modelling results. As a result, a third international round robin test was started in 1999 (Bork, 2002). This time the object of modelling was chosen to be a simple, largely rectangular, small studio room, approximately 8 m by 10 m by 5 m in size. The test was conducted in three phases in an attempt to reduce uncertainties: phase 1 was with the room empty with uniform absorption, phase 2 was with real absorption and phase 3 was with all actual geometry, absorption and scattering. The result is a lot better this time, as can be seen in Figure 9.19. The sound level result is particularly good. This is largely because of the simple room used for the testing, but also due to the maturity of both the models and the users. The good accuracy on the sound level in particular can be explained by the fact that the sound level is calculated by summing all the energy arrivals in the impulse response. It is therefore the most likely to be able to smooth out statistically the inaccuracies in individual arrivals. A more stringent test can be seen on the parameter *lateral energy fraction, LF*, which is a measure of the fraction of energy arriving to the listener within 100 ms of the

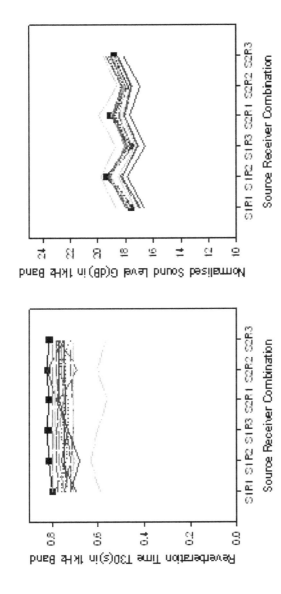

Figure 9.19 Reverberation time and sound level results from the third international round robin test of room acoustics computer models. The black squares are measured data and the black bars indicate the standard deviation. The other lines represent the modelled results from individual participated models.

direct sound from either side of the head. It has been found that this parameter has a very strong influence on the subjective impression of the apparent width of the source, and is often related to the spatial impression of the room, which in turn has been found to be one of the most important determining factors of subjective room acoustic quality. Within this short time period, the energy arrivals are relatively few and the models will also need to predict the direction of the arrivals with good accuracy. Figure 9.20 shows the result of the third round robin test on this parameter. As one can see, the discrepancies are significantly larger than those observed on reverberation time and sound level, and raise the question of whether the models arc good enough for predicting details of a small group of reflections.

The development of room acoustics computer modelling is an ongoing effort. The problems to be solved are not only on the modelling theory and algorithm, but also on the uncertainties in the input data and the reliance on user experience to represent the room geometry correctly. On the topic of input data, there are questions on whether the current database of absorption coefficients, which are measured in reverberant rooms with diffuse fields, should be used in a computer program that model reflections individually at different incident angles. The definition of the scattering coefficient has only recently been standardised (2004) and there are still very few data available on surfaces commonly encountered in real halls. In the ongoing effort to resolve these problems, a fourth international round robin test is being planned (2005). It is likely that auralisation will be included as part of the test in this round. There is no doubt that this ongoing independent round robin testing has provided much-needed guidance to both

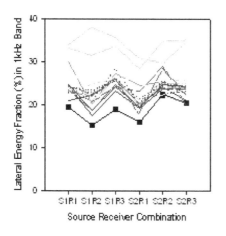 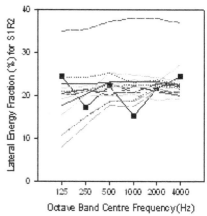

Figure 9.20 Lateral energy results from the third international round robin test of room acoustics computer models. The black squares are measured data and the black bars indicate the standard deviation. The other lines represent the modelled results from individual participated models.

developers and users of the computer models and has greatly helped the acceptance of these models in practice.

Advanced room acoustics modelling methods

Over the last 10–15 years, great improvements have been made on the geometrical room acoustics modelling method. The introduction of diffuse reflections to approximate energy scattering has been particularly useful. The geometrical modelling method is however ultimately limited by its geometrical approximation which cannot adequately represent the full wave nature of the sound. This is particularly important at low frequencies and for small rooms where the wavelength is not small compared with the size of the reflecting objects. In such cases the reflection and scattering are far from geometrical. There are several numerical methods available that take the full wave nature of the sound into account and are more accurate than the geometrical models.

The finite element method (FEM) is a well-established method in many disciplines of physics and engineering. Its application to acoustics is fairly well established especially in the study of acoustic devices such as loudspeakers, microphones and noise silencers. In room acoustics it is fairly straightforward to apply the method to model the internal space of a room. The volume of the room is divided into a mesh of finite elements. The physical equation that governs the behaviour of sound in the frequency domain is approximated within each element. Provided the elements are small compared with the wavelength, then the approximation is simple and accurate. The elements are then assembled together and the source is represented by some prescribed terms in the element at the source location. The wall surfaces are modelled as boundary conditions derived from the walls' complex impedance or alternatively elastic walls can be modelled by the usual structural finite elements to take full account of the interaction between the sound field and the wall structures. As proven in many other applications, the accuracy of FEM is very good provided that the mesh is adequate and the source and boundary conditions are represented correctly. The biggest problem with FEM in room acoustics is the size of the problem. The entire 3D volume has to be meshed by elements that are small compared with a wavelength. The number of elements required is rather large at high frequencies. It is therefore only realistic to use the method in small rooms at low frequencies. A useful application of the method is to study the coupling of two rooms (living room size) by a partition to work out the low-frequency sound insulation performance of the partition. This has obvious importance for the specification of low-frequency sound insulation requirements since the current standard method of measurement has serious limitations at frequencies below 100 Hz. In larger rooms or at higher frequencies the application of FEM is too expensive in practice. Another serious limitation on the application of the method is that it requires correct data on the complex impedance of the wall surfaces. Unfortunately in practice the acoustic properties of wall surfaces are currently only available in the form of the reverberant field measured absorption coefficient and perhaps also the scattering coefficient. Without proper input data, the usefulness of the method will be rather limited.

Another highly accurate numerical method that has found widespread use in acoustics is the boundary element method (BEM). The governing sound wave equation is transformed into an integral equation over the boundary surface of the room. In so doing the numerical solution only requires the meshing of the boundary and not the volume of the room. This reduces the size of the problem by one order of magnitude and makes the method much more attractive than the FEM for room acoustics calculations. The interaction with the wall surfaces is prescribed in turns of the complex impedance of the surfaces, similar to the FEM. It is also possible to couple the BEM mesh with an FEM mesh of an elastic wall surface to account for the coupling effect, although in such cases the calculation will be much more demanding. There is no approximation of the physical equations except the discretisation of the surfaces into boundary, elements and so the theoretical accuracy is better than the FEM. The discretisation also requires the elements to be small compared with the wavelength (about 1/6) but because the mesh only covers the boundary, the number of elements required is considerably less than that required by the FEM for the same problem. This allows the application of BEM to higher frequencies and larger rooms. There are reports of application to frequencies as high as 1 kHz and to medium-size auditoria. It is a promising numerical technique that may yet replace the geometrical models in the future as computing power increases. However, just like the FEM, it also suffers from the problem of requiring complex wall impedance data that currently do not exist without specialist measurements.

A third numerical method which is currently enjoying a lot of attention is the finite different time domain (FDTD) method. In room acoustics, the time domain impulse response carries many of the most important acoustic characteristics of the room. It is therefore useful to be able to perform the calculation directly in the time domain to obtain the impulse response. The BEM can also be applied in the time domain but the FDTD has the advantage that one can easily visualise graphically the propagation of the sound wave over time. The size of the finite difference step is limited by the highest frequency content of the impulse response and a smaller time/spatial step is required at higher frequencies. The demand on calculation time is therefore comparable to that of the other numerical methods, but the calculation can be cut short if the result is only required over a small time period. This is often acceptable since most of the important features of the impulse response that needs accurate modelling are within a short time (100 ms) after the arrival of the direct sound. The biggest problem facing the FDTD in room acoustics is the modelling of the boundary condition. A time domain representation of the wall reflection is required and this has to be convoluted with the FDTD propagated solution. The time domain representation of a reflection can conceivably be obtained by a transform of the frequency domain solution, but it is not easy. Furthermore, if the impedance of the wall is frequency dependent, which is usually the case, the time domain representation will be quite lengthy in time and the convolution with the FDTD propagated solution will result in a very large demand on computation resources. Until this problem is solved, the FDTD's main use is limited to the demonstration of the propagation and scattering of the sound in a room.

Concluding words

Acoustics is an important feature in the built environment that, the author feels, has been neglected in the main stream of built environment science. It is often the case that a building is designed with full considerations on architectural, visual, thermal and energy-efficiency features but not acoustics. Acoustic consultants are only called in when a problem appears. We all have had experience with bad acoustics in train stations and bad speech intelligibility even in some lecture theatres and communal halls, and yet good acoustics can be easily achieved with proper design in the first place. Many of the modern airport terminals, for example, have very good acoustics and sound systems, and not to mention modern concert halls that can combine good musical acoustics quality with visual beauty and efficiency in their operations. The building regulations in the United Kingdom, now extended to schools, have a positive influence on the industry, so that consideration of sound insulation in residential and school buildings is taken early in the design stage. For other spaces it is still very much up to the designer to add acoustics into the design and modelling process. It is hoped that this Chapter will serve to raise the awareness of acoustics in the built environment sector and the practitioners will find useful knowledge and concepts that would allow them to appreciate some common problems and solutions, and ultimately make better decisions regarding the acoustics of their buildings.

References

Barron, M. (1993) *Auditorium Acoustics and Architectural Design*, E & FN SPON, Oxford.

Beranek, L.L. (1996) Concert and opera halls: how they sound, Published for the Acoustical Society of America through the American Institute of Physics.

Bies, D.A. and Hansen, C.H. (1996) *Engineering Noise Control – Theory and Practice, Second edition*, E & FN SPON, Oxford.

Bork, I. 'A comparison of room simulation software – the 2nd round Robin on room acoustical computer simulation', *Acta Acustica* 86, 943 (2000), http://www.ptb.de/en/org/1/17/173/roundrob2 1.htm

Bork, I. 'Simulation and measurement of auditorium acoustics – the round robins on room acoustical simulation', Proc. of the Inst. of Acous. 24 (2002) http://www.ptb.de/en/org/1/17/173/roundrob3 1.htm

BS 8233:1999, Sound insulation and noise reduction for buildings – Code of practice, British Standards Institute, 1999.

BS EN 12354-1:2000, Building acoustics – Estimation of acoustic performance of buildings from the performance of elements – Part 1: airborne sound insulation between rooms, British Standards Institute, 2000.

BS EN 12354-2:2000, Building acoustics – Estimation of acoustic performance of buildings from the performance of elements – Part 2: impact sound insulation between rooms, British Standards Institute, 2000.

BS EN 61672-1:2003, Electroacoustics – Sound level meters Part 1: specifications, British Standards Institute, 2003.

BS EN ISO 140-3:1995, Acoustics – Measurement of sound insulation in buildings and of building elements – Part 3: laboratory measurement of airborne sound insulation of building elements, British Standards Institute, 1995.

BS EN ISO 717-1:1997, Acoustics – Rating of sound insulation in buildings and of building elements – Part 1: airborne sound insulation, British Standards Institute, 1997.

BS EN ISO 717-2:1997, Acoustics – Rating of sound insulation in buildings and of building elements – Part 1: impact sound insulation, British Standards Institute, 1997.

BS EN ISO 11654:1997, Acoustics – Sound absorbers for use in buildings Rating of sound absorption, British Standards Institute, 1997.

BS EN ISO 140-4:1998, Acoustics – Measurement of sound insulation in buildings and of building elements – Part 4: field measurements of airborne sound insulation between rooms, British Standards Institute, 1998.

BS EN ISO 140-7:1998, Acoustics – Measurement of sound insulation in buildings and of building elements – Part 7. field measurements of impact sound insulation of floors, British Standards Institute, 1998.

BS EN ISO 1793-3:1998, Road traffic noise reducing devices – Test method for determining the acoustic performance, Part 3. Normalized traffic noise spectrum, British Standards Institute, 1998.

BS EN ISO 3382:2000, Acoustics – Measurement of the reverberation time of rooms with reference to other acoustical parameters, British Standards Institute, 2000.

BS EN ISO 354:2003, Acoustics – Measurement of sound absorption in a reverberation room, British Standards Institute, 2003.

Building Bulletin 93, Acoustic Design of Schools – A Design Guide, Department for Education and Skills (UK), 2003.

The Building Regulations 2000, Approved Document (2003 Edition), Resistance to the passage of sound, Office of the Deputy Prime Minister (UK), 2003.

ISO 9613–1:1993, Acoustics – Attenuation of sound during propagation out-doors – Part 1: calculation of the absorption of sound by the atmosphere, International Organisation for Standardization, 1993.

ISO 266:1997, Acoustics – Preferred frequencies, International Organisation for Standardization, 1997.

ISO 226:2003, Acoustics – Normal equal-loudness-level contours, International Organisation for Standardization, 2003.

ISO 17497-1:2004, Acoustics – Sound-scattering properties of surfaces – Part 1: measurement of the random-incidence scattering coefficient in a reverberation room, International Organisation for Standardization, 2004.

Kuttruff, H. (1991) *Room Acoustics*, Third edition, Elsevier Applied Science, Cambridge.

Lam, Y.W. (1996) A comparison of three diffuse reflection modelling methods used in room acoustics computer models, *Journal of the Acoustical Society of America*, 100, 2181.

Lam, Y.W. and Windle, R.M. (1996) The noise transmission through profiled metal cladding, part iii: double skin SRI prediction, *Journal of Building Acoustics*, 2, 403.

Planning Policy Guidance 24, Planning and noise, Department for Environment, Food and Rural Affairs (UK), 1994.

Vorländer, M. (1995) 'International round Robin on room acoustical computer simulations', Proc. of the 15th Int. Congress Acous., 689.

Wu, T., Cox, T.J. and Lam, Y.W. (2000) From a profiled diffuser to an optimized absorber, *Journal of the Acoustical Society of America*, 108, 643.

Chapter 10

nD modelling to facilitate crime-reduction 'thinking' within the project briefing process

David Hands and Rachel Cooper

Introduction

Designing out crime presents many complex challenges for the design development team; these may include access to specialist crime prevention knowledge, utilisation of appropriate tools and techniques to enable crime misuse scenarios within the front-end stages (in particular during the design-briefing stages which is the primary focus of this section) and design concept development and construction phases.

Furthermore, the complexity of attendant design issues related to crime requires access to current research and data which is either unavailable or difficult to obtain. This chapter investigates how nD modelling can facilitate and focus the design development team to develop a crime-aware consciousness at the initial stages of the design briefing process. It discusses how designing against crime is an emergent issue within design, and how the designer can contribute valuable skills and approaches to reducing crime throughout the development programme. The chapter also provides an overview of design briefing and why it is so important to undertake this process effectively and efficiently; if executed successfully, the final solution should be tailored to its specific context, addressing crime in both an innovative and subtle way. Finally, the concept of DAC (design against crime) for nD modelling is introduced, outlining the key issues of what constitutes a holistic nD-enabled construction project.

Design against crime – an overview

Designing against crime: an emergent issue

Designing against crime has featured highly in the disciplines of spatial design and urban planning. Architects and planners have become more understanding of the public's perceived fear of crime, and have now started to develop new ways of addressing these issues in the design process. Although the British Crime Survey (Home Office, 2004) identifies a downward trend in crime activity, the public's perception of crime and fear of being a victim of crime is high. A recent BBC News feature (BBC, 2004) attempts to explain this phenomenon by suggesting that 'crimes are now being committed in more affluent areas . . . making the crimes more visible'. It further adds that '. . . if the media report even a small number of cases, readers may see themselves as being indirect victims.'

Public transport and travel is another area that has received considerable attention in reducing criminal activity through a more enlightened approach to designing-out crime. Mature industries such as the automotive manufacturing sector have taken crime prevention into account, focusing heavily on securing their vehicles or making it harder to steal from the vehicles. Vehicle crime accounts for just less than one-fifth of all crime recorded by the police (Home Office, 2004). Although the number of crimes has been falling steadily since 1992, around 340,000 vehicles are still stolen every year (HOCD, 2004).

The role of design and designers, can make an effective contribution to reducing the impact of crime. Designing against crime can be utilised throughout many different contexts that include the built environment, product design, packaging design, new media and textiles. Ekblom and Tilley (1998) argue that '. . . design can be used as a tool to aid crime prevention by incorporating features into potential targets which make criminal activity less appealing for the criminal and disrupting the causes of a criminal event.'

Erol *et al.* (2000) further add that '. . . the influences on design decision making are wide ranging with technological, environmental and market developments all having a significant effect. One role of the designer is to integrate all these factors to produce the best design solution.'

Crime-reduction thinking is already considered in certain disciplines of design, in particular the built environment – where crime reduction is an integral part of the design process. Environmental design, most notably architecture, can significantly contribute to situational crime prevention. Situational Crime Prevention (SCP) has developed considerably since the late 1970s, in a further attempt to minimise the occurrence of crime. SCP is based on the premise that crime is context specific, predominantly opportunist in nature and influenced by the predisposition of the offender. To reduce opportunist crime, measures can be taken to reduce the likelihood of success for the criminal by implementing 'control' measures within a given situation.

In a research study on situational crime prevention and economic crime, Lehtola and Paksula, suggest that

> the strategy of situational crime prevention consists of measures designed to decrease the opportunity for crime; measures that are directed at specific and closely identified crime types. 'Measures' refer to control, planning and influencing of the immediate environment in as systematic and permanent a manner as possible. The measures are designed to induce a wide range of potential offenders to realise that the effort required to commit crime and the risks involved have increased, and that the rewards that can be secured through crime have decreased.
>
> (1997)

Situational crime prevention measures are predominantly applied to the built environment, in particular to reduce the likelihood of burglary, vandalism and shoplifting. Although 'SCP can apply to every kind of crime, not to just 'opportunistic' or acquisitive property offences; there are examples of successful applications to more calculated or deeply motivated crimes, as well as ones committed by hardened offenders e.g. hijacking, homicide and sexual harassment' (Ekblom, 1999).

Crime-prevention features are most often, considered retrospectively, applied to both products and built environments after the event of crime, rather than at the initial stages of their design development. Consequently, crime prevention attributes are often 'added-on' or considered as an afterthought rather than 'embedded' within the product or system at the initial stages. To overcome this perception of thinking in relation to crime, the designer needs to view the way they develop design proposals from an alternative perspective. This will involve looking at the final design as 'misuser-unfriendly' as opposed to 'user-friendly' thus requiring a paradigm shift in the designer's understanding and involvement in the way that they identify and solve design problems.

Ekblom and Tilley (2000) argue that '. . . We must start "thinking thief," anticipating criminals' actions, researching the tools, knowledge and skills available to them now and in the near future and incorporating attack-testing into the design process.' Professor Ken Pease further adds that

> This by no means involves designing heavy and ugly objects and places. Indeed, windows, which thwart burglars, are likely to be less ugly than windows that are insecurely designed and need protection from bars. Designers are trained to anticipate many things: the needs and desires of users, environmental impacts, ergonomics and so on. It is they who are best placed to anticipate the crime consequences of products and services, and to gain the upper hand in the technological race against crime.
>
> (Design Council, 2001)

Changing the designer's approach to the way that they design, taking into account crime-resistant measures at the initial stages of the design programme requires an understanding of how the final product or environment could be misused by the criminal (Design Policy Partnership, 2002). Not only is the designer required to take fully into account the user's experience of the product/environment but also how it could be misused or abused in support of criminal activity or being a target of crime itself.

The impact of crime: trends and perceptions

The most recent and sophisticated analysis of crime in England and Wales puts the total annual cost at a staggering £60 billion, with, on average, burglaries costing £2,300 each, vehicle thefts £4,700 and robberies £5,000 (Brand and Price, 2000).

Although recorded crime has witnessed a slight decrease, more than half of the population in England and Wales considers crime as the number one problem facing the country (ICPC, 1997). The British Crime Survey (BCS) (Home Office, 2004) shows a fall between 1997 and 1999 in nearly all the offences it measures. Burglary fell by 21 per cent and vehicle-related theft by 15 per cent Robbery increased by 14 per cent and theft from the person by 4 per cent. If one measures the importance of property offences in terms of value, rather than the quantity of incidents, fraud is of far greater significance (University of Cambridge, 2000). By contrast, the combined costs of prolific offences of 'auto-crime' and burglary for 1990 were estimated by the Association of British Insurers to be just under £1.3 billion (Maguire, 1997).

Although recent statistics from the BCS (Home Office, 2004) Report highlight a downturn in incidents of recorded crime, to fully understand its effects on society as a whole we would need to move beyond mere statistics. The perception of crime is heightened by high-profile court cases in the media having a profound effect on society at large.

The general public's concern about crime in the BCS survey (Home Office, 2004) highlighted the following key issues:

- They are pessimistic about the problem of crime, with over one-third believing that national crime levels had risen considerably.
- A substantial minority (29 per cent) thought that it was likely that they would have their vehicle stolen in the next year, that they would have items (such as a radio, camera, bags etc.) stolen from a car (32 per cent) or their home burgled (20 per cent). One-tenth thought it was likely that they would be a victim of mugging or attacked by a stranger, although in reality the average risks of victimisation are far lower.
- Around one-fifth of the people were 'very worried' about burglary, car crime, mugging and physical attack by a stranger and rape.
- Concern about crime will be linked to both peoples' beliefs about their chances of being victimised and what they feel about the consequences of victimisation. Levels of worry are higher among those living in high-crime areas, recent victims of crime, those who consider it likely that they will be victimised and those who are socially or economically vulnerable (e.g. the elderly or single mothers).

The level of anxiety of being a victim of crime can be a mild concern ranging through to high anxiety. The survey also highlighted that

- 8 per cent of adults in the United Kingdom say that they never walk in their local areas after dark, at least in part because of the fear of attack. Amongst the elderly and women this figure rises to nineteen per cent.
- 13 per cent are 'very' or 'fairly' worried about their home being burgled all or most of the time.

- six per cent consider that the fear of crime greatly impacts on their quality of life, slightly more than those (4 per cent) who said the same about crime itself.

Designing against crime in the built environment

Crime Prevention through Environmental Design (CPTED) is a well-recognised and suitably mature approach to reducing the opportunity for crime under the broader framework of Situational Crime Prevention (SCP). McKay (1996) argues that '. . . it is defined as a proactive crime prevention technique in which proper design and effective use of a building and its surroundings, leads to a reduction in crime as well as an improvement in the quality of life.' Ekblom and Tilley (1998) add that '. . . offenders can only exploit potential crime opportunities if they have the resources to take advantage of them', suggesting that '. . . situational crime prevention must also consider offender resources and their distribution and social-technical change.' CPTED aims to reduce the opportunity for crime by focusing on the specific contextual situation as opposed to hardening the crime target or product; in essence it is a context-specific approach to crime reduction. Ekblom (1991) details how 'situational crime prevention rests on the observation that people's behaviour is not simply influenced by their fundamental personality, [but] by the physical and social situation they find themselves in at a given moment in time for a crime to happen'. By its very nature, environmental design is a broad multi-disciplinary aspect of design practice, requiring the input of architects, planners, statutory and legal professionals, and key stakeholders in the use and functioning of the designed space. Hence, CPTED requires the close involvement and collaboration of all the various stakeholders for it to be effective in reducing the opportunities for crime (Ekblom and Tilley, 1998). Due to the flexibility of CPTED principles, they may be embedded into the design early within the design programme, or alternatively, applied retrospectively after the design has been completed (Ekblom and Tilley, 1998). Crowe (1991) defines the distinction between CPTED principles and commonplace crime-reduction techniques, suggesting that

- '. . . where CPTED differs from traditional target hardening strategies is that the techniques employed seek to use environmental factors to affect the perceptions of all users of a given space – addressing not only the opportunity for the crime but also perceptions of fear on the part of those who may otherwise be victims.'
- CPTED draws heavily on behavioural psychology, with its key strategies and concepts taking advantage on the relationships between people and their immediate environment.

Crowe (Ekblom and Tilley, 1998) further explains this relationship by highlighting the 'softer' side of CPTED, arguing that

'. . . the way we react to an environment is more often than not determined by the cues we are picking up from that environment. Those things which make normal or legitimate users of a space feel safe (such as good lighting), make abnormal or illegitimate users of the same space feel unsafe in pursuing undesirable behaviours (such as stealing from motor vehicles).'

Three core principles are fundamental to CPTED for it to be an effective strategy to reduce criminal activity (Newman, 1972; Mayhew, 1979)

- Surveillance: the ability of the inhabitants to observe their own territorial areas. By creating areas of open space, it dramatically increases the numbers of people who could observe acts of crime being committed. Therefore, the probability of the offender being caught in the act is high. Built environments that restrict the ability of the residents to see what is taking place outside and around them will suffer from crime and anti-social behaviour.
- Territoriality: the feeling of possession and belonging by the inhabitants. Murray (1983) strongly argues for a fundamental review of approach towards designing communal residential dwellings. He argues that '. . . a family that has a sense of territoriality about the entryway to its apartment will more likely defend it against intruders than a family who does not have this sense of territoriality.' This concept first emerged from the work of Ardrey (1997) who investigated the concept of territoriality and animals and how it can be applied to human behaviour. Newman (1972) places great emphasis on the aspect of territoriality within 'defensible space', which should have a 'clearly defined sense of ownership, purpose and role'.
- Access Controls: Beavon *et al.* (1994) advocate the reduction of access points to built developments in order to restrict the movement of unauthorised visitors or potential offenders. If access was not controlled, potential offenders could freely pass through the area, developing a familiarity with the estate, whilst also having a legitimate reason to be there. Therefore, reducing access to a minimum lessens the opportunity for crime. SBD principles highlight the need to reduce unauthorised access, suggesting that '. . . this problem tends to be worst where fully public space directly abuts private space, with no intermediate "buffer zone" spaces in between. In public spaces, everyone has a legitimate excuse to be there, and wrongdoers become indistinguishable from legitimate users.'

CPTED: mechanisms and applications

Secured by Design (SBD) is an accreditation scheme based upon CPTED principles intended to reduce criminal activity in the built environment. The key designed elements it advocates in the early planning stages are as follows:

- Defensible space: positioning and detailing of boundary treatments, thus preventing unauthorised access. Boundaries will provide the first line of defence in securing the development from both the opportunist and dedicated criminal.

- Improved surveillance: natural surveillance is enhanced thus allowing a high degree of 'social control' to minimise criminal activity. The guidelines argue that '... optimum natural surveillance should be incorporated, whereby residents can see and be seen.'
- Promotion of territoriality: designing-in a clear demarcation between public and private space. Rachel Armitage (2000) talks about territoriality in the context that '... if space has a clearly defined ownership, purpose and role, it is evident to residents within the neighbourhood who should and more importantly who should not be in a given area.'
- Community interaction: by creating a sense of community through mixed use developments encouraging social interaction and a human scale in the architecture, the community will exercise a high degree of control over their environment.
- Circulation management: reducing the number of 'unnecessary' footpaths and access points within the development. Armitage (2000) argues that '... through routes and footpaths provide the opportunity for offenders to attach to an area for criminal intent and purposes.'

The following are typical examples where CPTED principles are often applied:

- Hospitals
- Schools and Colleges
- Public transport networks
- Vehicles parks
- Residential, commercial and industrial developments
- Hotels, bars and restaurants
- City centre developments.

Design briefing – an overview

The importance of effective briefing

The design brief acts as a central point of reference from which dialogue and the sharing of values can depart, and which provides a firm basis for all subsequent design decisions within and throughout the design programme (Philips, 2004). The design brief is to get everyone started with a common understanding of what is to be accomplished. It gives direction and provides clarity to designers as well as to clients. According to the AIGA (American Institute of Graphic Arts), '... a brief can be as valuable internally as it is externally.'

Often, briefing encapsulates the findings and conclusions of a whole process that precede the creation, review and approval of the document.

In an ideal situation, the client is aware of the value of discussing the brief with the designers or designers' representatives, and so, incorporate designers into the group at the initial stages of the design programme. However, on a day-to-day basis, this practice is often the exception rather than the norm among the briefing makers. In general, once briefing is considered finished, it is often given to the design team to act upon (Topalian, 1994).

Depending on many factors, the designer's response can vary, for instance, it can

- start right away generating the design solution: This may be the case of a small project, or a typical design competition, which does not allow the design firm to thoroughly understand its client in order to give valid advice. It could be the case that to protect their investment in the design competition, the competing design firms will play it safe, not questioning the brief and simply providing the client with what they are expecting. When discussing about how to get creative results, the AIGA highlight that design competitions, in which the client receives artwork at a cost below market value, owns the intellectual or creative property and can exploit the work without the involvement from its creator and are a threat to the designer, the client and the profession. 'The designer gives up creative property without a fair level of control or compensation. The client fails to get the full benefit of the designer's talent and guidance. The profession is misrepresented, indeed compromised, by speculative commercial art'.
- discuss the brief and generate a 'contra-brief': The Design Council (2002) suggests, 'an outline brief can be developed [from the beginning or] further with the designer to help form a common understanding of objectives throughout the project'. Alternatively, the process of debriefing the client is extremely important for the designer, who can often complete or even discuss the information contained in the briefing.

As the Italian designer Michele DeLucchi learned, even the perfect translation of a client's brief cannot guarantee that designer and client mean the same things when they talk about design and its subtle qualities. So, he always produces a contra-brief in response to the owner's brief, to clarify his terms and his vision.

The Design Council suggests that the brief for a designer should be a comprehensive document. It should contain all of the information a designer needs to fulfil the design objectives of the project. The AIGA further add that '. . . the brief should not tell the designers how to do the work.' To decide what makes sense in terms of how to structure the brief is a process of thinking, not just completing a standardised form.

Cooper and Press (1995) describe a design brief as something '. . . that defines the nature of the problem to be solved' whilst Ken Allinson (1998) describes it as '. . . relating to problem framing and problem finding as opposed to problem solving.' Hymans (2001) succinctly argues that '. . . systematic briefing defines the problem to which design is the answer.' In a report for the Construction Industry Board (1997) 'Briefing the Team', Frank Duffy defines briefing

Table 10.1 RIBA Plan of Work: Stages A – B – Briefing

Stage	Purpose of work and decisions to be reached	Tasks to be done	People directly involved	Commonly used terminology
A	To prepare general outline of requirements and plan future action	Set up client organization for briefing	All client interests, architects	Briefing
B	To provide the client with an appraisal in order that he may determine the form in which the project is to succeed	Carry out studies of user requirements etc.	Clients' representatives, architects, engineers, according to the nature of the project	
C	Begins when the architect's brief has been determined in sufficient detail			

Source: RIBA, 2000.

as '. . . the process by which a client informs others of his or her needs, aspirations and desires, either formally or informally and a brief is a formal document which sets out a client's requirements in detail.'

The new edition of the RIBA Plan of Work (2000) has now placed greater emphasis on briefing, moving it to a more central position within the design programme (Table 10.1). This follows on from major recommendations highlighted in the Latham report (1994) which attributed poor briefing as a significant factor in the poor performance of the UK building industry in terms of producing buildings that were actually needed. Although, both 'Briefing the Team' (1997) and the RIBA Plan of Work (2000) still do not see the extent to which briefing is a continuous process requiring much iteration.

Worthington (1994) describes the process of briefing as a multi-layered iterative process '. . . spread over the complete design and construction process, which is layered succeeding through continual feedback.' Barrett (1995) follows the same argument as Worthington, pointing out the common dilemma in briefing – how early it should be fixed within the design and construction programme.

The Design Council have developed guidelines to enable organisations to effectively brief designers, highlighting what information is required and how it could greatly assist the designed outcome. It states that the initial brief should include the following:

- a description of the issues to be addressed, highlighting specific problem areas;
- project objectives, expressed in terms of desired outcomes;
- information about the target users, particularly detailing any specific or unusual requirements;
- guidelines about the desired positioning of the product or service.

Then, having appointed a suitable designer or design consultancy, the second brief should capture more detailed strategic issues that include the following:

* agreed budgets, milestones and deadlines;
* the business strategy behind the project;
* the brand strategy within which the project will operate;
* the activities and deliverables the organisation expect the project to involve.

The guidelines tentatively raise the important aspect of aligning the proposed designed outcome to corporate strategy, but they do not provide detailed advice or suggestions as to how the organisation can translate and embed their 'values' within the brief itself.

Michael Slade (1990) talks about how quick decisions early on, prior to briefing, can actually jeopardise new projects, offering advice to overcome the common pitfalls in pre-project activity. He argues that: '. . . An attempt in instant solutions often places the projects team in a false position. Either the initial response is anodyne – key problems have not been exposed, but everyone is happy. Or the response does not meet with approval and inevitably more detailed requirements begin to emerge from the subsequent debate. Especially in this latter case, valuable credibility is lost and the project begins to fall in the classic pattern of continually moving goalposts.'

nD modelling to facilitate crime-reduction 'thinking'

The opportunity for 'prevention' of criminal activity can and does embrace a vast array of activity, from environmental design (CPTED principals) to social control and criminal justice system institutions. However, it is time architects and designers take some responsibility for preventing crime and play a significant role when developing built environments.

However, designing out crime presents many complex challenges for the designer; these may include access to specialist crime prevention knowledge, utilisation of appropriate tools and techniques to enable crime misuse scenarios within the front-end stages and design concept development and manufacturing phases. Neglecting to do this could result in the failure of the designer(s) and construction team to anticipate the vulnerability of the proposed new building to crime. However, organisations and design professionals lack effective tools, techniques and knowledge resources to assist them in the initial stages of the building programme leading up to design brief development.

nD modelling allows the multi-disciplinary team to envision and anticipate potential crime misuse scenarios within the initial stages of the design development programme. The nD knowledge base is a fluid and highly responsive tool enabling information from multi-perspectives and knowledge domains to be incorporated within the briefing process. Crime tensions and trade-offs can be fully explored allowing decision makers to prioritise crime issues early within the process. Often, crime issues are not generally raised and anticipated within the briefing process, and if they are, they are not fully articulated to all the stakeholders. The success of designing against crime lies in its focus on engaging and embracing the user's culture with designed solutions closely aligned to all their specific requirements.

Key stakeholder involvement

In developing an effective project brief, the views of all 'stakeholders' both internal and external to the organisation must be identified and represented within the document. Wootton *et al.* (1998) define 'stakeholders' as '[a]ny party (both internal and external to the company) who has a significant influence over the design, development, manufacturing, distribution and use of a product.' This will ensure that needs, preferences and requirements of all individuals or groups who have an interest in the designed outcome will be represented within the brief. Internal stakeholders could include the following, engineering, sales, finance, research and development, legal, marketing and production.

External stakeholders could include the following: customers, end-users, suppliers, subcontractors and distributors.

When identifying relevant stakeholders, the significance of their contribution to the success of the proposed designed outcome must be taken into account. Within a 'crime' context, stakeholders to a consumer product susceptible to theft could include the following:

- police;
- insurance agencies;
- reformed offenders, who could contribute specialist knowledge regarding how to steal, misuse or dispose of the product.

On a larger scale, not only the end-users or customers of the products, services and environments can play a vital part in contributing specialist knowledge, but manufacturers and suppliers as well. In a report to the Home Office (Design against Crime, 2000) the authors argue that

> motor cars for example, are designed with the consumer in mind rather than the potential offender. The 'crime-free' car would have secure locks, an effective immobiliser, a speed limiter and a device to render drunken driving impossible; at present no such vehicle exists, although the technology is almost certainly available.
>
> (Design against Crime, 2000)

Through increased and effective collaboration between industry partners, suppliers and crime-reduction agencies, a multi-stakeholder approach could contribute to the reduction in criminal activity connected to high-value and susceptible products. The report (Ekblom and Tilley, 1998) suggests that the 'role of the government may be to promote and encourage their [industry] contribution to detect or anticipate emergent crime problems in product design.'

An approach to adopting greater stakeholder involvement within the crime reduction process could be to draw upon tools and techniques commonly used within the development of 'green' products and 'strategic environmental thinking'. Prentis and Bird (1999) discussing the holistic inclusion of stakeholders within the sustainable design process argue that 'customer data is a crucial part of any market dynamics research; however, the process of first engaging customers needs to be carefully managed.'

nD Modelling: towards a holistic approach to DAC briefing

Incorporating the nD modelling tool within the initial stages of the briefing process will allow the development team to further enhance their approach and understanding of potential crime misuse scenarios in the latter design development stages. Table 10.2 provides an overview of how a design against crime nD-enabled project can anticipate potential misuse scenarios and unify the team towards a crime-aware focus.

- Shared vision: The design and construction team should have a shared vision throughout all stages of the design and construction process ensuring that the integrity of a crime-focused and crime-aware vision remains throughout the project process. Through demonstrating excellence and adopting a socially responsible attitude towards crime-reduced construction, the benefits are significant to all stakeholders.
- Stakeholder involvement: Having identified the relevant stakeholder partners, it is crucial that their views are incorporated within the briefing process. Barrett and Stanley (1999) argues that 'without engaging end-users in the process, an important stimulus to the creative process of design is lost

Table 10.2 Design against crime nD-enabled construction project

Shared vision	All major project 'actors' work towards a common goal of providing a crime-aware solution
Stakeholder involvement	Incorporating all key stakeholders in the decision-making process
User-centred approach	Positing the core values and culture of the end-user/ client at the heart of the process
Clear objectives	Establish clear crime-reduction objectives at the initial stages of the project
Effective communication	Ensure all stakeholders contribute and have access to specific crime-reduction information
Specialist expertise	Identify and source specialist crime-prevention expertise

and the client is likely to be left with long term dissatisfaction amongst the very people they would wish to support through the built solution being developed.'

- User-centred approach: A user-centred approach to briefing is more focused on the client/end-user culture in preventing crime and feelings of insecurity (Town *et al.*, 2003). Solutions are tailored to their particular context with the user culture providing the core to innovative and subtle design against crime interventions.

- Clear objectives: Set clear crime-reduction objectives at the initial briefing stages and then develop a subsequent crime reduction strategy aimed to develop a design solution. This could be through identifying the different types of crime that impact upon the built solution through consultation with stakeholders and end users.

- Effective communication: Maintain an effective dialogue with all stakeholders. Information can be incorporated within the nD tool accessed at critical stages of the design development and indeed construction phases of the project. Information can be both quantitative and qualitative in nature allowing the user to make informed and intelligent decisions that could decrease crime risk associated with the design.

- Specialist expertise: Crime-prevention expertise and other professionals engaged in crime reduction, have specialist insights and understanding of design issues from a crime perspective. By signalling critical points within the briefing process, the close involvement of Architectural Liaison Officers (ALO) and local police crime-prevention advisors can contribute their knowledge to guide the focus and direction of the brief.

Summary

In developing crime-resistant built environments, the task of design briefing is an important and highly complex part of the whole design process. Failure to effectively execute this process can often lead to incorrect, incomplete or ambiguous expression of the design requirement, which in turn leads to a designed solution that fails to prevent or reduce, or actually increases criminal activity. In essence, the solution fails.

The process of arriving at a design brief that encapsulates design against crime thinking and attributes is a complex one, characterised by the application of (often external) expertise in circumstances that vary about a multitude of sometime conflicting tensions. This in turn makes the whole process of effectively managing the briefing process and anticipating criminal activity difficult. As increasing demands upon the organisation and design function become increasingly more complex and the penalties for failure in design increase, so too does the need to ensure that the crime-focused design brief is developed more effectively.

nD modelling actively encourages and enables key stakeholders to identify and engage with (often unexpected) crime 'misuse' scenarios in relation to their work. Through the multi-layering of information and its instant point of access, nD modelling offers the users immediate access to relevant information, allowing them to consider alternative solutions and designed responses. Currently, few professional bodies provide advice and practical guidance to designers, thus making the decision-making process difficult and often time consuming. nD modelling obviates this problem by raising greater emphasis of crime misuse scenarios and encourages crime-reduction 'thinking' throughout the entire design and construction phases. As a consequence, nD modelling aims to encourage the construction industry to adopt crime prevention measures throughout the entire construction process by providing a flexible and responsive simulation tool offering holistic crime-centred solutions to create safe and secure urban environments.

References

Allinson, K. (1998) *Getting there by Design: An Architect's Guide to Project and Design Management*. Architectural Press, London.

Ardrey, R. (1997) *The Territorial Imperative: A Personal Inquiry into the Animal Origins of Property and Nations*, Kodansha Globe, New York.

Armitage, R. (2000) An Evaluation of Secured by Design Housing within West Yorkshire. Briefing Note 7/00 to the Home Office p. 1.

Barrett, P. (1995) *Facilities Management: Towards Best Practice*. Blackwell Science, London.

Barrett, P. and Stanley, C. (1999) *Better Construction Briefing*. Blackwell Science, Oxford, pp. 59–70.

BBC (2004) BBC News: Panic on the Streets, http://news.bbc.co.uk (Accessed January).

Beavon, D.J.K., Brantingham, P.L. and Brantingham, P.J. (1994) in Clarke, R.V. (ed.), *Crime Prevention Studies* Vol. 2, Criminal Justice Press, New York.

Brand, S. and Price, R. (2000) The Economic and Social Costs of Crime. Home Office Research Study 217. Home Office: London.

Construction Industry Board (1997) *Briefing the Team: Working Group 1*. Thomas Telford, London.

Cooper, R. and Press, M. (1995) *The Design Agenda: A Guide to Successful Design Management*. John Wiley & Sons, Chichester, pp. 51–97.

Crowe, T. (1991) *Crime Prevention through Environmental Design: Applications of Architectural Design and Space Management Concepts*. Butterworth: Stoneham, MA.

Design against Crime (2000) Design against Crime Report. University of Cambridge, Sheffield Hallam University and University of Salford.

Design Council (2001) Cracking Crime through Design Design Council Publications, London. p. 27.

Design Council (2002) Frequently Asked Questions: 'How to brief a designer.' Design Council Publications, London.

Design Policy Partnership (2002) Design against Crime: Guidance for Professional Designers. University of Salford.

Ekblom, P. (1991) Talking to offenders: practical lessons for local crime prevention. In conference proceedings of Urban Crime: Statistical Approaches and Analyses, Barcelona Institiut d'Estudis Metropoitans de Barcelona, p. 1.

Ekblom, P. (1999) Situational crime prevention: effectiveness of local initiatives. In Reducing Offending: An Assessment of Research Evidence on Ways of Dealing with Offending Behaviour.

Ekblom, P. and Tilley, N. (1998) What works database for community safety/crime reduction practitioners: Towards a specification for an ideal template. Policing and Reducing Crime Unit, Home Office RDS and Nottingham Trent University.

Ekblom, P. and Tilley, N. (2000) Going equipped: criminology, situational crime prevention and the resourceful offender. *British Journal of Criminology*, 40, 376–398.

Erol, R., Press, M., Cooper, R. and Thomas, M. (2000) Designing out crime: raising awareness of crime reduction in the design industry. In conference proceedings of the British Society of Criminology, Leicester, July 2000.

HOCD Car Theft Index 2001 (2001).

Home Office (2004) The 2002 British Crime Survey, England and Wales 2002–03, HMSO, London.

Hymans, D. (2001) *Construction Companion – Briefing*. RIBA Publications, London.

ICPC (1997) Crime Prevention Digest, International Centre for the Prevention of Crime. (ICPC), Montreal.

Lehtola, M. and Paksula, K. (1997) Situational Crime Prevention and Economic Crime. National Research Institute of Legal Policy. Publication number 142.

McKay, T. (1996) The right design for securing crime. *Security Management*, 40(4), 30–37.

Maguire, M. (1997) Crime Statistics, Patterns, and Trends: Changing Perceptions and their Implications, in R. Morgan and P. Reiner (eds), *The Oxford Handbook of Criminology*, Clarendon Press, Oxford.

Mayhew, P. (1979) Defensible space: the current status of a crime prevention theory. *Howard Journal*, 17, 150–159.

Murray, C. (1983) The physical environment and community control of crime. Cited, in J. Wilson (ed.), *Crime and Public Policy*. ICS Press, San Francisco, CA, pp. 349–361.

Newman, O. (1972) *Defensible Space*. Macmillan, New York, pp. 87–90.

Philips, P.L. (2004) *Creating the Perfect Design Brief*. Design Management Institute, Boston, MA.

Prentis, E. and Bird, H. (1999) 'Customers – the forgotten stakeholders'. *The Journal of Sustainable Product Design*, January, 52–56.

RIBA (2000) *Architects Job Book, Seventh edition*, RIBA Publications, London.

Slade, M. (1990) Understanding the brief. In *Engineering Journal*, March, p. 21.

Topalian, A. (1994) *The Alto Design Management Workbook,* Spon Press, London.

Town, S., Davey, C.L. and Wootton, A.B. (2003) Design against Crime: secure urban environments by design – guidance for the design of residential areas. Pelican Press, Manchester.

University of Cambridge (2000) Design against Crime Report. Sheffield Hallam University and University of Salford.

URL <http://news.bbc.co.uk/hi/english/uk/newsid.>

URL <http://www.design-council.org.uk> (Accessed 28.11.2002).

Wootton, A., Cooper, R. and Bruce, M. (1998) *A Generic Guide to Requirements Capture*. University of Salford. ISBN 0902896 16 4.

Worthington, J. (1994) Effective project management results from establishing the optimum brief. *Property Review*, November, 182–185.

Part 3

nD modelling

The application

So far, 'Constructing the Future: nD Modelling' has defined the concept and scope of the new construction ICT phenomenon. This section describes the application of nD modelling, concentrating on a number of pertinent issues that impinge on the global widespread uptake of nD-enabled construction, such as data management, visualisation and education.

In the first chapter, Wix comprehensively describes how data can be identified and classified within an nD model. He indicatively covers issues such as type, occurrence, symbols, relationships, dictionary, thesauri and so on. The need for classification grows ever greater as the AEC/FM industry moves from a drawing-based approach to using a model-based approach, and thus is pertinent and timely.

For a comprehensive nD model, project requirements should also be integrated. Bacon builds from Wix's chapter and uses his vast experience of knowledge management in both the public and private sector organisations to comprehensively describe and depict how to successfully achieve this. Starting from the Brief, he systematically and comprehensively outlines the purpose and structure of the information so that it can be subsequently managed in an nD model. Finally, he highlights the limitations of this approach before describing how he has tackled the management of construction information in his own organisation, ARK e-management.

Continuing on the theme of data management, Froese adds to the debate by exploring the requirements of project information management for the uptake of nD modelling. Current practices for managing project information and ICT need to evolve and current skills need to improve. The chapter discusses aspects of such a strategy, describing underlying conceptual models, an overall framework for project information management and technical and organizational considerations. The chapter also touches on the relationship between project information management and project management in general. Bacon and Froese both fuel the need for structured project data to be included in the nD model.

Karlshøj espouses the visualisation benefits of nD. In his chapter, he describes the varying forms of computer visualisation, ranging from 1D (graphs) through to 3D (augmented reality, geometric models, etc.). The advantages and limitations of each method are clearly discussed before the need for effective nD modelling visualisation is emphasised. nD visualisation techniques, as applied in Danish projects, are described to demonstrate the clear need for nD and the subsequent benefits. Finally, Karlshøj proposes future visualisation research areas.

Hassan and Ren's chapter seeks to start the timely debate on the legal issues arising from working in an nD modelling environment. It critically highlights topical areas for concern, including access rights, liability, ownership and so on of the information embedded in the building information model. It benevolently highlights the relevant European Laws and Directives that impinge on working with nD tools, and clearly sets out a timely and topical area for research before nD can be applied globally.

Interactive graphics and technologies are one of the most important developments that are affecting ICTs. Kähkönen and Rönkkö's chapter explores the possibilities of using interactive computer avatars for testing and understanding the performance of buildings. It is considered that avatar technology can provide new multi-physical dynamic simulations, which are interactive and communicative to both professional and non-professional users. With avatars, building performance can be analysed interactively from a human viewpoint, and can be designed as a set of interactive graphics for facilitating client's decision making.

In the technology transfer chapter, Sexton identifies key enablers and obstacles to the effective adoption and use of nD modelling technology. He presents a technology transfer framework with a supported review and synthesis of relevant literature, before discussing and describing the findings of case study research. Conclusions and implications are drawn on the potential use of nD.

Horne's chapter addresses the role of education in nD modelling implementation. The chapter highlights the timely need to move from typically 2D curricula and onto 3D, 4D and nD modelling to reflect the needs and advancement of the AEC/FM industry. Using Northumbria University as an example, Horne illustrates how the use of product modelling and VR is, and should be, embedded in various Built Environment courses. The introduction of nD modelling technology will require a change in the classic ways of teaching if it is to support the requirements of different disciplines, and facilitate and encourage a multi-discipline culture.

In the final chapter, on the nD game, of the 'application' of nD modelling, Aouad *et al.*, continue Horne's contention for the need to include nD in the curricula for global nD-enabled construction. Aouad *et al.* present the development of the nD game, a computer game for school kids that pro-actively teaches various design parameters, highlights that some criteria may conflict with each other and promotes the concept of nD design. The project is a continuum of the successful '3D to nD modelling' project at Salford.

Chapter 11

Data classification

Jeffrey Wix

Introduction

Building Information Modelling (BIM) is progressively taking over from Computer Aided Design and Drafting (CAD) as the way of representing information in building construction. Leading software applications are evolving from a drawing-based approach to using a model-based approach. In this chapter, key ideas about the organisation identification and classification of data within models are presented and their importance outlined.

Building information modelling

Moving from CAD to BIM involves much more than changing the name of the technology. A BIM provides a much more comprehensive, complex and realistic representation of the real world than a CAD drawing. In CAD, a shape is defined and this is used to represent a real world concept (e.g. a space, wall, air duct, etc.). The shape can be assigned a meaning through placing it on a 'layer' which is named or classified as containing only shapes used for a particular purpose (e.g. all shapes on a particular layer are used to represent external walls). If a stored shape (or symbol) is used, some further meaning can be defined through the symbol name.

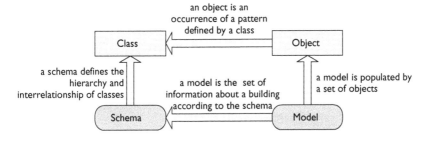

Figure 11.1 Object concepts.

BIM works in a different way as illustrated in Figure 11.1 (Stephens *et al.*, 2005):

- First, it establishes templates for the data that characterise a concept such as wall, space and so on. Data within a template can include allowed shape representations. These templates are called 'classes'.
- A class has a name that is indicative of what it represents (e.g. Door, Window, Task, etc.).
- Classes define the attributes of an object but do not set the values that may be assigned to the attributes (these define the 'state' of the object). Attributes are named for the information that they convey (e.g. Width, Height, Weight, etc.).
- Attributes define the type of properties that are associated to an object (e.g. the attributes' value could be a real number, integer number, alphanumeric text string or another class, etc.).
- Classes are collected into schemas which also specify the relationships that are allowed between the classes (such as a valve can be attached to a pipeline but not to a window!).
- Objects are occurrences of classes where each occurrence has values associated with data items (or attributes). Attributes with values specify the identity and state of an object.
- A building information model (the 'Model') is the complete population of objects that are used to describe a building.

Type and occurrence distinction

Within a schema, a distinction may be made between classes that define a particular pattern of attribute values that characterise a 'type' class and an occurrence of a class within a project having identity and location (Wix and Liebich, 2004) (Figure 11.2). A type can be used as the specification of the common data that is used by all occurrences that conform to that type. Thus, it has value as the

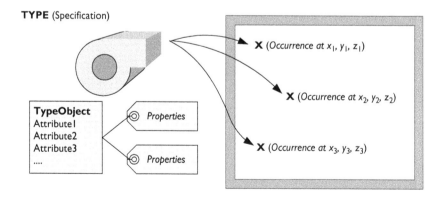

Figure 11.2 Type/occurrence differentiation.

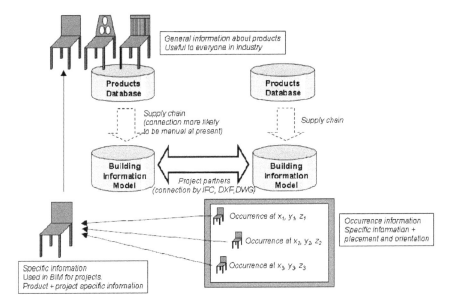

Figure 11.3 Type specifications in product databases.

specification of a product or other specification set (such as specification of design performance) that can be stored within an external catalogue or product library.

Whilst the type class allows the common data to be specified, each occurrence of a type needs only to add identification and placement (position) data plus any other data that is particular to the occurrence (Figure 11.3).

Identification

In BIM each object should be identified. However, there are several forms of identification that may be applied and it is important to understand the difference between them, how they may affect working and actions that should be undertaken to ensure their appropriate use. Wherever possible, use should be made of existing international, national and industry standards. Examples include the following:

- BS EN ISO 4157-1:1999 Construction drawings – Designation systems – Part 1: Buildings and parts of buildings,
- BS EN ISO 4157-2:1999 Construction drawings – Designation systems – Part 2: Room names and numbers,
- BS EN ISO 4157-3:1999 Construction drawings – Designation systems – Part 3: Room identifiers.

Globally unique identifier

A globally unique identifier (GUID) is an identifier that may be given to an object to make it unique. A GUID is typically generated from an algorithm that ensures that the value is, as far as can be imagined, completely different to any other value that ever has been or could be created. Generally a GUID value is dependent on the identity of the machine and the date and time of its generation.

The importance of a GUID is that it ensures that an object can be tracked throughout its entire life cycle. It gives an absolute assurance as to which object is being referenced. Systems that work with life-cycle databases and that recognise the need for information sharing will apply a GUID to an object that they create. In the longer term, it is likely that GUID assignment will become normal practice.

Handle

A handle is an identifier or token applied by a particular system which gives uniqueness in the context of that system but that does not guarantee universal uniqueness. As with a GUID, a handle is automatically generated by a system.

Instance identifiers

There are a number of instance identifiers that can be applied. Some of these are pre-assigned by product manufacturers, others can be applied in the context of the project. They include the following:

- serial numbers of components,
- asset identifiers,
- radio frequency identifiers,
- bar codes,
- order numbers,
- tag numbers for component schedules.

Type identifiers

Similarly, there are a number of type identifiers that can be applied. Again, some may be pre-assigned whilst others may be selected or assigned in the context of the project. They include the following:

- product code,
- classification.

Assignment of identifiers

Where the assignment of identification is within the context of the project, strategies for development of both the 'instance' and 'type' identifiers should be determined. The identifier should

- indicate the context (asset, order, schedule identity, etc.),
- provide some structured information that relates to the context (e.g. location information for asset, departmental information for order, etc.),
- give a number in sequence that is unique for the context and the project.

Naming strategies

In working with BIM, the aim is to work with 'classes' that define the characteristic attributes (properties) of an object. A well-designed object-based system will provide enough classes to fulfil the intention of a software system but not so many as to make software implementation unnecessarily difficult. Thus, classes tend to define general concepts. For instance, a space may be defined generally as 'representing an area or volume bounded actually or theoretically'. This recognises that spaces are areas or volumes that provide for particular functions within a building without being specific as to what those functions are or may be. Definitive attributes can be applied, such as size, capacity and so on.

The specific function of a particular class is often determined by applying a further parameter. In the example of a space, this parameter is often a type of identification or classification that establishes the functional usage. Taking the example of a space, the parameter might define a specific space as a living room, office or kitchen. The use of such a 'name' parameter is widespread in BIM applications. Without this ability to specify the function of certain classes, the number of classes required would make software implementation very difficult.

Names provide a means of creating 'types' of classes without having to actually create new classes. Names for 'types' of classes should be defined according to an established convention such as a classification or thesaurus or from a dictionary. This is so that they can be understood and applied in a consistent way.

Naming of types by purpose

A 'type' is the specification of a pattern of data that is always applied to objects in a given set of circumstances. For instance, every space in which cooking is to take place may be specified as being of type 'kitchen' whereas a sleeping space is specified as a 'bedroom'. Purpose names that are to be used should be defined and agreed between users. This should be both for users within the same organisation and for the users from separate organisations working on the same project. Examples of some purpose types for which names should be defined and agreed are indicated in Table 11.1.

Table 11.1 Examples where naming of a 'type' is required

Example	Naming method
Buildings	By building type or purpose group
Components	By 'types' of components, for example for a valve, isolating, regulating, check, pressure reducing, pressure relief, commissioning, control and so on
Materials	By name of material
Spaces	By space or zone function for example living room, kitchen, office, storage, circulation
Systems	By system function, for example low temperature hot water heating, cold water service, chilled water, compressed air and so on

Naming of types by construction

Construction objects such as walls, floors, ceilings and roofs on a project usually follow patterns of construction. These patterns identify common constructions with consistent attributes/properties. They can usually be identified and given a name. For example, a wall comprising multiple layers of 102 mm common brick, 75 mm cavity, and 100 mm concrete block could be named as a '275 mm cavity wall'.

The different names for construction types should be recorded so that other software systems within an organisation (and by a project team) can recognise them.

Naming of materials

Materials from which objects are constructed should be consistently identified and named. This name may apply to a basic material such as copper, iron, brick, glass and so on, or it may be a composite name such as common brick, engineering brick, 18:8:1 stainless steel, timber studding with mineral wool inserts and so on. The names given to materials should be recorded so that other software systems can consistently use the same material names.

Naming of attributes/properties

Consistent names should be defined for attributes/properties that may be added to a class. These names should be taken from a dictionary of established property names (which should also define units used) (Figure 11.4). By specifying approved property names, consistency can be defined between different users of the same application whilst the meaning of the property is defined between different software applications.

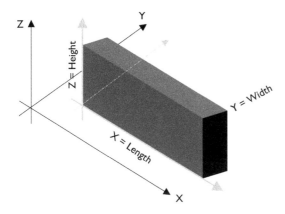

Figure 11.4 Consistent naming of properties is critical.

This is particularly important where common words are being used to define an attribute, or where a common word is used as part of the attribute name. For example, consider the word 'length' which might be used to define an attribute that represents the length of a wall. A general view is that 'length' is the longest dimension of an object. However, to be unambiguous, it is necessary to be more precise and to define the length as a distance measured along the X axis of the local coordinate system of the object.

Shape

Representation

The shape of an object may be represented in several different forms. In BIM, an object may have several different representations associated with it.

Presentation

Shapes in BIM also need to know how to 'present' themselves according to the context in which they are viewed. This is done either by ensuring that there are sufficient different representations defined to deal with all viewing contexts or by defining behaviours of objects that respond to context. Essentially, a context-responsive behaviour is a set of rules that determine how the shape will appear. Primarily, context concerns the level of detail required. Table 11.2 lists two of the many different presentation contexts that can be used to represent some examples of objects.

Level of detail is broadly equivalent to scale concepts in drawing. A level of general presentation might be considered as equivalent to 1:50 scale or less whilst detailed presentation might be equivalent to 1:20 scale or more. The level of detail

Table 11.2 Presentation contexts

Object	General presentation context	Detailed presentation context
Door	A door needs to show only that it fits into an opening and the direction in which it opens	A door needs to show its frame, sill, panel and ironmongery in detail. For greater detail, material textures may also be included to enable visualisation
Space (plant room)	A 2D plan of a plant room shows only single line representations of pipe work and ductwork	The detail layout of a plant room needs pipes and ducts to be shown as twin lines
Electrical components	It is generally sufficient to show electrical switches and sockets as schematic symbols	For detail purposes, electrical sockets and switches need to be represented realistically for dimensional and visualization purposes

for shapes to present themselves in different viewing contexts (scales) needs to be defined so that users within the same organisation and users on the same project from different organisations

- understand the shapes that are being presented,
- understand the level of detail being given.

Note that as systems become more sophisticated, it may also be possible to control which person can view what level of detail through 'access control lists' directly on the object. Whilst this is not possible on most systems at present, provision should be made to understand which person can view and change (read/write) information for future needs.

Symbols

Traditionally a 'symbol' is simply a shape representation that can be saved and re-used. In object-based working, it is the multiple shape representations that can be saved and re-used for attachment to an object. That is, it is equivalent to a multi-view symbol in CAD.

Principles of symbol naming in BIM are the same as for CAD (BS 1192-3, BS 1192-5).

Shape naming

Within BIM, an object may have multiple shape representations and there is a need to identify each according to the purpose that it is fulfilling and according to the type of representation used in a consistent way. Strategies for identification depend upon the capabilities of the system concerned. BIM systems generally

Table 11.3 Types of shape representation

Shape representation	Description
Curve 2D	2 dimensional curves
Geometric set	Unconnected set of points, curves, surfaces (2 or 3 dimensional)
Geometric curve set	Unconnected set of points, curves (2 or 3 dimensional)
Surface model	Face-based and shell-based surface model
Solid model	General solid including swept solid, Boolean results and Brep bodies
Swept solid	Specific solid model using swept area solids, by extrusion and revolution
Brep	Specific solid model using faceted boundary representations with and without voids
CSG	Specific solid model using Boolean results of operations between solid models, half spaces and Boolean results
Clipping	Boolean differences between swept area solids, half spaces and Boolean results
Bounding box	A simple representation defined by a bounding box
Sectioned spine	Cross-Section-Based representation of a spine curve and planar cross sections. It can represent a surface or a solid and the interpolations between the cross sections is not defined
Mapped representation	Representation based on a mapped item, referring to a representation map. Note: Mapped Representation is the representation used in IFC for to define the geometry of a symbol/block/cell

allow for meaningful names to be given directly to an object. Table 11.3 provides a system neutral list of names for types of shape representations (IFC, 2005).

Relationships

Groups

Many BIM applications are able to group sets of objects together to form 'group' objects. This allows users to work with both groups of objects and individual objects in the group (for instance, moving a single object does not change its relationship with the group). The use of groups can add significant value to a building model and such relationships can be created to form the following:

- a group 'systematic' object, for example bringing individual components into systems and sub-systems (for building services),
- a group 'action' object, for example collecting objects into assets (for facilities management) or determining objects that are to be acted upon by a task (for planning/scheduling),

- a convenient 'arbitrary' grouping of objects, for example identifying the items of furniture belonging to a particular space.

An individual object can be related to more than one group. For instance, an object may belong to one group for costing purposes and to another group for fabrication. When working with groups of objects it is important to ensure that

- 'systematic' groups are named according to an established convention for the particular systematic type,
- 'action' objects are identified according to the established identification strategy for the particular action type,
- 'arbitrary' groups are named and/or identified in such a manner that they can be easily recovered at a later time; practically, this means recording the groups created,
- care is taken to ensure that a group does not contain more than one copy of the same object; generally, this should be taken care of by software applications but care may need to be exercised when a group also contains other groups.

Layers

A particular example of a group object is a layer. Layers are as relevant to BIM as they are to CAD. For BIM however, layers are simply one way of defining a 'systematic' group. Generally, layer names in existing layer conventions should be treated as identifiers (since they are generally defined from some form of classification). The name of the layer group should be the equivalent of the 'alias' value normally associated with a layer. This is because the alias name is more semantically meaningful than the conventional identifying name. Standards that have been defined for layers within CAD remain applicable to BIM including (BS 1192-5, ISO 13567-1, ISO 13567-2).

Whole/part relationships

Whole/part relationships are used to handle assemblies so that whilst the whole is visible as the assembly, the individual parts remain separately visible. For example, stairs are assembled from parts that include steps, landings, stringers, balustrades and so on. The process of assembly from parts is often known as aggregation and it is a powerful part of BIM working. Some viewing context (scale dependence) may need to be handled using aggregation relationships (Figure 11.5). This will be the case when the nature of the information to be viewed changes. For instance, utility services at small scale do not need to show detailed connections; a straight run of pipe or cable may be viewed as a single object.

Cost

Another example is a group object to which a cost can be associated. Cost schedules such as 'bills of quantities' or 'cost plans' are usually developed according to

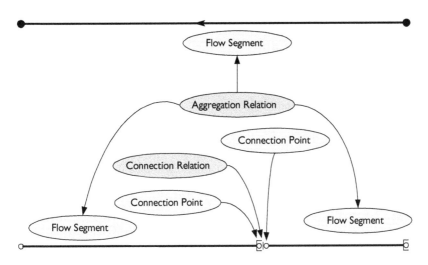

Figure 11.5 Level of detail using aggregation relationships.

measurement rules that require objects to be grouped in particular ways. Whilst an individual object may have a specific cost, the group object may be expected to have the cost that relates to labour, plant and materials cost items that are related to group object – an item in a bill of quantities.

Note that for cost purposes, several different groupings may be appropriate according to circumstances. This is because there are many different ways in which the cost of an object (or group of objects) can be represented (cost to buy, cost in tender, replacement cost, life-cycle cost, etc.).

Task

Another group is a set of objects that are subject to a single task in a work schedule. Each task should be named and may be assigned an identity in a work breakdown structure. The grouping of objects by task and the collection of tasks into a work schedule will provide support for the development of object-based 4D simulations. Such simulations can be seen for a project as a whole or, depending on the detail of the task breakdown, may be able to be seen at the level of an individual space.

Glossary/thesaurus

Glossaries are important components in schema and ontology definition since they provide good indication of high-level content (by subject) and also the meaning of the terms used. However, in developing schemas to be supported by BIM use (such as IFC), the extent of the content of the standard has been found to be limited covering only a small proportion of requirements. The standard glossary for building and civil engineering is BS 6100 (and its international equivalent ISO 6707, BS 6100-0, BS 6100-1).

Thesauri

An alternative to glossaries are thesauri. A thesaurus is a common way of presenting a controlled language index so that it may be effectively and efficiently used in information retrieval (Cann, 1977). There are many thesauri relating to building construction terms and these are generally well organised into hierarchical structures; there are also explicit translations between terms. These are potentially very valuable at the level of ontology (since they can prescribe the terminological content). A thesaurus term without a semantic definition is of limited use for schema development however.

Some key examples of English language thesauri for building construction include the Construction Industry Thesaurus (Roberts, 1976) covering general terms and the Building Services Thesaurus (Beale, 1993) covering terms more specific to building services practice and use.

Classification

Classification provides an essentially taxonomic (hierarchical) structure that enables the organisation and grouping of things into specific families. Traditionally, classification systems have taken one particular view of the world and how it should be organised and this has led to a proliferation of such systems becoming available. For instance, the CI/SfB system used in UK practice (Ray-Jones and Clegg, 1991) has an essentially element-based view of the world (e.g. as doors, windows, etc.) whilst the Common Arrangement of Work Sections (CAWS) has a system-based view. These have been useful in the development of layer conventions within CAD and will continue to have value within BIM (since layers provide one grouping mechanism that is valuable for viewing purposes).

More recent classification system developments including the Uniclass system applied in the United Kingdom (Uniclass, 1997) and the Omniclass system for US (Omniclass, 2005) use are based on a framework structure defined within ISO 12006 part 2 which can be broadly viewed as a schema (Figure 11.6). Such classification systems provide a multi-facetted view of the world and enable a classification to be applied with various classification facets.

Classification provides a further strategy for identifying objects. Some systems allow classification to be handled as a separate object that can be related to an object of interest. Allowing classification as a separate object enables the possibility of classifying an object in multiple classification systems. This is particularly useful where classification systems in use provide a singular world view. In systems that do not have this capability, classification needs to be applied as an attribute/property.

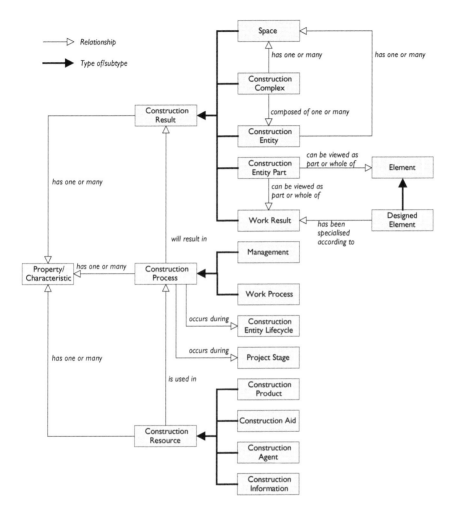

Figure 11.6 Simplified view of ISO 12006 part 2 metaschema.

Dictionaries

A dictionary (in this sense of its usage, a dictionary is correctly an ontology since it defines not only terms and meanings but also the relationships between terms and how they may be instantiated with units and values) provides a comprehensive reference to terms and their meanings and can deal with alternatives (synonyms and homonyms) as well as defining the context in which the terminology is used and the units on which values relating to the terms can be expressed.

Dictionaries are of critical importance in the development of BIM usage because they provide the means by which the names that we use to describe concepts, facts, ideas and relationships can be captured and shared.

Dictionaries are also a powerful tool in enabling concepts, facts, ideas and relationships to be expressed appropriately for a local usage and yet be consistent with other local usages. This can be done by creating a 'master dictionary' in which each term has its own globally unique identifier and then mapping the local usage (in whatever language is appropriate) to this identifier. This is the principle behind the development of ISO 12006 part 3 which provides an object-oriented framework for dictionaries.

ISO 12006 part 3 (which is sometimes called 'International Framework for Dictionaries' or abbreviated to IFD) defines a schema for a dictionary that provides the ability to (Figure 11.7)

- define concepts by means of properties,
- group concepts,
- define relationships between concepts.

Objects, collections and relationships are the basic entities of the schema. The set of properties associated with an object provide the formal definition of the object as well as its typical behaviour. Properties have values, optionally expressed in units.

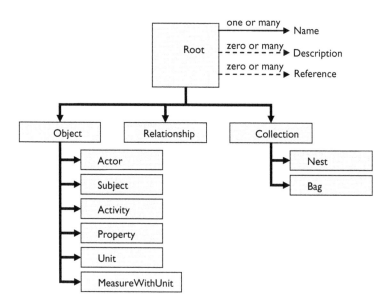

Figure 11.7 Top level structure of ISO 12006 part 3.

The role that an object is intended to play can be designated through the schema (shown as subtypes of the object class in (Figure 11.7) and this provides the capability to define the context within which the object is used.

Collections are either all of the same type of object (class) in which case a 'nest' collection method is used or of various types of objects in which case a 'bag' collection method is used. Various types of relationships between objects are allowed including acts upon, aggregation, collection, composition, grouping, measure assignment, property assignment, sequencing and specialisation. These allow complex references to be developed within a dictionary.

Each object may have multiple names and this allows for its expression in terms of synonyms or in multiple languages. The language name of each object must always be given in English (the default language). An object may also be named in terms of the language of the location in which it is determined or used. Objects may be related to formal classification systems through the provision of references.

There are several implementations of dictionaries that are compliant with the schema given in ISO 12006 part 3. These include developments in Norway (BARBi, 2005), Netherlands (Lexicon, 2005) as well as efforts currently in France, Germany and elsewhere (IFD):

- BARBi is a Norwegian based project that is looking to define reference data approaches that are relevant both to the construction industry (through the IFC schema) and to the process engineering/offshore industries (through the implementation of the ISO 15926 schema). It is implemented within an object-oriented database and has a browser that enables web access.
- LexiCon is used nationally within the Netherlands in support of the STABU specification system and provides definitions and specifications of concepts that are of interest for the construction industry (Figure 11.8).

Future development

Building Information Modelling is coming into use within the building construction industry and its benefit is being reported. However, its use does require changes in approach. This is being realised through the development of guidance approaches such as the UK Avanti program (Stephens *et al.*, 2005) and similar initiatives elsewhere.

Building Information Modelling is progressively being accepted as an approach rather than simply being a type of software application. This is leading to the realisation that information exchange is a key component not only within the project but also within the supply chain. The techniques used to specify data exchange structures are also being expanded to handle knowledge representations and this will enable knowledge provision and application through the building

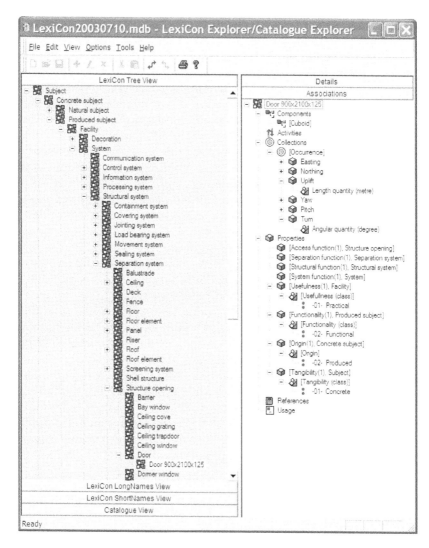

Figure 11.8 Lexicon screen showing dictionary terms in English.

information model. The dictionary developments that are presently emerging will play a key role in supporting knowledge representations.

Building Information Modelling is also emphasising the impact that data has on processes within building construction. This is likely to lead to the identification of data specifications that support particular processes. These are termed 'business objects' and will be the next key to unlocking the potential of this technology.

References

BARBi (2005) Building and construction reference data library, The Norwegian Building and Construction Reference Data Library, http://www.barbi.no

Beale, G. (1993) *The Building Services Thesaurus, Fifth edition*, BSRIA, Oxford.

BS 1192-3 (1987) Construction drawing practice. Recommendations for symbols and other graphic conventions, British Standards Institution.

BS 1192-5 (1998) Construction drawing practice. Guide for structuring and exchange of CAD data, British Standards Institution, British Standards Institution.

BS 6100-0 (2002) Glossary of building terms. Introduction, British Standards Institution.

BS 6100-1 (2004) BS ISO 6707-1:2004, Glossary of building and civil engineering terms. General terms, British Standards Institution.

BS EN ISO 4157-1 (1999) Construction drawings – Designation systems – Part 1: Buildings and parts of buildings, British Standards Institution.

BS EN ISO 4157-2 (1999) Construction drawings – Designation systems – Part 2: Room names and numbers, British Standards Institution.

BS EN ISO 4157-3 (1999) Construction drawings – Designation systems – Part 3: Room identifiers, British Standards Institution.

Cann, J. (1977) Principles of classification: suggestions for a procedure to be used by ICIS in developing international classification tables for the construction industry, ICIS.

CEN (2004) CWA15141 Wix, J. and Liebich, T., European eConstruction Meta-Schema (EeM), European Committee for Standardization, http://www.cenorm.be/cenorm/businessdomains/businessdomains/isss/cwa/econstruct.asp

IFC (2005) Industry Foundation Classes Releases http://www.iai-international.org

ISO 13567-1 (1998) Technical product documentation. Organization and naming of layers for CAD. Overview and principles, International Standards Organization.

ISO 13567-2 (1998) Technical product documentation. Organization and naming of layers for CAD. Concepts, format and codes used in construction documentation, International Standards Organization.

Lexicon (2005) Lexicon: A Common International Construction Language, STABU, http://www.constructionlexicon.com/

Omniclass (2005) Construction Specifications Institute, http://www.omniclass.org

Ray-Jones, A. and Clegg, D. (1991) CI/SfB Construction Indexing Manual, RIBA Publishing.

Roberts, M.J. (1976) *Construction Industry Thesaurus, Second edition*. London: Property Services Agency, Department of the Environment, UK.

Stephens, J., Wix, J. and Liebich, T. (2005) Object Modelling Guide, Avanti Toolkit 3 Version 1.0, The Avanti Programme: c/o Constructing Excellence in the Built Environment, June 2005.

Unified Classification for the Construction Industry (UniClass, 1997) National Building Specification (NBS), UK.

Chapter 12

Management of requirements information in the nD model

Matthew Bacon

Proposition

At the outset let me set out a clear objective for the nD model:

> The substantial body of written information that comprises the Brief and which exists external to the nD model should be integrated into it, so the model integrates *all* project information.

With current briefing practice, such an aspiration is largely unattainable. But this is not to say that current practice cannot be changed. The challenge is in fact both one of business practice and technology. Current practice needs to be changed in order that briefing information can be an integrated part of the nD model, which means that the way in which briefing teams use technology to document the Brief also has to change.

This is the premise behind this chapter and in it I will discuss both the technological opportunity as well as the process change required to embrace it, and so deliver briefing information that can form an intrinsic part of the nD model. First, I will discuss the need for integration and the benefits that arise from it. Clearly, just because integration is possible, it does not mean that it is necessary. I will then discuss the process of creating the Brief. It is important to consider the process perspective because it is important to understand the complexities of information management in current working practice. This section of the chapter will highlight the need for structured information, if we are to achieve the integration that is required. I will also explain how this can be achieved. With information managed in a structured environment, I will then explain the opportunity of being able to manage information over the entire project life cycle, and support the different perspectives of the team working at each stage of that life cycle. Finally I will illustrate how we have achieved some of these objectives in our own business.

Why the need for integration?

Without a Brief, the professional team would have little basis on which to start the Design Process. The Brief should be, indeed must be, the driver behind the design. If the Brief fails to adequately communicate the customers' requirements, then inevitably the resulting design will fail to deliver what the customers believe they have requested. To avoid this situation, the Brief must not only be understandable, but as importantly the information it contains must be *accessible*. It is too trite to argue that the Brief only has to be read for this to be achieved. On large complex projects the Brief can accumulate into hundreds, if not thousands of documents. Many of these documents are inter-related, and as such, to extrapolate specific requirements is a significant task. A major justification for integration is to make the briefing information accessible to the designers at each stage of the Design Process in order to make the process much more efficient and effective.

The way in which briefing information is documented, made accessible and then retrieved by the professional team is fundamental to the success of the project. It is on complex projects where these needs become paramount. It follows that the designers who are creating and modelling the graphical information (which become the instructions to the construction members of the team) need to have the briefing information (their instructions from the briefing team) in a form that is useful to them. This is another justification for integration – improved quality of information.

The reality is that on many occasions the process is rarely efficient and often it is not particularly effective. The expensive and well-publicised major projects of recent times are testament to this, where dissatisfied users of new facilities complain that the professional team did not listen to them and ignored their requirements. On some occasions these failings can even delay the opening of a new facility, until the work is rectified. The reality is often not so much that the professional team did not listen, or set out to ignore requirements, but that that they had been overlooked in the substantial body of information that can comprise a Brief.

But if a key need for integration is to make information accessible, another is to enable the design information to be systematically checked for compliance. In other words, if the briefing information is not integrated into the nD model, the team is denied the ability to systematically test the design against the briefing requirements. If one contrasts architectural design and software design, then in software design we design tests and document performance criteria in the specification, and we design test routines to automate the testing process to conclusively prove compliance with the specification. Furthermore, we are able to 'scope code test' as well, so that we are able to determine how much of the software code has been systematically tested. I am not aware of this possibility in architectural or engineering design.

In my experience as an architect and later as professional client, I never witnessed (or indeed recognised the need for) any programmatic testing from multiple use-case scenarios, other than those performed by engineering disciplines. I would suggest that the closest that the majority of architectural design gets to programmatic compliance testing (using computer-generated scripts) is through clash detection routines in some software applications. However, this is a practice far removed from systematic testing of the design against the requirements stipulated in the Brief. I believe that this situation stems from the lack of structured briefing data for the designers to test their designs in this way. The third justification for integration is the ability to programmatically test design options and decisions against the briefing information.

Documenting the Brief

The Briefing Process is essentially one of communication between a customer and the professional team. The customer (or their representatives) is required to communicate their business needs, and more specifically, their requirements for the works. The professional team is required to interpret those requirements into terms that are useful to them, and in doing so will create a technical requirements brief, which is to be read alongside the customers' own requirements. Often the Brief will either contain or reference other documents, which will provide additional information, such as references to standards, legislative requirements, reports and commonly, minutes of meetings or workshop findings. Some teams have even used paintings and poetry to help communicate the essential characteristics of the customers' requirements.

This body of documentation is the primary means of recording requirements, and is the primary output of the Briefing Process. During this process the professional team will elicit those requirements through a series of meetings and workshops, in order to discover the issues that will need to be interpreted for the Brief into specific requirements. During the Briefing Process, the team will start to create exploratory sketches and schedules, the objective of which will be to tease out further requirements and articulate conceptual issues. These sketches will require yet more meetings in order to discuss the requirements arising from them. Manual appraisal of design information against specific requirements, such as room data sheets, will be carried out. This process can take some months, and on larger projects, it can take years. As explained earlier, the body of information that is accumulated through this process can be substantial. The written information will form a major part of the Brief, alongside the sketches, diagrams, photographs and even paintings or prose.

As the process develops, the team will commence the development of a technical brief, and this will document how the team will achieve the business requirements. It too will reference other documents, such as planning legislation, building code legislation, design principles documentation, technical standards and so forth.

The Briefing process for software design is very similar to that for architectural design. However the major difference is the ability to systematically record requirements as structured digital information. With such an information resource we are able to configure a simulation engine to automate sequences of highly complex routines and identify areas of non-compliance in the process of checking compliance of software code with the specification. This process is far more thorough, systematic and efficient than any manual process could ever be.

To return to the briefing process, the Briefing team will attempt to co-ordinate all of the information and typically will assemble it into five types of documents:

1   Statement of need
2   Business case
3   Statement of requirements
4   Statement of technical requirements
5   Specification.

These documents will reference other documents, such as reports, and meeting minutes, as described earlier. It will be from these documents that designers will source the requirements and then aim to reflect those requirements in the design – in other words the graphical information. The process is completely reliant on the diligence of the design team to assimilate the requirements into the design. However in order to do so they will need to develop a thorough understanding of all parts of the primary documents listed as well as the referenced information. To understand the complexity of this task, we need to consider the content of these documents:

Statement of need   This will establish the overall objectives and the key needs that have to be met in the project. It will often reference board minutes and reports, and it will reference the preliminary business case. The information will be factual as well as prescriptive.

Business case   This will document the business justification and the parameters of cost versus revenue and or service affordability of the proposed facility. It will often contain a risk assessment and may even explore different business case scenarios with an evaluation of the risk issues associated with each. A systematic risk assessment may even consider what risks would be best managed at each stage of the project.

Statement of requirements (SOR)   This will introduce the broad requirements concerning the nature of the new work, as well as specific accommodation requirements, functional requirements, spatial relationships and user requirements. The information will be both prescriptive (e.g. floor area requirements) and performance related (the building must be light and spacious).

The SOR will set out detailed schedules of accommodation, and a well-documented Brief will contain room schedules, which set out details of equipment, finishes and fittings.

Statement of technical requirements (SOTR)   This will document the requirements inferred by the professional team, and will be as concerned with standards (ranging from those that determine spatial planning, to those specifying finishes and equipment) as well as design, system and construction strategies. Standards may well be arranged by accommodation types (particularly the case in public buildings), by element type, such as floors, walls ceilings and even by service type, such as electrical, mechanical, safety, security and communications.

Legislative requirements that control building use, access, security, environmental constraints and so forth will be referenced by the SOR.

Specification   This document will be arranged using a standard structure, such as the National Building Specification (NBS). It will describe the standard of work and the specific standards that the work must conform to. It will contain schedules and in this regard there will be a substantial body of very prescriptive information. However there may also be performance standards and criteria will be documented that will define how compliance with these standards must be demonstrated.

During the briefing process, all of these requirements should be co-ordinated. This co-ordination function is essential if there are not to be conflicts between one requirement and another. The co-ordination task can be highly complex, because the effect of one requirement on another may not always be easily recognised.

The ideal situation would be to have the ability to interrogate the body of documentation with a question such as 'What are the requirements for building access?' To get an answer from the accumulated body of information along the lines in the following page would be ideal for a briefing team. The answer could be displayed in a report set out as follows:

Statement of Requirements
Section 1.1.2
The building must display a gravitas commensurate with the importance of its public function . . . .. Section 3.2.7
Access requirements are to be planned for the following categorises of user. For each category refer to . . . .

Hard landscaping design standard
Scope
Where public access is to be provided then the following considerations will apply . . . ..
Section 3.1
Ramps. These are to be designed in accordance with the following requirements . . . ..

Local Planning Authority – Meeting with Highways Department on . . . .. at . . . .
Minute item 2.1.2
The requirement for vehicular access will be . . . .

Report
Space syntax study. Author . . . .. Date . . . .
Section 1.3 – Executive summary
The report suggests that the access routes across . . . . and that the following primary flows must be incorporated into the site layout.

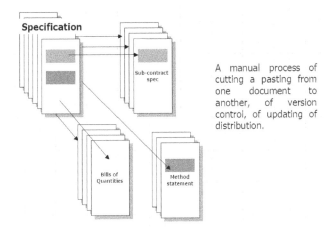

Specification

Sub-contract spec

Bills of Quantities

Method statement

A manual process of cutting a pasting from one document to another, of version control, of updating of distribution.

Figure 12.1 Information re-use (storing only one instance of the truth) is not achievable using word-processed documents.

This illustrates in the context of just one issue, the diversity of information that could exist in a Brief and the associated body of documentation. Concerning the issue referred to, there will undoubtedly be many more references. Performing a search of the information (interrogating it) as described in the previous paragraph is simply not possible with conventional practice where the information is created and stored in documents. This is because the information is *unstructured* and means that it is unable to be interrogated using standard query methods. Manual referencing and copy typing are the traditional means of carrying out this co-ordination function, but it is both very time consuming and inefficient. This is illustrated in Figure 12.1. It is also highly prone to error because it relies on individuals possessing a thorough knowledge of the whole body of documentation, which is almost impossible on major projects.

Consequently, it is the traditional method of creating information and storing briefing information in documents that leads to the need for manual processing of information. The greater majority of briefing information will have been created using word-processing and spreadsheet applications and in this form I refer to it as *unstructured information*.

Structured information

So in what form should the briefing information be to enable co-ordination of requirements and integration with design data to be achieved? Fundamentally it must be structured, by this I mean that it should be stored, in a database format in which the information can be uniquely identified and managed.

Briefing information stored in a database means that it is able to be stored at a 'granular', some might say 'atomic' level of abstraction. In comparative terms, a word-processed document could be thought of as a 'blob' of information. The advantage of creating and storing information at the 'atomic' level means that every 'atomic particle' can be uniquely identified, and in doing so it can be used to inform the property set of the nD objects in the project model, which will have been defined at an identical level of abstraction. In other words the data model for the briefing information would be defined in the same terms as that for the nD model, thus providing the integration that I believe we need to be aiming to achieve. The illustration in Figure 12.2 explains the opportunity for integration when both textual and graphical information are modelled at the same atomic level.

As documented in previous chapters of this book, work has already been under-taken where information pertaining to building codes is stored in such a way as to make integration between briefing information and design information much more explicit. This is one area where programmatic testing of the design for com-pliance with codes is possible. However, building code information forms only part of a Brief, and indeed there is a substantial body of briefing information in the Statement of Requirements, which drives the requirements for these codes.

In the 3D to nD model, described in Chapter 1, we can see the opportunity to integrate briefing and design information. Objects in the model have the poten-tial to be referenced to briefing data, but only if the briefing data is created and stored in such a way as to make this possible, as illustrated in Figure 12.2. Having referenced the briefing information to the model in this way, it means that the properties of the data objects within the model can be populated by values defined by the briefing data. Furthermore, there is no reason why the rule base that gov-erning object behaviours cannot be expressed as design constraints should not also be informed by the briefing requirements. To achieve these outcomes will

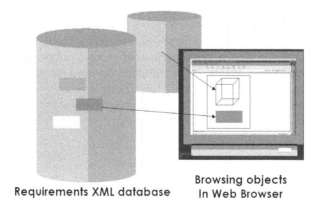

Requirements XML database   Browsing objects In Web Browser

Figure 12.2 A Requirements XML database enables briefing information to be stored at an 'atomic level' so that it can be referenced by objects at the same atomic level in the nD model.

mean changing the briefing process and specifically the way in which briefing information is created and stored. To explain the changes required, Figure 12.3 describes the essential requirements of a web browser supported process.

The example illustrated in Figure 12.3 shows an input screen on the left of the diagram connected to two different types of database. The input screen can either read from or write to, each of the databases. A manager of the briefing process would use the input screen. As information content is saved from the input screen (form) to the database, meta-information about the context within which the information is created is also saved.

On the right hand side of the diagram is a report view of the data. This 'virtual document' assembles the information dynamically from either or both of the databases depending on what is required in the 'virtual document'.

The 'virtual document', which is a web-browser-based representation of the information, can present the information within any database, in any way that is required by the process. This means that different views of the information can be constructed to suit the needs of different users. For example, the Statement of Requirements will present one view of the information, as much as the Statement of Technical Requirements will present another view. Inevitably, the same information will be re-used from one view (document) to another. Through a managed process, whenever that information changes, both documents will read from the same data source and present the updated information in the 'virtual document'.

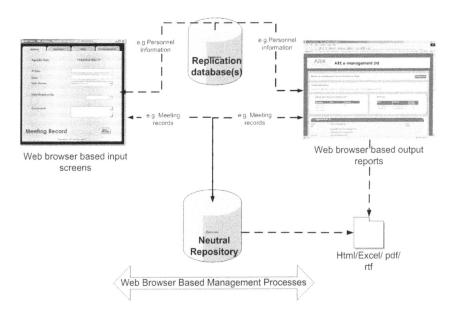

Figure 12.3 Schematic of web-browser-based tool enabling the creation of structured briefing information.

The implication of my proposition is that at each stage of the Briefing process for each different activity in it, the briefing team will require access 'tools' to execute that will enable structured information content to be created. The types of tools (software applications) that will be required to achieve this are as follows:

1   Briefing management tool
2   Meeting management tool
3   Specification management tool.

Briefing management tool

Figure 12.4 illustrates another type of 'virtual document' – the Briefing management tool. Such a tool enables the briefing team to work in the briefing process in a collaborative environment. It obviates the need for multiple versions of the same document to be managed by the team and provides complete traceability for each part of it. Indeed it provides total traceability of every contributor, information source, date and time of every change. Unlike traditional word-processing applications, it automates versioning of every atomic element of information.

A significant advantage of this approach to creating structured information is that it enables the briefing team members to work on the same 'virtual document' at the same time. They are able to work from any location and as such collaborative authoring in a managed process is achieved.

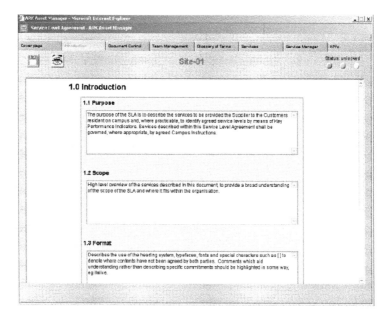

Figure 12.4 A 'virtual document template' viewed with a web browser and through which structured briefing information can be created.

As explained earlier, the Briefing management tool represents the Brief as a virtual document by an html rendering in a web browser, with fields (information place-holders) in which information content is created by the user (Brief writer). Each section of the document and each of the fields within it will be uniquely identified in the database, in what I have described before as the 'atomic level' of information. As information content is changed, it is automatically version controlled. Key meta-information will include the authors' details, the date and time when information was created. The system captures the information automatically, without user intervention.

The tabs represent the document sections, and clicking on each with a mouse pointer opens up that section of the document in the web browser. Document sections can also be controlled by role profiles. In this way, each member of the briefing team will have access rights to different parts of the document. I refer to these users as 'Section Editors'. Another type of user would be the 'Document Controller' and their primary role might be called something like 'Design Brief Manager'. With overall responsibility for the briefing process this person will manage the inputs of the other members of the briefing team.

I hope that it is now clear how it is possible to make queries into the database, such as the one I cited – 'What are the requirements for building access?' Storing information at an atomic level and being able to categorise it at that level is the means for achieving this.

Meeting management application

The briefing process is also characterised by many meetings and workshops, where different user groups are able to articulate their requirements. Often such meetings will challenge previously documented requirements in the Brief. On other occasions they will add to them, by providing clarification. Whatever happens, a co-ordination issue immediately arises, which has to ensure that requirements do not conflict, or are at least are acknowledged and reconciled one way or another.

Creating structured meeting information content is essential if this co-ordination and integration process is to be managed efficiently and accurately. To do so means also changing the way in which briefing information is created and stored. Using a word-processing application is not an effective tool, because it is unable to create structured information.

Figure 12.5 illustrates an on-line agenda, which ensures a consistent approach to the management of each meeting. By using a standard agenda for each meeting type, we can ensure that the meeting content associated with each agenda item is stored consistently in the database. By this means we are able to query the database for all information content relating to a specific agenda item in a specific context. On a typical project there may be a number of standard meeting types, such as

1   User Study Group meetings
2   User Study Group workshops
3   Operational review meetings

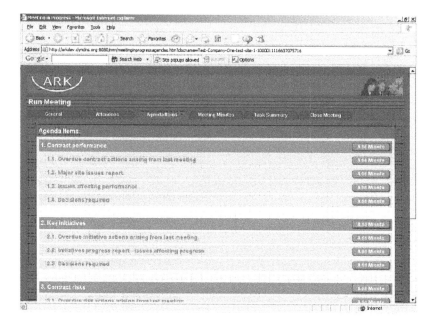

Figure 12.5 An online meeting agenda accessed though a web browser.

4   Business Case development meetings
5   Project strategy meetings.

All of these meetings will generate information, which in one form or another will impact on the Brief.

For example, in terms of a User Study Group meeting, a briefing team could organise meeting information as follows:

1.0   Facility

    1.1   Department

        1.1.1   User functional requirements

            1.1.1.1   Standards
            1.1.1.2   Operations
            1.1.1.3   Proximity requirements
            1.1.1.4   Assumptions

        1.1.2   User performance requirements

            1.1.2.1   Standards
            1.1.2.2   Objectives
            1.1.2.3   Constraints

1.2   Common areas

   1.2.1   User functional requirements

      1.2.1.1   Standards
      1.2.1.2   Operations
      1.2.1.3   Proximity requirements
      1.2.1.4   Assumptions

   1.2.2   User performance requirements

      1.2.2.1   Standards
      1.2.2.2   Objectives
      1.2.2.3   Constraints

Whatever hierarchy is used, it must be consistently deployed and as such the Brief should also recognise this structure because it enables substantial information re-use. As mentioned earlier, having such a method ensures that information can be readily retrieved and assembled with other like information.

Within a meeting context – such as a User Study Group meeting for example – discussion about requirements can be recorded, and in the context of that discussion, decisions might be made. These decisions need to be integrated into the Statement of Requirements. In a conventional process, the briefing team will have to 'cut and paste' meeting information – or at least re-key it into that document. In the approach that I advocate, the Briefing management application would assemble this information dynamically from the meeting decision content.

The way in which the meeting information is committed to a database is through the form-based approach illustrated in Figure 12.6. When a user activates an agenda item from the online agenda, it activates this form and the user selects the appropriate tab: Discussion – Decision – Task – Risk. Whichever tab is selected the information content will be documented in the database accordingly.

Within each form it would be possible to tag content (i.e. assign a further property) in order that it could be readily retrieved and used elsewhere in the nD model. For example, if within the User Study Group meeting a requirement for a space is stipulated, then it could be tagged as <spatial attribute property>. Or if there is a generic requirement, then it might be tagged as <generic spatial attribute property>.

With this approach, every atomic fragment of meeting information is stored with essential meta-information, such as meeting date, time and location; attendees, and then the agenda items and the associated content type as explained earlier. It is this traceability of briefing information content which is essential for the effective management of the briefing process.

Specification management tool

The Specification can be constructed in exactly the same way that I propose for the Brief, using the Briefing management tool referred to earlier. The organisation of the Specification is likely to follow an industry standard, such as

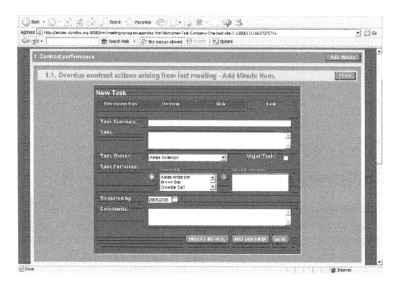

Figure 12.6 An online meeting content form through which structured briefing information can be created.

National Building Specification. Using a database of clauses such as that provided by NBS users will need the ability to tag meta-data to the existing NBS data
in order to be able make much more flexible use of it. The need for flexibility
arises from the need to manage information across the whole life cycle of a
facility.

Technologies such as Java and XML enable tools such as a Specification
management tool to be developed as interfaces to existing databases. Furthermore,
they can also be configured to query the briefing information database content
and extract any information concerning standards for example. Indeed the
Specification Management tool could query any relational database, and if a
manufacturers database is exposed to the Internet, it too can be queried. In doing
so, product data can also be referenced into the specification. Other specification
technologies such as Code Book (see www.codebook.co.uk) provide a database
of structured requirements, which would also provide a data source by which
information can be structured and referenced into the nD model.

Some might argue that the complexity of achieving this outweighs the benefits.
However, this is where standards are important in terms of how specification data
should be structured. Standards such as RFD (Resource Definition Framework)
offer huge potential and can be applied post-processing of manufacturers data.
Put simply, RDF offers the opportunity to create a simple but powerful meta-
information model, by which standards information can be categorised, searched
and retrieved *in context.* I emphasis the context issue because standards are

designed to be used *in context*. If the context in which they are to be used is expressed clearly in the RFD vocabulary (the means by which properties and their associated values are matched), then referencing the data appropriately is readily achievable.

The power that RFD provides is that it is able to manage meta-information external to the database to which it is referencing. This means that it does not affect the integrity of the original data wherever it is stored, NBS, Code Book or any other specification database. RFD browsers are able to query any of the connected data sources and return values associated with meta-information being searched, which in turn references the database source.

It is when information is considered in the context of the whole life cycle of a project and facility that the need for specification meta-information becomes obvious. Without this capability project teams have to re-process existing information for the specific needs of each stage of the life cycle. This has often been the criticism of project extranets in that if they are set up in a design phase, they are inappropriate at a construction phase, because the information needs to be organised differently.

Summary

In order to be able to integrate all information into the nD model, all of it that needs to be managed in it needs to be structured. Conventional office type applications are inappropriate because they are unable to produce structured information. To structure the information requires specific 'tools' (software applications), as well as the ability to create and manage meta-information. Finally, these tools need to be able to work over existing data sources such as those often found with specification databases.

Management of information over the project life cycle

A common objection to the creation of structured project information is that the information needs to be structured differently at each stage of the project life cycle. The effort required in restructuring far outweighs the benefits, it is argued. In terms of briefing information, this potential complexity affects the arrangement of the Specification. As we have seen, design teams will tend to organise specification information using the structure of the National Building Specification (NBS). However the Cost Planners may wish to organise the Cost Plan information using the Standard Method of Measurement (SMM). This will organise trades of work differently than in the NBS, but is designed to suit their methods of working. Equally valid, a contractor may wish to organise information by work packages, and as such the trades of work might be arranged differently to the SMM. Finally, the operational teams will require information to be structured to suit the way in which they will wish to operate the facility. In this case they may wish to organise information not by work package or by trade, but by 'service line', which is the means by which the Facility Management teams manage the delivery of services to the facility occupiers.

It is through the attempts of the various members of the team, performing different roles at different stages of the project, to 'cut and dice' the information to suit their own work processes, that the complexities of different information structures become self-evident. If the nD model is to serve the whole information life cycle, it must embrace the complexities of the needs of the different roles of the organisations in the project process, and provide views into the model appropriate to those roles. The illustration in Figure 12.7 shows how it is possible, using meta-information and data views, to create different information structures to suite the needs of different phases of the life cycle.

To illustrate this issue we need to consider the contract context in which a new facility is being briefed. In the briefing process, the type of contract that will be used to deliver the facility will be discussed. The type of contract will determine who is responsible for producing the final information for the specification. For example, in public finance contracts team members, who were not part of the original briefing team, often produce this information. During the Briefing Phase, the briefing team may well have specified requirements in terms of performance standards. It is in the 'downstream' Construction Phase that team members are required to specify products and materials that conform to the performance requirement.

At this stage of the process, specifications are likely to be developed in the context of a 'work package', and not by facility, unit of accommodation or building element, which the briefing team may have used to organise the

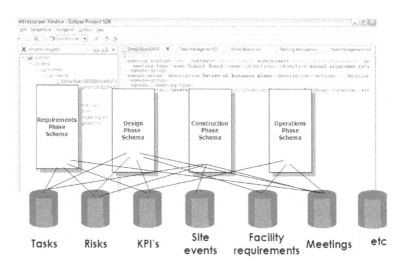

Figure 12.7 The concepts of data views is now well understood in the world of XML database design.

specification information. This can be illustrated by an example of where the slip resistance of floor finishes in public spaces may have been specified by the briefing team as a performance standard. They may have organised the Brief in terms of Building Element <Floors>; Space Type <Public Space>; Finishes <Floor finishes>. However a Construction Phase team may well wish to organise information in terms of: Construction Phase <Phase 1>; Level <Level 1>; Work Package Type <Hard floor finishes>. In the Operations Phase, the Facility Management team however has a different need, because they will organise cleaning and maintenance contracts. For them, the information would be more usefully organised as: Facility <Facility Name>; Service Line <Cleaning>; Finishes <Floor finishes>; Finishes type <Hard floor finishes>.

The use of relational databases in order to manage this information is far from ideal. The rigidity of the relational database means that changing relationships between different data sets is difficult (but not impossible) to manage. However with the advent of XML databases, and the standards that support them, it is possible to design highly flexible and adaptable information structures (called ontology's, that is, sets of representational terms) to represent the different perspectives required by each process phase. If project-based, or auditing teams are to be able to systematically and programmatically test the design and specifications for compliance with the Brief, and collaborating teams are to be able to access information in the context that it is required, then the need to be able to manage and model different information structures within the nD model is paramount.

Making it happen

Whilst I hope that I have demonstrated the need for integration of briefing information into the nD, and I have suggested how this can be achieved, I am sceptical that my proposition will be realised because

1   There are too many vested interests that will prevent it from happening.
2   Lack of ownership and responsibility of the whole building life cycle.

Where ownership of the whole life cycle does exist, for example with Public – Private Finance projects or large facility owners, then I can envisage substantial opportunities existing to invest in integration. However for the wider industry, where there are few positive incentives for professional teams to share data and move from a document-based paradigm to a content-based paradigm, then change will be slow to take place.

The vested interests that I refer to are as much to do with established working practices, and almost slavish use of existing technologies, as they are to do with the attitude that 'structured information' per se is not often seen as a necessary deliverable alongside the facility outcome.

However this status quo can be successfully challenged if there is a tangible business case that provides clear commercial reasons why conventional practice needs to undergo structural change. The most compelling reason will be significantly improved efficiency in all process phases from Briefing, through Design and Construction into Operations. Businesses need only to understand the waste (often described as non-value-added work, or work that should not be required to be carried out) in current practice to see the need for change. Ultimately greater efficiency has to lead to teams being able to be more productive with less effort. The value-add is that the quality of the output should also improve, and consequently customer satisfaction.

Waste in the current process includes the following:

1   Manual processing of information rather than automated processing of information.

    a   Manual creation of documents rather than using automated document generation from database drive technologies.
    b   Manual correction and editing of documents, rather than automated updating when changes are required.

2   Correcting errors because of working with incorrect information.
3   Manual checking of information for compliance, rather than automated compliance checking.

In my business, our own studies indicate that substantial efficiency improvements are achievable, but we do not have any data for the better management of the briefing and design process in construction. This is where research can play its part in order to help define the metrics and benchmarks and establish pilot projects that will collect the data that will underpin the business case.

However my business does have substantial evidence of the gains that can be made by adopting the integrative approach that I propose. We develop software and we have invested substantial effort in building a technology infrastructure to support our briefing, design and development process, where data is being integrated across these processes. The integrative approach enables us to achieve substantial data re-use, where data captured in the briefing process is then used in the design process and later in the configuration of our technology. This approach enables us to automate information processing, and substantially reduce the need for manual processing. It is this experience that has led me to believe that the same kind of approach advocated here could be applied to the management of briefing in construction projects.

Our approach is in Figure 12.8. Essentially there are four parts to the architecture.

1   Specification Portal
2   Object-based business analysis software
3   ARK Smart Information system
4   The Customers Business System (the product of our work).

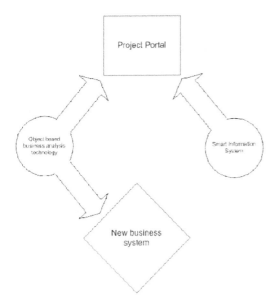

Figure 12.8 The key elements of the ARK e-management Ltd project documentation system.

This Project Portal is a database-driven system (developed using the open source software called Plone, see http://plone.org) operating over the Internet from a web server. It is the means by which we present information over the Internet in a web browser to the members of the team. Requirements and specification information can be created directly in this portal using a server side text editor. Data is stored as xhtml content in a relational database. It is also able to receive file uploads from conventional office-type applications. A key feature of the portal is that it is able to read dynamically from our object-based business analysis software, which also stores all business requirements in a relational database. The business analysis technology enables us to model the business processes, information flows and business rules associated with them. It is this data that is then processed into the Project Portal, so that team members are able to view the results of briefing meetings and requirements capture workshops as soon as the data has been processed by the team using the business analysis technology.

The ARK Smart Information System (see http://www.ark-online.net) is a suite of online technologies that enables remotely located teams to collaborate over the Internet. It comprises a Business Process Engine and an Information Processing Engine. Over this are deployed applications such as a Meeting management tool, as illustrated earlier. This enables the creation of structured meeting content (e.g. notably customer requirements arising from briefing meetings and user study group meetings). The Project Portal will integrate this data into the relevant part of the business requirements.

Finally, data from the Business Analysis technology (ARK has developed and configured its own business meta-model over a proprietary product called System Architect (see http://www.telelogic.com) is used to programme the 'New Business System' being developed for the customer. By this means the customers can see what they will be delivered through the Project Portal, and then witness the system functioning as specified. Changing the business rules, for example will be reflected in both the Project Portal and ultimately in the operation of the customers' system.

With all of our requirements and specification information structured in this way, we are then able to carry out substantial automated testing of software as it is developed. This is a significant benefit of structured information, because test scripts are developed from Use Case Scenarios, which are the means by which we determine the exact functional requirements for the software. In essence these Use Cases are no different from a Brief that describes how accommodation layouts in a facility must enable key processes to perform. These Test Scripts form part of the project documentation and are used to configure the simulation engine. Reports from this process inform the test team where software faults are occurring.

I hope that the correlation between this approach and that which I have set out in my proposition is established. Whilst a major product of our business is software and supporting information but not a building model, I believe that briefing disciplines are identical in many respects. Just as we have adopted the integrated approach that I advocate, and automate many processes, so I believe that the nD model offers the same opportunities.

Chapter 13

Information management in nD

Thomas Froese

Introduction: information management as a sub-discipline of project management

This book has so far explored how an nD-modeling approach can support the design and management activities of building construction projects. However, if nD is adopted, then the nD information and ICT (information and communication technologies) must itself be the focus of information systems design and management activities. This chapter looks at the requirements of Project Information Management for nD approaches.

Information and information management have always been recognized as being an important aspect of project management. But to date, they have not been well-formalized – there are a wide range of perspectives and approaches to address information management, such as the following:

- One perspective is that project management is inherently all about information and communication. Taken to the extreme, information management is not a sub-discipline of project management, but rather, information management is project management. We see an element of truth in this perspective: certainly, information is a critical, inseparable part of project management. Yet other significant aspects of project management, like leadership, are not fundamentally forms of information management. Further, information management practices are likely to benefit if they are treated as a formal function within the overall project management process. Information management, then, should be seen as an essential, explicit sub-discipline of project management.
- Another perspective on information management is that it should be treated largely as an area of technical support to the project management function. This view would be typified by a project management team that thinks of their computer support technicians as providing the project's information management. It is true that information management can involve a large and ever-growing amount of technology and that adequate technical support of the information systems is vital. Yet at its core, information management deals with the management processes of the information – perhaps the

project's most vital resource and coordination mechanism. This is not a function that technical support staff (often relatively junior and off-line positions) is able to contribute to major projects.

- Some consider information management to be a corporate, rather than project, responsibility. Again, there is some justification for this. There are many issues around policies, work practices, and information systems that generally are – and should be – addressed by a corporate "head-office" group that has direct input to information management and ICT issues on individual projects. Yet the very close interplay between the project information practices and the overall project management objectives and procedures demands that information management be treated at a project management function.

In summary, information management should be treated as a critical, explicit function within the overall project management process. This project information management function may have close ties with corporate information management functions, with ICT technical support functions, and certainly with the overall project management function – yet it exists as a distinct role within the project management organizational structure.

This could be considered as very analogous to functions such as safety management, risk management, or quality management. Safety, risk, and quality have always been important to project management in the construction industry, yet over time, these areas have evolved from loosely defined project management objectives to distinct sub-disciplines with well-understood requirements, procedures, bodies of knowledge, and roles within the overall project management process. The same can be said for information management. For example, one chapter of the Project Management Institute's Project Management Body of Knowledge (PMI, 2000) defines a communications planning framework (defining requirements and technologies, analyzing stakeholder issues, and producing a communications plan) and then three sub-issues of information distribution, performance reporting, and administrative closure. Yet this falls well short of a comprehensive approach to information management. Information management seems far behind the areas of safety, risk, or quality as a well-defined and understood sub-discipline of project management.

As projects move into the realm of nD, the requirements for ICT and information management become all the more significant. The use of nD tools and techniques, then, require a well-developed project information management strategy. This chapter discusses aspects of such a strategy, describing underlying conceptual models, an overall framework for project information management, and technical and organizational considerations. The chapter also touches on the relationship between project information management and project management in general.

Modeling the context for ICT in construction

To understand the practice of project information management, one must first have a clear understanding of the role that information and ICT plays with respect to construction projects. We have found a number of conceptual models useful in better understanding this role. In particular, we find it valuable to consider a simple project processes model for exploring the role of information and ICT in construction projects (Aouad *et al.*, 1999 use a more elaborate project process model – the Process Protocol model 1 – to analyze project ICT). This model adopts a process perspective of construction projects, and views projects in terms of the following elements (illustrated in Figure 13.1):

- a collection of tasks carried out by project participants (all tasks required to design and construct the facility);
- a collection of transactions involving the exchange of goods or communication of information between tasks;
- a collection of integration issues – issues relating to the interactions between the tasks and transactions as a whole rather than as a set of individual elements. This also includes issues relating to integration across organizational boundaries, integration over time (such as integration with legacy systems or future systems), and so on.

The model considers these elements across all project participants.

Given this process view of a construction project, it can be seen that construction projects are heavily information based. Design and management tasks involve the processing of information rather than physical goods and even the actual construction operations involve critical information-based aspects in addition to the physical processing of the building components. Similarly, many of the transactions involve the communication of information rather than (or in

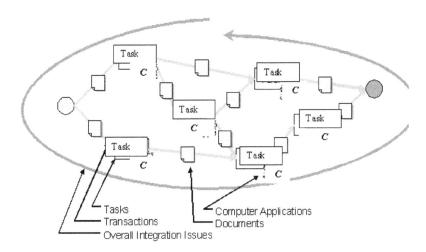

Figure 13.1 A model of project processes that considers projects in terms of tasks, transactions, and overall integration issues. From an information perspective, tasks are associated with computer applications and transactions are associated with documents.

addition to) a physical exchange of goods. Finally, there are many overall integration issues that relate specifically to information, like providing appropriate access to the total body of project information for any of the project participants.

From an ICT perspective, the model provides a categorization of the important ICT elements of a project, and a basic understanding of the main roles of ICT in supporting the construction process:

- The project tasks correspond to the individual tools or computer applications used to help carry out the task.
- The transactions correspond to the documents or communication technologies that are used to convey the information.
- The overall integration issues correspond to ICT integration and interoperability issues.

It is discussed later in this chapter how this breakdown helps to structure project information management. nD technologies, of course, can impact all three of these areas, and particularly the integration issues.

Other perspectives or models that we have found useful include views of the various human-to-human, human-to-computer, and computer-to-computer communication channels that make up construction projects; views of the ongoing development of distinct classes of software (comparing, for example, mature areas such as CAD or financial software with younger areas such as web-based project collaboration tools, or recently emerging technologies found in nD approaches); and views of the overlapping range of different ICT issues (e.g. ICT related to individual tasks versus overall projects versus across corporations versus across entire industry segments).

A framework for project information management

A comprehensive list of all of the issues involved in the management of information systems for construction can grow very long indeed. To provide some structure to these issues, we propose that project information management be defined as the management of information systems to meet project objectives. Though simple, this definition suggests a breakdown of project information management into four main topic dimensions: a management process, project elements, information system elements, and objectives. The following sections examine each of these topic areas. (In the following sections, we have revised our earlier approach with issues suggested by Mourshed, 2005.)

A management process for information management

The management of information systems should follow general management processes:

- Plan all aspects of information system. This includes analyzing the requirements and alternatives, designing a suitable solution taking into account all objectives and constraints, and adequately documenting the plan so that it can be communicated to all. Some of the analytical tools that can be used include cost/benefit analysis (though this may not be a straightforward process since the costs involved in improving information management elements may be incurred by parties that are different from those receiving the resulting benefits), and a consideration of life-cycle issues in assessing costs and benefits (e.g. future information compatibility or hardware obsolescence issues).

- Implementation of the plan, including issues such as securing the necessary authority and resources for the plan, implementing communication and training, and so on.
- Monitoring the results, including appropriate data collection relative to established performance measures and taking necessary corrective action.

Other generic management processes such as scope definition, initiating and closing the project, iterating through increasingly detailed cycles of the plan-implementation-monitoring sequence, and so on. are all equally applicable.

Project elements

The information management actions of planning, implementing and monitoring an information system should be applied to all parts of a project. This can involve the same project work breakdown structures used for other aspects of project management (e.g. breaking the project down by discipline, work package, etc.). However, there are perspectives on decomposing the work that are of particular relevance to information systems. We adopt the project processes model (shown in Figure 13.1) as a basis for structuring an information management approach. Information management should address the three primary elements in the model: project tasks, information transactions, and overall integration issues.

First, the process should define each task, transaction, or integration issue, including identifying participants, project phase, and so on. This should correspond largely to an overall project plan and schedule, and thus it may not need to be done as a distinct activity.

For each of these elements, the information management process must analyze information requirements, design information management solutions, and produce specific information management deliverables. The level of detail required for the breakdown of project tasks and transactions should reflect the detail needed to achieve an effective overall project information management system. In general, this will be at a level where distinct work packages interact with each other, not a finer level at which work is carried out within the work packages themselves (e.g. it will address the type and form of design information that must be sent to the general contractor, but not the way that individual designers must carry out their design tasks).

The model considers these elements across all project participants (spanning all participating companies, not just internal to one company), and the information management tasks should be carried out for each of these project elements.

Also, the project should be considered to be made up of not only the physical elements of the facility to be constructed, but certain information artifacts should be considered to be project elements in their own right, with their own value distinct from the physical facility. For example, an nD building information model resulting from the design and construction, which may be used as the basis for a facilities management system, is a significant project element.

Information system elements

For each of the project elements to which we are applying our information management processes, here are a number of different elements of an information system that must be considered:

- Information: foremost, we must consider the information involved in each of the project elements. First, the process should assess the significant information input requirements for each element, determining the type of information required for carrying out the tasks, the information communicated in the transactions, or the requirements for integration issues. With traditional information technologies, information requirements generally correspond to specific paper or electronic documents. With nD and other newer information technologies, however, information requirements can involve access to specific data sources (such as specific application data files or shared databases) that do not correspond to traditional documents. Second, we must assess tool requirements by determining the key software applications used in carrying out tasks, communication technologies used for transactions, or standards used to support integration. Third, we must assess the significant information outputs produced by each task. This typically corresponds to information required as inputs to other tasks. After analysis, these results should be formalized in the information systems plan as the information required as inputs for each task, and the information that each task must commit to producing.
- Resources: the information management process should analyze the requirements, investigate alternatives, and design specific solutions for all related resources. These include hardware, software, networking and other infrastructure, human resources, authority, and third-party (contracted) resources.
- Work methods and roles: the solution must focus not only on technical solutions, but equally on the corresponding work processes, roles, and responsibilities to put the information system to proper use.
- Performance metrics, specified objectives, and quality of service standards: the information systems plan should include the specification of specific performance metrics that can be assessed during the project and used to specify and monitor information systems objectives and standards of service quality.

- Knowledge and training: the information systems solution will require certain levels of expertise and know-how of people within the project organization. This may well require training of project personnel.
- Communications: implementing the information systems plan will require various communications relating to the information system itself, such as making people aware of the plan, training opportunities, procedures, and so on.
- Support: information system solutions often have high support requirements, which should be incorporated as part of the information management plan.
- Change: the information management plan should include explicit consideration of change – how to minimize its impact, how to address un-authorized changes by individual parties, and so on.

Information systems objectives

The previous sections outline a number of information system elements to be developed for all of the project elements as part of the information management process. Solutions for these should be sought, which meet the general project objectives of cost, time, scope, and so on. However, there are a number of objectives that are more specific to the information system that should be taken into account:

- System performance is of primary concern, including issues such as efficiency, capacity, functionality, scalability, and so on.
- Reliability, security, and risks form critical objectives for information systems.
- Satisfaction of external constraints: here, we have placed the emphasis on the project perspective, but the information management must also be responsive to a number of external influences. Of particular significance is alignment with organization strategies and information management solutions, including appropriate degrees of centralized vs. decentralized information management. Other external influence include client or regulatory requirements, industry standards and so on.
- Life-cycle issues should be considered. These include both the life cycle of the information (how to ensure adequate longevity to the project data), and of the information system (e.g. the life-cycle cost analysis of hardware and software solutions).
- Interoperability is key objective for many aspects of the information system.

The technical body of knowledge: project systems and areas of expertise

The previous section outlines a very generic framework for information management. While this focus on the conceptual frameworks and management processes provides one leg to the practice of project information management, the other leg consists of the technical body of knowledge that underpins the information systems used throughout the construction industry. Ideally, there would be a well-developed and widely understood body of knowledge for this discipline, but this does not seem to exist. At present, technical expertise is built up mainly through extensive industry experience with little in the way of unifying underlying theory or frameworks. Recent developments such as Master degree programs focusing on construction ICT (e.g. the European Masters program in Construction ICT, Rebolj and Menzel, 2004), or initiatives such as this text on nD are helping to contribute to a more formalized body of knowledge for both traditional and emerging construction ICT. As one perspective on the range of technologies and issues involved, this section outlines a range of systems that might be considered to represent a "best practice" tool set for a comprehensive project information management solution:

- Project document management and collaboration web site: a web site should be established for the project to act as the central document management and collaboration vehicle for the project. This will include user accounts for all project participants, access control for project information, online forms and workflows, messaging, contact lists, and so on. A commercial service would generally be used to create and host the site.
- Classification systems, project breakdowns structures and codes, and folder structures: much of the project information will be organized according to various forms of classification systems. These range from the use of industry-standard numbering schemes for specification documents, to the use of a project work breakdown structure, to the creation of a hierarchical folder structure for documents placed on the project web site. The information management process must consider relevant industry classification systems such as OCCS (OCCS Development Committee, 2004), and establish appropriate project classification systems.
- Model-based interoperability: many of the systems described later work with model-based project data, and have the potential to exchange this data with other types of systems. The project should adopt a model-based interoperability approach for data exchange for the lifecycle of the project. The information management process must consider relevant data exchange standards, in particular the IFCs (IAI, 2003), and must establish specific requirements and policies for project data interoperability. It must also establish a central repository for the project model-based data (a model server).
- Requirements management system: a requirements management tool may be used to capture significant project requirements through all phases of the project and to assure that these requirements are satisfied during the design and execution of the work.
- Model-based architectural design: the architectural design for the building should be carried out using model-based design tools (e.g. object-based CAD). Although this improves the effectiveness of the architectural design process, the primary motivation here is the use of the resulting building information model as input to many of the downstream activities and systems.

- Visualization: using the building information model, which includes full 3D geometry, there can be extensive use of visualization to capture requirements and identify issues with the users, designers, and builders. This may include high-end virtual reality environments (e.g. immersive 3D visualization), on-site visualization facilities, and so on.
- Model-based engineering analysis and design: the building information model is used as a preliminary input for a number of specialized engineering analysis and design tools for structural, building systems, sustainability, and so on.
- Project costs and value engineering: the building information model can also be used as input to cost estimating and value engineering systems. These will be used at numerous points through the lifecycle of the project (with varying degrees of accuracy).
- Construction planning and control: the project should use systems for effective schedule planning and control, short interval planning and production engineering, operation simulation, resource planning, and so on. Again, the systems will make use of the building information model and will link into other project information for purposes such as 4D simulation.
- E-procurement: project participants will make use of on-line electronic systems to support all aspects of procurement, including E-bidding/tendering, project plans rooms, and so on.
- E-transactions: on-line systems should be available for most common project transactions, such as requests for information, progress payments claims, and so on. These will be available through the project web site.
- E-legal strategy: project policies and agreements will be in place to address legal issues relating to the electronic project transactions.
- Handoff of project information to facilities management and project archives: systems and procedures will be in place to ensure that complete and efficient package of project information is handed off from design and construction to ongoing facilities operation and management, as well as maintained as archives of the project.

This provides a breakdown of ICT areas of expertise from the perspective of the major systems that might be used on construction projects. This is a useful approach in considering the required areas of expertise for ICT. However, it does not provide the best way of organizing a comprehensive "body of knowledge" for construction ICT.

Organizational roles: the project information officer

Organizational issues for information management

The previous sections have argued that nD and emerging ICT could significantly impact construction project processes. The magnitude of this potential for ICT to improve project processes depends upon the degree to which these processes evolve to fully embrace and exploit the ICT. With ICT playing a critical central role in the work processes, the information management becomes correspondingly critical to the overall project management processes – managing the project will be just as much about managing the information and ICT as it is about managing people, managing costs, managing risks, and so on. With information management becoming an increasingly important element of overall project management, the following challenging criteria must be considered in defining the organizational responsibility for information management:

- Project focus: information management should be project-focused and organized as a project management function, as opposed to centralized within a corporate ICT department. The information management process, as described earlier, is tightly coupled to the project processes and, inversely, the project processes should be strongly influenced by the ICT perspective. Furthermore, the information management must be responsive to project objectives and the needs of all project participants, rather than being driven by the corporate objectives and the needs of one company alone. This does not imply that a centralized ICT group is not needed: the depth of ICT expertise and resources required may be well served through some centralized resources. Thus, a matrix organizational structure may be suitable, with primary organizational responsibility for information management residing in a project position supported by a centralized information management group (although matrix organizational structures are generally not ideal, their use here would be similar to other common applications in the construction industry such as estimating or field engineering services).
- High level: since information management is central to the overall project management, it should not be relegated to a low level within the project organizational structure (e.g. as might be found with typical ICT support personnel), but should be the primary responsibility of someone within the senior project management team.
- Separate function: Although the responsibility for information management should lie within the senior project management team, it would often be a poor fit with current senior project management staff. It requires a depth of specialized knowledge in areas of technology that are rapidly evolving. It may also be overshadowed by traditional practices if it is added as a new, additional responsibility to someone that already handles other aspects of the project management, such as a contracts manager, a project controls engineer, or the overall project manager.

These criteria suggest that, where possible, information management requires a new, senior-level position with the project management team. We call such a position the Project Information Officer (PIO). The overall responsibility of the PIO is to implement the information management as described previously. The following sections outline additional issues relating to the PIO position.

Organizational role

The PIO may be an employee of the project owner, lead designer, or lead contractor organizations, or may work as an independent consultant/contractor. Regardless of employer, the PIO should be considered to be a resource to the project as a whole, not to an individual project participant organization. The PIO should be a senior management-level position within the project organization (i.e. not a junior technology support position). The PIO should report to the owner's project representative and work with an information management committee consisting of project managers and information specialists from key project participants. Depending upon the size of the project, the PIO may have an independent staff. In addition to the information management committee, liaison positions should be assigned within each project participant organization.

Skills and qualifications

Candidates for the position of PIO must have a thorough understanding of the AECFM industry, information management and organizational issues, data interoperability issues, and best practices for software tools and procedures for all of the major project systems described previously. Preference would be for candidates with a master's degree relating to construction ICT and experience with information management on at least one similar project.

Compensation and evaluation

Advanced construction ICT offers great promise for improving the project effectiveness and efficiency while reducing risk. Not all of these benefits directly reduce costs, yet the overall assumption is that the costs of the PIO position will be fully realized through project cost savings. This will not be a direct measure, but will be assessed on an overall qualitative basis through an information management review process that examines the following questions of the information management and technology for the project:

- To what degree was waste (any non-value-adding activity) reduced?
- What new functionality was available?
- How efficient and problem-free was the information management and technology relative to projects with similar levels of ICT in the past?
- What was the level of service and management effectiveness offered by the PIO?
- What is the potential for future improvements gained by the information management practices on this project (i.e. recognizing the long learning curve that may be associated with new ICT)?

There is a need for the development of good metrics and data about industry norms related to these issues.

Validation: critical success factors

In current practice, many aspects of the foregoing description of project information management are routinely performed, and project management practitioners exist that readily fit the given description of a PIO. Yet project information management is not well formalized or recognized as a distinct and critical function within project management practice and project organizational structures. Moreover, advances in nD technologies impose significant new demands on any existing information management practices. The description of project information management given here, then, is intended to represent a substantial evolution of current practice. It is a description of a conceptual framework, not a well-tested and proven methodology – thorough validation of the approach has yet to be carried out, but would likely involve studies of existing industry best practices in information management. A relevant investigation of best practices for ICT in construction has been carried out by the Cooperative Research Centre (CRC) for Construction Innovation in Australia. The analysis of the best practices led to the definition of a series of critical success factors, which must be addressed to maximize successful outcomes. These critical success factors are summarized as follows:

Organizational commitment

- The commitment of an organization's employees is vital to the successful adoption and use of ICT.
- An organization's continuous and conspicuous investment in staff development and training is vital to its successful adoption and use of ICT.
- Successful ICT implementation requires the commitment of a firm's senior management.
- Transparency and trust among project team participants is vital for the successful adoption of ICT across a project team.

Organizational attitude

- The availability of standard conditions of contracts that specifically accommodate the issues raised by the use of ICT will encourage the use of these technologies for project communications.
- When using ICT for project communications, an organization needs to be prepared to engage in long-term collaborative relationships, such as partnering.

Support and assurance

- The security of information is vital in an ICT-enabled project environment.
- A "champion" should support all new technology that is to be used across a project team within a firm.

Rights and duties

- The identification of the ownership of the intellectual property generated during a project is a significant issue affecting the adoption of ICT across a project team.
- Project teams require a powerful ICT "champion" to support the technologically weaker organizations in order to ensure that communication processes continue to function as planned.
- Project team members must acknowledge the sensitivity and confidentiality of other participants' information.

Investment drive

- Organizations commit to ICT as a long-term, strategic decision.

Communication structure

- A fragmented project team will lead to the ineffective performance of ICT-enabled operations.
- Organizations try to limit their use of multiple online systems promoted by different project participants' (Brewer, 2004, p. 9).

There is no mention of an explicit project information management discipline or practitioners here. However, many of the critical success factors represent fairly significant policy issues or management tasks, and we contend that the likelihood of succeeding in these factors would be greatly increased if the issues are treated as explicit, high-level, high-priority, well-understood functions within project organizations, as suggested by the approach outlines in this chapter.

The impact of nD on project management as a whole

The premise of this chapter has been that project management must include an explicit information management component, and that nD approaches increase the need for formal project information management practices. That said, nD also impacts project management at an even broader and more fundamental level. This section briefly discusses ways in which nD technologies impact project management as a whole.

nD modeling of project management information

At the level of data modeling technology, it is important to recognize that integrated nD project models can fully represent and inter-connect all types of project management data in addition to the product design information. nD models or building information models are generally described as being comprehensive models of all project data, yet the vast majority of work with building information models use the data models to either exchange product design information between different model-based CAD systems, or to use these CAD models as input to downstream tools such as detailed design or cost estimating. In contrast, nD techniques can and should be used to integrate all project management information and processes. An example of this is the use of 4D simulation which links building product geometry with scheduling processes.

Integrated models such as the Industry Foundation Classes (IFCs) include data structures that allow information about products, processes, costs, resources, information sources, and so on to all be represented, linked, and exchanged. These could be used, for example, to exchange cost data from one estimating system to another; to use a construction plan from a scheduling program as input to a cost estimating program, or to link documents in a project collaboration web site with the objects in a CAD-based building model. These forms of interoperability are all fully supported by the IFC model, but almost no software has yet been commercialized with these types of data exchange capabilities. Further technical detail of these capabilities is beyond the scope of this chapter, but it is important for the industry to recognize that nD-based integration can extend to all aspects of project management, and to create demand for such products from the construction ICT software industry.

nD-enabled virtual design and construction

The ability for nD technologies to support comprehensive, semantically rich, computational models of construction projects creates an opportunity for these models to become the primary medium for all project collaboration, coordination, communication, and information sharing. The nD models act as the project prototype; they become the canvas upon which all project team members can make their contribution to the project. This goes well beyond the use of new ICT tools to support current project practice – it introduces a new paradigm for the way that projects are designed, managed, and constructed. Many of the chapters in this book illustrate aspects of this new style of virtual design and construction. Project management practices will need to evolve to match this new approach. The focus of all coordination activity must shift to the production of the nD model as the precursor to the physical facility, much of the communication and information flows will take place via the model, and the model must become a major element of the way that the key participants envision their role in the project. However, these are substantial changes to current practice that will take a long time to emerge.

Conclusion

In summary, nD techniques offer great potential to improve project outcomes, but they come at a cost. They involve complex systems with high technical and organizational requirements. Current practices for managing project information and ICT need to evolve and current skills need to improve. This chapter has laid out a framework for project information management as a distinct sub-discipline of project management, and defined the role of the PIO as the central organizational focal point for the information management function. Without these management and organization changes, it will be very difficult for projects to realize the enormous potential of the technical aspects of nD approaches.

References

Aouad, G., Cooper, R., Kaglioglou, M. and Sexton, M. (1999) 'An IT-supported New Process', in M. Betts (ed.), *Strategic Management of I.T. in Construction*, Blackwell Science, London.

Brewer, G. (2004) Best Practice Guide: Clients. Technical Report of Project 2001–016-A, Critical Success Factors for Organisations in Information and Communication Technology-Mediated Supply Chains, CRC for Construction Innovation, Brisbane, Australia. October 6 2004.

IAI (2003) Industry Foundation Classes, IFC2x Edition 2, International Alliance for Interoperability.

Mourshed, M. (2005) Online Survey: Management of Architectural IT, web page at http://www.ecaad.com/survey/ (accessed May 16, 2005).

OCCS Development Committee (2004) 'OCCS Net, The Omniclass Construction Classification System', web page at: http://www.occsnet.org/ (accessed June 24, 2004).

PMI (2000) A Guide to the Project Management Body of Knowledge (PMBOK Guide), 2000 Edition, Project Management Institute: Newtown Square, PA, USA.

Rebolj, D. and Menzel, K. (2004) 'Another step towards a virtual university in construction IT', *ITcon* 9, 257–266, http://www.itcon.org/2004/17

Chapter 14

Data visualization

A Danish example

Jan Karlshøj

Introduction

This chapter describes different forms of visualization that illustrate the results of computer-generated simulations. The simulations can, for example, demonstrate the appearance of buildings before construction or analyze their technical mode of operation under different conditions. The need to handle large amounts of data increases concurrently with greater usage of ICT (Information Communication Technologies) tools to handle and simulate more aspects of buildings (hence, nD modeling), be it an imaginary mode of operation planned during the design phase, planning of the production, or the use and operation of the building. Since man possesses the ability to visually handle large amounts of information, it is obvious and expected that different visualization techniques are increasingly used among actors, authorities, users, and decision makers within the building industry.

Historically the use of computer-generated visualizations has developed considerably for the past 15 years. Research and development organizations and the gaming industry have, for a number of years, invested considerable resources in improving computer-generated visualization by developing new algorithms, optimization methods, and hardware components. At the same time, computer-based visualization in the AEC/FM industry has benefited from the general development of the ICT arena. The development has now reached a level where computer-generated photo-realistic pictures are included in many sales brochures, and video sequences form a natural part of documentation for an architectural competition. This chapter presents some suggestions as to where computer visualization is expected to be headed in the future.

Visualization

In this context visualization is to be conceived as creating or evoking a picture of something. Visualization is not only about photo-realistic presentations but also about forming a picture of something. Hence, graphs and thematic map presentations apply as examples of visualization, which can contribute to forming an overview of large amounts of data. An example of different types of visualization is given in Table 14.1.

Table 14.1 Types of visualization divided according to dimensions that they represent

Types of visualization	May be supplemented by	Examples
Table		
1D geometry		Graphs
2D geometry		Drawings
	Thematic contents	Maps
	Free navigation and scale	Digital prints
	Textures	Pictures
3D geometry		Geometric models
	Textures	Pictures and video
	Free navigation and scale	Virtual reality
	Animation	Animated objects morphing
	Overlay	Augmented reality
	Thematic contents	Simulation results

In this context simple bar charts and graphs are considered to be 1D. The display of figures in graphs and bar charts is much easier and quicker to understand than tables with infinite numbers of figures.

Ordinary paper drawings or maps are typical samples of 2D visualization. For centuries this technique has been used as a communication media both between small and large groups of people. Within this context of communication, the sender who wants to transfer the description of a building to a recipient has typically used façade elevations, plans, and sections to produce an image of the building in the recipient's mind. Plans and sections provide clear and simple sections of the building, which are not "disturbed" by the third dimension. On the other hand, 2D visualization poses large demands on the recipient's ability to visualize 3D models based on plane presentations. Today it is possible to make digital presentations where the degree of detailing can be varied according to demand, and where the scales are practically suspended, because it is possible to seamlessly move from one overall level to the smallest detail. In GIS (graphical information system), representations based on 2D geometric elements are widely used for combining geographic data with other information to form thematic maps.

In 3D visualizations it can be possible to move freely in a 3D environment. The visualization can be shown with a simple color scheme and surface structure, or with symbolic or realistic surfaces. The most advanced forms of visualization permit animation of the whole model or part of it. This may be doors opening, or people or cars moving.

Pictures, showing 2D or 3D elements, are almost always designed as plane subjects without stereo effect, be it a photo, a photomontage, or a computer-generated picture.

It is possible to supplement the visualization of a 3D geometric model with more dimensions, representing, for example time, air current, activities, and so on. When combining the building model with the building activities and the expected duration of the execution, it is made possible to take in a considerable amount of data that otherwise would have been difficult for the recipient to compare. In practice this means that a vast number of logistic problems, geometrical conflicts, and inconveniences will be discovered before being met in practice. During construction, it is expensive and virtually impossible to compensate for the faults and inconveniences without influencing quality negatively. In this case, visualization will enable the recipient to take in large amount of data, thus reducing the risk of faults and clashes at a later stage.

Visualization media

The next section describes the various types of media used for forming or capturing visualization.

Paper

For centuries paper has been the primary medium for presenting real or imaginary issues to be passed from one person to another. Paper is inexpensive and permits a high resolution of the visualization. A further advantage in using paper is that it is easy and inexpensive to reproduce. The ability of the paper to contain information is both its strength and its flaw. Its static form makes it unsuitable for showing issues containing a dynamic aspect. Paper cannot show a space of time, or give the recipient the freedom to choose his own center of projection or scale.

Digital paper

Digital paper is not yet within reach commercially, but research is in progress to make inexpensive and stable solutions. Some suppliers expect that the digital paper is going to be produced in a few years (Fujitsu, 2005).

Physical models

Physical models have been used for illustration of the appearance of a final product for many decades. For many people, physical models are the most natural way of getting an idea of an imagined object. Since the scale of the physical models is normally not the same as the scale of the finished product, it may be necessary to change the proportions of parts of a building in order to be able to produce the model. In the mechanical industry, however, it is often possible to make true scale models, but in another material than the finished product for cost-efficiency reasons. Physical models have always been, and continue to be widely used in the AEC/FM industry. More often than not, these models have been produced through analogous processes, but lately the use of digitally produced models has increased. 3D printers producing 3D physical models can produce physical models at competitive prices. The fact that it is possible to produce new models without a large amount of manual work makes it possible to currently produce new models, and hence ensure compliance, between physical models and other types of documentation. This enables use of physical models which are currently replaced by new, updated models, which can be used in a continuous dialogue with the client. Physical models made of gypsum are inexpensive (3D printer, 2005).

Monitors

For more than 25 years, monitors have been widely used for visualization in the AEC/FM industry. They are suitable for dynamic presentations and for visualizations of a relatively high graphic quality. However, the size of the monitor limits the number of recipients who can have a good view. The flat panels developed in recent years make it possible to produce large monitors, but up to now the resolution has been too poor to show high-quality visualizations.

Multiple-monitor technology has never really become common but a few application suppliers have utilized the technology. Only few users have experienced true 3D by viewing the image on monitors through active or passive 3D glasses. The active 3D glasses are expensive and have never really become popular. A general weakness of many monitors is that the colors of the monitor are not realistic.

Holograms

Holograms can be used to form visualizations, where more pictures can be stored in the same hologram thus permitting the recipient to see the building from several sides, in spite of the fact that the hologram itself is flat. This technology has been used for a number of years outside the building industry, and a number of companies have specialized in making holograms based on CAD models (Zebra Imaging, 2005). A hologram can provide a relatively good 3D view and give the recipient the possibility to choose his center of projection along points defined in advance. Compared to monitors, the advantage of hardcopy holograms is that they are passive and do not require power to work.

Projectors

Ordinary projectors are often used for presentations, and in contrast to monitors they permit a larger group of recipients to view at the same time. As is the case for monitors, the colors of most projectors are not realistic. Furthermore, like monitors, they rarely provide a real 3D experience. This would require use of active or passive 3D glasses.

Furthermore, only a few users will be able to get surround visualization or just a view of a large part of the picture from a projector, thus limiting the recipient's impression of the visualization. It is now possible to simultaneously use more projectors working parallel and thus covering a large field of vision.

Caves

Visualization caves enable powerful graphic computers and projectors to work in parallel together, thus making the user experience that most of his field of vision is covered. In some caves both vertical and horizontal surfaces are projected, and in this case the user is fully surrounded with the graphics. Another advantage of caves is the possibility of working in true scale, thus making the recipient's experience more realistic. By using 3D glasses the user is not only surrounded with graphics but is experiencing a true 3D vision. The price for cave equipment has followed a downward tendency, but previous high prices have prevented wide commercial use of caves.

Augmented reality

Augmented reality is a type of presentation where the recipient views the existing physical world overlaid by a digital model, which is correctly scaled and oriented in relation to the physical world. The overlay can be made through furnishing the monitor with a video camera to show the surroundings, on which the monitor is throwing a shadow. Another technical solution is to use transparent screens built into glasses. It is, for example, possible to overlay a wall with a digital model of the installations in the wall.

Heliodisplay

A Heliodisplay is a display that can visualize 3D models directly in the air. Heliodisplay is not dependent on using ordinary monitors and projectors to show 3D models, thus providing a more free method of presentation. Heliodisplay or similar technologies are hardware solutions which are expected to free the visualization from the rigid frames of the computer when developed (IO2Technology LLC, 2005).

Examples of visualizations

Virtual reality

The VR Media Lab of Aalborg University has worked with visualization of buildings in their cave, primarily with support of decision making in architectural competitions and urban development projects. The experiences gained have proved that it is easier for laymen to understand and evaluate alternative projects if the projects are presented in full-scale with a possibility of navigating in a 3D environment (Kjems, 2004).

In addition to using 3D models to illustrate building texture, the cave has been used in connection with analyses of large amounts of data (Svidt *et al.*, 2001). The cave has been used to present and evaluate the results from CFD (computerized fluid dynamics) – calculations containing information on velocity, direction, and temperature of air flow. In the cave it is possible to see air flow as colored vectors, where the color depicts the velocity of the particle. Furthermore, it is possible for the user to send out particles using the mouse, which can be moved freely in the space, following which the particles are conducted with the virtual air flow for a few seconds before disappearing. Unfortunately it is not possible to carry out the CFD-calculations in real time, making it possible dynamically to change the model and immediately calculate the derived change of the air flow.

The Centre for Advanced Visualization and Interaction, (CAVI) of Aarhus University has utilized the cave to show alternative city models in connection with town-planning projects (Oadsen and Halvskov, 2003). They have demonstrated how virtual reality technology can be used for improvement of decisions to be taken in a city region that is continually undergoing development. Virtual Urban Planning (VUP) has to function as a combined tool for the benefit of politicians, public administrations, building constructors and architects, private businesses, and the town's citizens. It has been possible, interactively in the cave, to place large blocks of flats, storage buildings, a sports stadium, and other significant buildings in the port of Aarhus, permitting the audience simultaneously to experience how the different buildings influence the look of the port, seen from arbitrary positions.

Uni-C has through the Virtual Reality Centre of the Technical University of Denmark helped the national Danish Broadcasting Company to build up a visionarium where several thousand recipients have been able to view the coming new headquarter of Danish Broadcasting Company in a new developed part of Copenhagen called Ørestad (Henrik, 2005).

Augmented reality

Like several other institutes, the Alexandra Institute has worked with augmented reality, where computer models are combined with reality (Grønbæk et al., 2003). Augmented reality has been used in connection with building surveys where the engineer has a PC (personal computer) which currently updates the position in the computer model corresponding to the engineer's position. The engineer can post yellow labels in the building model with his recordings and does not have to waste time navigating the model to find the construction section in question, since the PC automatically finds the position. The system can also show "invisible" installations, which may have caused damage.

Interactive floor

Likewise, the Alexandra Institute has worked with integration of various types of ICT in everyday life: a combined floor surface shown on a transparent screen with a video camera that makes it possible for people to use the floor as a dashboard (Grønbæk, 2005). The floor has been demonstrated in a library where the visitors can send information to the floor via a computer or via SMS (short messaging service). The messages were shown on the floor, and the visitors could subsequently navigate and read the messages on the floor. If the visitors posed a question on the floor and the question was not answered by the time the visitor had left the library, the system could offer to send the answer as an SMS. The solution was awarded the Danish design prize in 2004, and has been nominated for a German design prize in 2006. The system is still being developed and is used in connection with school projects.

RFID

Furthermore the Alexandra Institute has worked with RFID (radio frequency identification device) tags for combining digital work places with analogous material. By furnishing analogous material with RFID tags, it has been possible to integrate them in the digital world. When traditional drawings are furnished with a tag, it is possible by placing the drawing on the digital desktop to retrieve a digital edition of the drawing and project the drawing onto the table or the wall (Grønbæk, 2005).

Visualization of nD

Modeling more than three dimensions, hence nD modeling, has been described in previous chapters. The previously described examples of visualizations have mainly focused on the geometric element.

The use of visualization for presentation of nD modeling is expected to increase considerably in the future. Correspondingly, the use of visualization of geometric elements will make it easier for the recipient to understand a 3D building. Through presentation of a 3D model, it will be easier for the recipient to understand the indoor climate in buildings, where the maintenance costs are going to be higher than average, and how an evacuation situation will be based on the latest design changes, and so on. When we are able to simulate and handle more factors in terms of calculations in real time, visualization of the results will be essential. From the moment that we are able to develop an alternative solution and the consequences derived – such as changed energy consumption, escape routes, and indoor climate – it will be possible to quickly interpret the results. In some cases key figures will be sufficient, but in many situations the recipient does not wish exclusively to be informed about savings and a change of costs of 20 per cent, but wishes visually to be able to see the changes in relation to a building model.

Visualization is also expected to play a large role in operation situation. In Norway, fire-fighting authorities are working on using computer games for simulation of rescue operations in buildings (International Alliance for Interoperability, 2005). This makes it possible for the fire station to prepare for accidents in buildings, and during the transportation to the scene of accident, to visualize the building to plan the rescue operation. It is natural to combine the geometry of the building with its fire resistance and if possible, with information from the monitoring and safety systems about the number of persons present in the building, along with information of particularly inflammable equipment, location of the nearest fire hydrant, and so on. At the scene of accident, the building model is combined with feedback from the fire fighters' RFID system, which guides the rescue team in the building. Computer aid for surveillance of the injured is also being developed.

Public client requirements

In 2003, the National Agency of Commerce and Building implemented "Det Digitale Byggeri" (Digital Construction) which originates from the Government's competitiveness package "Vækst med vilje" (Growth on Purpose), which was launched in January 2002. The objective of Digital Construction is to promote the use of IT in the building sector with the underlying objective of improving quality, increasing productivity, and also lowering prices.

The National Agency of Commerce and Building invests 20 million DKK, and the Realdania Foundation 10 million DKK, whereas the building industry contributes with 10 million DKK through self-financing from companies and organizations participating actively in the program. "Digital Construction" focuses on new building, where the State is the client, and where the client's requirements will influence the building industry's use of ICT. Hence architects, consulting engineers, contractors, and suppliers will be the first to encounter the requirements of "Digital Construction," in addition to state clients such as the Palaces and Property Agency, the Danish Defense Construction Service, and the Danish Research and Education buildings. It is the goal to implement client requirements in 2007, and it is hoped that it will spread to other public clients and the private market. It is also possible to spread the results from "Digital Construction" to authorities and other actors, for whom building is only one among more fields that they deal with. Digital Construction is divided into three main areas:

- client requirements;
- a digital base;
- best in building (examples of best practice).

All the areas have been put out to tender and have been won by different consortiums, each mainly consisting of enterprises within the line of business and at least one knowledge institution. A learning network has been established in Denmark, consisting of these consortiums, persons connected with the development of building, and representatives from the building industry. The network is an open network, where mutual communication is possible via www.detdigitale-byggeri.dk and in frequent workshops.

The clients have requirements within the following areas:

- digital tender;
- 3D-models;

- projectweb;
- digital information delivery.

Digital tender focuses on retrieval of information on quantities from 3D models, which the contractors can use for calculation of their tender price (BANK, 2005). It is the objective that the consultants shall be in charge of retrieval of information on quantities, and that the retrieved information is gained as automatically as possible from 3D object-oriented building models. The building models must be documented as drawings in pdf format and be transferred to the contractors in their original format and in IFC (Industry Foundation Classes). Subsequently it will be possible for contractors to digitally fill in the bill of quantities (BOQ) and submit the tender in digital form. Another objective is that the contractors shall use the building models in their in-house production planning and calculations, thus enabling them to use the building models for simulation of the execution of the project. The requirements of 3D models are dealt with in detail in the following paragraphs (B3D konsortiet, 2005).

In client requirements to projectweb, directions for how to set up projectweb shall be scheduled together with directions for how to use projectweb in various situations (Projektwebskonsortiet, 2005). Projectweb will primarily be used for handling traditional documents such as specifications, minutes of meetings, and drawings. Client requirements are directed towards the various parties in a building project and towards the suppliers of projectwebs. It is required that a few metadata shall be filled in per document and that these data shall follow the ISO 82045 standard.

As for digital information delivery, the client requirement's are about how the client wishes to receive digital information on the building and its operation and maintenance (Dacapo, 2005). Digital information delivery is distinguished between different types of presentations and extent of documents to be handed over. It is mandatory to submit a data model of the building in XML (extendable mark-up language) format. The digital handing over is primarily used in connection with operation and maintenance, but the availability of digital data increases the client's possibility of reusing the same data for simulation or 'what-if' scenarios.

Client requirements: 3D models

The B3D Consortium composed of Arkitema, NCC, Rambøll, and Aalborg University has published the second issue of client requirements to visualization, simulation and 3D models. The requirements are directed toward competitions and the design phase in new buildings.

The B3D Consortium's requirements are based on the so-called object-oriented software programs, which besides forming the building as 3D geometric elements, builds up an inner logical correlation between the elements. Some software suppliers call this concept Building Information Models (BIM). In such systems a wall "knows" that it is a wall and that the window is located in an opening, which is again connected to the wall. Since information is created in the design phase, the requirements are based on a principle permitting a successive specification of the building. A solution was inspired by work executed in Finland under the ProIt project (ProIT, 2005). For data exchange between various software programs the B3D Consortium has chosen IFC (International Alliance for Interoperability, 2005), which is an open, neutral standard, which the large CAD suppliers have already implemented in their products. The Consortium finds this a proper solution, and the choice is furthermore based on the rule not to favor individual software products in the market.

The Consortium distinguishes between the contents of the 3D building model in the competition and design situation (Figure 14.1):

- Volume model
- Room model
- Unit model
- Building element model
- Construction model.

Besides, the client may have a model with requirements to existing conditions for the project, for example a site model.

The first edition of the requirement specification was tested in January and February 2005, in connection with a competition at Aalborg University, where the National Research and Education Building Agency was the client. The test

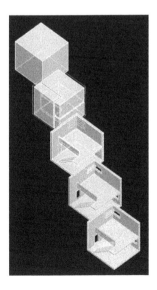

Figure 14.1 Five levels of content in the building model.

Figure 14.2 Proposal by Kjaer and Richter architects.

project at Aalborg University is characterized by being a revitalization of a part of the university, and is therefore an extension and rebuilding of the existing building than a new building. In the test project, the architects had access to 3D models in IFC format of the existing buildings. The experience gained from the project indicates that the 3D models helped the lay judges of the selection committee understand the spatial conditions (see Figures 14.2–14.5). However, there is no unambiguous answer as to how the winning architects experienced the requirements. One of them did not have any problems with the requirements, whereas the winning architect solved the task, but found that the requirements could be improved. It is still possible to adjust the requirements in the project, since the requirements will not apply until 2007, but some of the problems are expected to be solved by adjusting the building industry's tools and working procedures.

Advantage of 3D models

The state clients expect first and foremost to benefit from visualizations of buildings, both the interior and exterior, based on 3D models. Visualization of exteriors can contribute to the following:

- showing the location of buildings in their future environment;
- showing shades;
- giving the citizens an insight into the development of an area;
- clarifying the risk of people being able to look into other people's private rooms.

The 3D models can furthermore form part of 3D urban models, which some municipalities use. The 3D urban models have in various cities in the world replaced or supplemented existing physical urban models. A number of advantages are gained from using digital models, one of them being that the urban model is placed at the disposal of the citizens via the internet. The urban models can also be used in advanced Virtual Reality Caves offering real time showing in real 3D, using stereo effect.

In order to be able to utilize building models in an urban model, the degree of detailing has to be suitable or to have a system which can change presentation.

Figure 14.3 Proposal by Kjaer and Richter architects.

A digital site model is necessary to be able to place the building vertically and horizontally correct. Attention is drawn to the fact that the detailing of the building model and the site model shall be suitable.

Furthermore, 3D-models can be used in connection with analyses of light, wind and material conditions. Today it is possible to reasonably correctly simulate local wind conditions and thus define locations which are not suited for people to stay in. The possibilities of simulating the operation of the buildings are numerous, and in this field the B3D Consortium expects simulation tools and reuse of data from the building model to be more widely used in the future.

Figure 14.4 Proposal by PLOT architects, interior 1.

Future analyses

The Danish state requires a general economic evaluation of its own buildings and therefore has an immediate interest in evaluating both construction costs and expenses for operation and maintenance. From a theoretical point of view, the State ought to be interested in simulation and modeling of many aspects. Historically the expenses for operation and maintenance have not been taken into

Figure 14.5 Proposal by PLOT architects, interior 2.

account to the same extent as the construction costs, but this might change. Investigation of the correlation between construction costs and costs of operation and maintenance will probably be mandatory as the clients are being increasingly aware of this correlation. If it can be proved that the savings in using various forms of simulations exceeds the expenses, there is no reason for the clients not to make use of this possibility. Construction costs have traditionally been in focus when building new buildings, but the requirements for improved control will increase the focus on the construction costs during the design phase.

Not only are the expenses for heating and cleaning pertinent for the operation of buildings, but key issues such as the durability of materials and the structure of the building should also be considered. Software that examines access conditions for handicapped people already exists, and so do programs which can evaluate sight conditions which can be important for vandalism and for crime-prevention purposes.

Furthermore, an increasing number of analysis methods systematically evaluate environmental conditions based on the materials used in the buildings are expected to be developed in the near future. By using this type of analysis already in the design phase, it is possible to take into account both purchase price, life time and operation costs, as well as environmental impact when selecting materials for the building. A linking of geometric data, time and supplies and working processes during the production preparation phase can give an unprecedented overview. Commercial solutions covering large or small parts of the relevant modeling areas have already been developed, and even though statements have not been documented, the tendency is clear.

Rambøll Danmark's experience

Rambøll Danmark has gained experience in combining the results from different technical analyses. A number of areas have been subject to regulatory requirements based on the actual task, whereas other areas have been handled using empirical rules and preconceived solutions, for example, fire protection of structural components. The latest development has, however, made it possible to make analyses of the actual conditions and hence the regulations in Denmark concerning the fields of energy, fire, and indoor climate have been made more rigorous. The tendency of a higher degree of analyses of actual projects has therefore increased. Furthermore, Rambøll Danmark has carried out simulations of selected competition projects and has already in the preliminary phases analyzed, for example, the flow behavior around buildings, where special conditions so require. A concert hall in Iceland is a concrete example of a project where analyses of the wind conditions influenced the design of the competition project (Figure 14.6).

Conclusion

The change from the production of 2D drawings to 3D modeling in Denmark may lead to more dimensions or aspects to be added relatively quickly, since it would

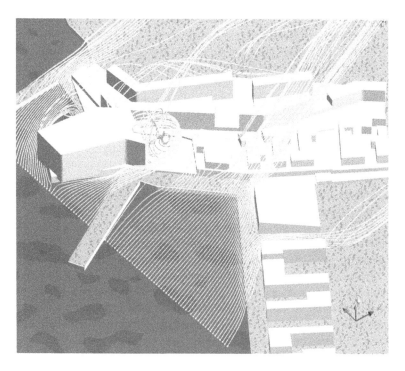

Figure 14.6 Music hall in Reykjavik, wind flow.

be obvious to include more conditions than the purely geometric items in the building models. In spite of the technical problems, it is possible to make many types of simulations and analyses to predict the operation of a building prior to its construction. It is difficult to foresee the speed of implementing this technology, but the conditions for an increased use of nD modeling are present, since the technology exists, and scientists have shown that the solutions work in laboratories.

Bibliography

3D printer (2005) Internet http://www.3dprinter.dk

B3D konsortiet (2005) Bygherrekrav – 3D modeller, kravspecifikation, version 2, 6 September.

BANK (2005) Bygherrekrav – Digitalt Udbud, revision 2, July, Internet http://www.det-digitalebyggeri.dk

Dacapo (2005) Bygherrekrav – Digital Aflevering, kravspecifikation, version 2, September.

Fujitsu (2005) Internet http://www.fujitsu.com/global/about/rd/200509epaper.html

Grønbæk, K. (2005) Internet http://www.interactivespaces.net/, Interactive Spaces, healthcare Informatics and Software development (ISIS).

Grønbæk, K., Ørbæk, P., Kristensen, J.F. and Eriksen, M.A. (2003) Physical Hypermedia: Augmenting Physical Material with Hypermedia Structures. In *New Review of Hypermedia and Multimedia (NRHM)*, 9, 5–34. Taylor & Francis, Abingdon, UK.

Henrik, S., (2005) http://www.proektweb.dk

International Alliance for Interoperability (2005) Industry Foundation Classes (IFC) 2x, Internet http://www.iai-international.org

IO2Technology LLC (2005) Internet http://www.io2technology.com/

Kjems, E. (2004) A Music House in Aalborg, Internet http://www.vrmedialab.dk/pr/activities/spatialmodeling/musikkenshus.html, VR Medielab, Aalborg University.

Madsen, K.H. (2003) Virtual Urban Planning, Internet http://www.pervasive.dk/projects/virtUrbPlan/virtUrbPlan_summary.htm, CAVI, Aarhus University.

Oadsen, A. and Halvskov, V.B (2003) Urban Modelling. PIT Press, Denmark.

Petersen, M.G., Krogh, P., Ludvigsen, M. and Lykke-Olesen, A. (2005) Floor interaction: HCI reaching new ground. CHI on April 2nd – Portland, Oregon, USA.

ProIT (2005) http://www.vtt.fi/rte/cmp/projects/proit_eng/indexe.htm

Projektwebskonsortiet (2005) Bygherrekrav – projektweb, manual, version 2, 1 October.

Søndergaard, H. (2005) Internet http://www.uni-c.dk/generelt/english/research/vr-c.html, Virtual reality Center, UNI C – Danish IT Centre for Education and Research.

Spohr, J. (2005) Digital Construction, National Agency for Enterprises and Construction, Internet http://www.detdigitalebyggeri.dk

Stangeland, B. (2005) Oslo Conference, Internet http://www.iai.no, Norwegian IAI Forum.

Svidt, K., Bjerg, B. and Nielsen, T.D. (2001) Initial studies on Virtual Reality Visualisation of 3D airflow in ventilated livestock buildings, AVR II and CONVR2001, Conference at Chalmers, Gothenburg, Sweden, 4–5 October.

Tónlistarhús (2005) Internet http://www.tonlistarhusid.is

Zebra Imaging (2005) Internet http://www.zebraimaging.com

Chapter 15

Legal issues of nD modelling

Tarek M. Hassan and Zhaomin Ren

Introduction

Lack of Intellectual Property Rights (IPR), litigation, sharing and reusing documents, contracts, data standards and trust are identified as some of the major barriers for the implementation of nD modelling (Lee *et al.*, 2005). While it is certainly true that ICT (Information and Communication Technologies) is an essential enabling factor for an nD model, the legal issues associated to the use of ICT in nD modelling will naturally appear to be an aspect to be addressed.

Information and Communication Technology Law – compared with all the other legal branches – is still quite an emerging area of research and is undergoing a continuous and rapid evolution. Despite prolific activity of the legislator at national, community and international levels, some issues still need to be addressed. ICT Law has a transversal nature, as it ranges across all legal sectors, from trade law to intellectual property rights, from data protection to criminal law, to commercial law, IPR, and so on.

Of these legal issues, this chapter mainly discusses the legal issues related to Industry Foundation Classes (IFCs) and IPRs. The former is believed as the way to move forward for the implementation of the nD modelling by national workshop participants (Lee *et al.*, 2003); the latter is particularly important for the innovative concepts and engineering practice like nD modelling. Addressing those legal issues will significantly contribute to eliminating the legal inhibitors to the full uptake of nD modelling.

Legal issues relevant to IFCs

Industry Foundation Classes (IFCs)

Industry Foundation Classes (IFCs) are being developed by the International Alliance for Interoperability (IAI), an organisation which aims to 'provide a universal basis for process improvement and information sharing in the construction and facilities management industries' (CIRSFID, 2001). The intention of the IAI is to specify how the 'things' that could occur in a constructed facility (including real things such as doors, walls, fans, etc. and abstract concepts such as space, organisation, process, etc.) should be represented electronically. These specifications represent a data structure supporting an electronic project model useful in sharing data across applications – without concern over which software was used to generate the data, or which software is reading and processing the data. Using a standard form of data representation, such as the IFCs or STEP, objects can be shared easily across the Internet. This will enable the design team to prepare object-based information such as design specification, performance and maintenance data.

IFCs provide a foundation for the shared project model, and specify classes of things in an agreed manner that enables the development of a common language for construction. The IFC Object Model provides a foundation for software authors to develop applications that read and write to the physical file format used to communicate with IFCs. These data then have to be exchanged with other users through their different applications. The IAI suggest three possible ways to share data using IFCs: physical files, shared databases and software interfaces.

Overview of legal issues in ICT

Owing to the heavy involvement of ICT in modern design, a range of legal issues are beginning to emerge which threaten to inhibit their growth and prosperity. This is mainly due to the lack of a solid contractual basis, which governs the electronic exchange of information and documentation within and between design participants. This results in a duplication of work rendering ICT to be an extra cost rather than an enabler. Examples of the emerging legal issues are the proof of receipt of electronic data, ownership of information, access rights, company v. project information and handling object-based information (Carter *et al.*, 2001, 2002; Hassan *et al.*, 2001; Hassan and McCaffer, 2003; Shelbourn and Hassan, 2004).

An investigation of common contracts used for construction projects was undertaken by the authors in the United Kingdom, Finland, Germany and Italy, searching for clauses that can be specifically related to the use of ICTs (Merz *et al.*, 2001). Generally, no mention of ICT-based communication methods was made in the contracts in Finland. The JCT98 (Joint Contracts Tribunal) contract (UK) makes no explicit reference to information technology or even, for example, to e-mail. This does not mean that it is therefore unsuitable for use on a project employing heavy use of ICT because the references in the clauses to, for example 'notices', are media neutral. However, without amendments, the contract is not suitable for use on a project seeking to be as 'paperless' as possible, because it still envisages important communications taking place 'in writing' and it is uncertain whether electronic communications will suffice as 'writing'. The position in Italy and Germany is only slightly better, with a nominal mention of ICT's being given. However, these references were typically the specification of a particular type of CAD software to be used on a project or that 'data has to be valid, secure, well organised and properly managed'. Even so, the underlying message was that the only method for achieving legal admissibility up to now is the use of a hand-written signature on a paper hard copy. Considering that it is the data that will be used for further processing in construction, rather than the paper copy, problems are bound to occur.

It is revealed that official documents (such as correspondence, drawings, specifications and raw data) are formally submitted solely on paper. The use of ICT seems to be only intended to speed up the transmission process, but effectively the parties did not attach to it any legal validity. The said picture changes with the implementation of the Directive 1999/93/EC. In particular, article 5, paragraph section 2 reads as follows:

> Member States shall ensure that an electronic signature is not denied legal effectiveness and admissibility as evidence in legal proceedings solely on the grounds that it is:
>
> - in electronic form, or
> - not based upon a qualified certificate, or
> - not based upon a qualified certificate issued by an accredited certification-service-provider, or
> - not created by a secure signature-creation device.

In addition, some recommendations on the said issues stem from other research projects. For instance, eLSEwise (Hassan *et al.*, 1999) identified the business and IT requirements and trends of the European large-scale engineering (LSE) industry, and proposed a 'road map' for the future exploitation of ICT solutions. This included the use of Euro-standards to support ICTs in LSE, and more particularly, the need to ensure the 'acceptance of legal accountability of ICT transactions'. Additional contracts are therefore required between key actors to assure proper communication within a design team.

In terms of the application of IFCs in nD modelling, some of the particular legal issues need to be addressed, such as the following:

- Access rights on objects: An object may have many attributes, but only specific ones are relevant to each discipline. Certain attributes may contain sensitive information that should be accessible only by certain parties. For example, a contractor may add the cost of installation of a system to a model. The client and the cost engineer would use this data. However, the contractor may not want other contractors or parties to view this specific data. Is it possible to restrict the access to particular attributes of a model by specific parties?
- Is the object legally binding: A drawing is just a view of a model containing objects – legal issues below may need to be considered:
  - How can an object and its attributes be verified and validated to ensure that the object is legally acceptable?
  - How can drawings generated from a model be time stamped, with an audit trail back to the model, to ensure that the drawing is a 'legally valid view' of the model?

- General liability for a model: There are potentially three distinct areas of a model where liability issues may arise:

  - Model software/database – who is responsible and liable for the safe storage of the model in a database, and the structure of the database?
  - Application software – when a model is read by someone using application software, who is liable for any problems relating to incorrect reading of the model or translation of attributes and so on?
  - Data provision – can the originators of objects be held responsible for the object data and their attributes, particularly if a manufacturer's catalogue changes. Also, if a designer uses another party's object in their design, who is responsible for ensuring the object is still a current product?

- Who is liable for derived attributes by conversion software: The issue of verification and validation of attributes for objects may require serious consideration, particularly when one considers the problems already faced during 2D data exchange: 'Even if one takes a drawing created using AutoCAD, saves it as a DXF file, and then re-imports it straight back into AutoCAD the result is different from the original. Different CAD systems create elements in different ways and the translator can easily change an element without the user being aware of what has happened' (Day, 1997).

However, the application of standards such as IFCs may influence this. If IFCs are being used to structure data in a database, then the DataBase Management Systems (DBMS) can perhaps support an audit of the attributes of an object. The data should be capable of being read by all compliant software applications. However, the integration of applications is still achieved by the translation and exchange of data between systems. If the appropriate application protocols are applied, and SDAI (Standard Data Access Interface, ISO 10303-22) and so on used, the legal concerns will be minimised.

Intellectual Property Rights (IPR)

To implement the nD modelling concept, the design team (either virtual or physically located closely) unavoidably faces problems such as accessing sensitive information and using (internal and external) know-how. Furthermore, issues such as how to protect the knowledge generated during a design process or by a design team are also crucial for the nD modelling practice. In other words, IPR (e.g. software license, copyright, trademark, know-how, etc.) is an important issue to be addressed. This section focuses on the copyright and know-how protection in a virtual design environment (VDE).

Copyright protection

Copyright protects the creations of human intelligence which are endowed with originality. The originality is usually intended in the objective sense of origination of the work from its author and not also in the subjective meaning of aesthetic pleasantness of the work itself. According to a long-established doctrine, there should co-exist two distinct legal contents: a moral facet, expressing the relationship between author and work, and an economic fact, consisting in the exploitation of the work for making profit. While the former is non-transmissible and cannot be prescribed, as it is a right of the person as a legal subject, the latter constitutes the very object of an ownership, and may be therefore subject to acts of disposal as any other goods.

In the case of the VDE, moral rights will generally be vested in the individuals who perform works for intellectual protection for each of the VDE members, while title to the economic rights will be attributed to the respective VDE members for whom those individuals work. Nevertheless in some legislations, priority may be given to the agreements reached by the parties, or the presumption may only operate for works of certain types. In this connection, it is particularly useful for the contracts with the workers to stipulate expressly that this assignment of the economic rights over the works created in the context of the employees' ordinary work in the organisation.

Computer software is presently regarded, at European level, as a literary work, although a patent protection for it might be considered foreseeable. The European Patent Office has granted patents on computer programmes on the basis of the elements of industrial novelty they presented. Such works, to be protected by copyright, must be original, which means that they must result from the creation of the author. It is remarkable that under Directive 250/91/EEC also a temporary storage of a programme and even a partial one are considered as covered by copyright. This rule is noteworthy because it is expressly tailored on computer programmes and creates a particularly strict regime.

Protection of copyright through formal procedures

Registration of the work on a public register is powerful evidence for the person registered to be presumed to be the author in the event of dispute although they are only necessary on certain occasions for the attribution of the rights over the works. For this reason, members are recommended to register the work in the country granting the most protection with regard to foreign countries, applying the principles of reciprocity. In the Interchange Agreement (the contract under which the VDE is to operate), it is possible to agree on the obligatory registration of the work on a specific register, and also on the choice of the State where its first publication will be made (Garrigues et al., 2002).

Members of the VDE must decide between registering the work as a collective work or as a work in collaboration. In the case of the collective work, the person who edits and publishes it under his/her name will be presumed to hold the rights and will thus become the only holder of the economic rights. Members should agree on the person who is to hold the exclusive rights over the work, who could preferably be the broker or any other key member in the VDE. If the work in collaboration is selected, all the members together will be recognised as joint holders of the work and they should therefore register the work together. For greater flexibility, the members are recommended to designate a representative authorised to register the work in the name of all the joint authors, who could be the architect or the broker.

IPR after termination of the VDE

After the dissolution of the VDE the joint authorship of partners should not terminate. Co-authors will keep their rights on the common works, following the rules set forth by the legislation indicated according to the system of international private law, or according to the regulation provided in the VDE framework agreement. However, the parties can decide to terminate any ownership in common at the moment in which the VDE is dissolved, by expressly specifying it in the VDE framework agreement or by joining a common agreement in that respect. Other ways to terminate the ownership in common can be found in legislation which private international law would refer to. Authorship held by a single natural person seems not to show particular difficulties.

Protection of the know-how

Protection of VDE's know-how

In the context of nD modelling, know-how is more concerned about a party's research, idea or development which may not yet be sufficiently expressed to acquire copyright, or which must be kept secret in order to qualify for patent when the application is ready to be submitted, and require to be protected from disclosure to third parties. So far as the VDE is concerned, there are two key categories of know-how which require to be protected:

- know-how brought to the VDE by the individual VDE partners;
- know-how which arises out of the VDE as a result of the collaboration between VDE partners.

Also, there are two key threats which know-how must be protected against:

- disclosure by a VDE partner to third parties not involved in the VDE/other inappropriate disclosure;
- violation of the confidentiality by third parties illegitimately acquiring the information.

These are not new problems in terms of commercial confidence; however, with the arrival of the computer revolution, and the internet especially, they are more pertinent than before. It is likely that VDE partners will store large quantities of confidential information in their computer networks, which will, inevitably, be connected to the internet. The information thus will require some form of technological protection, as with any communication between VDE partners. Furthermore, VDE partners' employees will require not only to be bound by duties of confidence, but also to be made aware of the need to care in communication of sensitive information.

While there are certain general principles in relation to know-how protection across many jurisdictions, the lack of a harmonised law specific to confidentiality agreements may lead to problems in the event of a dispute – for instance, what exactly is encompassed by the term 'confidential information'. In large part this will often be dependent upon the law of the jurisdiction in which the confidentiality agreements – most likely to be incorporated into the interchange agreement – are determined to have been concluded. It is thus desirable that such details are explicitly set out in the contract to begin with. Recommendations for contracting:

- Prior to the creation of the VDE: at the negotiation stage, the following should be in place:

    – Adequate trust between the parties – potential VDE partners should only consider participation in the VDE where trustworthiness of the other participants is a known quality.
    – A preliminary confidentiality agreement should be put in place in order to protect know-how which may be disclosed at the negotiation stage and before the final confidentiality agreement within the interchange agreement has been finalised. This agreement should encompass all relevant points as in the final version, albeit in a simpler form.
    – In particular, this should include obligations of the following:

        ▪ adequate security of potential partners' electronic communications which will be used during the course of negotiations;
        ▪ confidentiality in relation to know-how disclosed where negotiations conclude without the formation of a VDE. These should incorporate all the main points in relation to the information which has been disclosed during the unsuccessful negotiation.

- During the life-cycle of the VDE: the interchange agreement should incorporate a confidentiality agreement, with clauses relating to the following issues:

    – an agreement on what constitutes confidential information that is not to be divulged to parties outside the VDE membership;
    – a clear point in time at which the confidentiality agreement begins;
    – a clear agreement as to where the Interchange Agreement is concluded and the appropriate jurisdiction under which it will be regulated;
    – a requirement that the VDE partners introduce confidentiality agreements for individual employees of the VDE partner companies who will come into contact with the know-how, creating a duty of confidentiality

to the VDE, and an obligation to put in place procedural rules for employees in order to assist the maintenance of such confidentiality. Know-how should be disclosed to employees on a 'need-to-know' basis only;

- an obligation upon VDE partners to provide and maintain a certain minimum standard of data security for VDE know-how, including the following:

  - physical security of the premises in which information records (both physical and digital) are stored;
  - secure information disposal procedures;
  - adequate controls over systems access and provision of a recognised standard of communication security;
  - provisions for recourse to ADR where necessary in the event of a dispute;
  - agreed penalties in the event of a breach of confidentiality by one of the parties;
  - these should apply equally to both know-how brought to the VDE by individual VDE partners, and know-how created during the life cycle of the VDE.

- After termination of the VDE: the interchange agreement should incorporate provisions designed to protect confidence following the dissolution of the VDE, as follows:

  - a continued obligation not to disclose confidential information belonging to either the VDE in general, or to other VDE partners;
  - where appropriate, the agreement might also incorporate 'non-use' clauses, obliging VDE partners not only to avoid disclosure, but also not to use for their own purposes, know-how brought to the VDE by other individual VDE partners. In practice, however, it may be preferable to create an agreement which gives VDE partners, especially those who may be in competition following dissolution, the right to use know-how gained from others during the course of the VDE, on condition that they maintain secrecy regarding those outside the former VDE members. Disclosure of know-how amongst VDE members during the VDE would thus be regarded as an information exchange, which could benefit all; however, a certain level of trust, that there will be a balance in contribution to this exchange amongst partners who may otherwise be in competition, will be required.

Sources of law

The performance of a legal analysis necessarily requires the reference to the existing legal framework which can possibly be applied to regulate the subject of the research. Owing to the constantly evolving and still incomplete nature of ICT law,

Table 15.1 ICT relevant European laws and directives (CIRSFID, 2001)

EC directives and regulations	Directive 2000/31/EC on electronic commerce; Directive 1999/93/EC on electronic signatures; Directive 97/7/EC on distance contracts; Directive 95/46/EC on data protection; Directive 96/9/EC on database protection, and so on.
National constitutional laws	Provided that there is a national constitution
International conventions	1980 EEC Rome Convention on law applicable to contractual obligations; 1980 United Nations Vienna Convention on contracts for the international sale of goods; 1968 EEC Brussels Convention on jurisdiction and enforcement of judgements in civil and commercial matters; 1886 Berne Convention on copyright law; 1883 Paris Convention on patent protection.
National legislation	Italian law No. 218/95; the Italian Civil Code; the general principles of the law
Non-State rules, commercial usage, codes of conduct, standards and best practices	ICC INCOTERMS; UNCITRAL model laws; GUIDEC; EOCD rules; UETA (URL1); UCITA (URL2)

a very relevant role in the sources of law for parties of a virtual enterprise is the contract (Table 15.1).

The importance of informal rules

At present, in the absence of specific national or supranational legislation and case law on VDE, the legal framework in force is not yet in the position to provide a coherent and certain regulation of the VDE structure as a whole, nor of their actors, activities and complex interactions. This situation can be compared to the economic revolution which gave origin to the Lex Mercatoria, a set of rules developed by merchants who could not apply a law based on land property to their commercial exchanges. Similar to its historical antecedent, the Lex Electronica may include the following categories of norms:

- treaties and international conventions,
- model contracts,
- arbitration case law,
- trade usage,
- general principles of the law.

While the first three ones are regarded as 'institutional sources', the last two are commonly described as 'substantial law'. It is in particular the international character of VDE which stresses the relevance of non-State rules. Normative tools such as commercial usage, codes of conduct, standards and best practices may be particularly adequate for those entities, which involve a plurality of jurisdictions, as they would help overcome possible incompatibilities between the different national laws of the partners.

VDEs are dynamic, fluid subjects and this feature may clash with the rigidity and necessary stable character of formal legislation in force. This usually needs a very long time and a complex procedure for its adoption and subsequent implementation and may find it hard to take account of rapid developments in information and communication technologies, thus becoming obsolete over a very short time, despite the presence of re-examination mechanisms through subsequent technical implementation decrees.

The need for flexibility may instead be met by referring to non-institutional rules and practices. This necessarily implies the compliance with formal rules in force but can provide a valid support for those areas which still remain uncovered or blurred from the legal standpoint or are undergoing a legislative transition phase. Besides, it has to be remembered that non-State rules traditionally find a wide adoption in those fields which imply a high-level technological knowledge. The rules which derive from codes of conducts or best practices are actually developed by subjects who possess a deep experience in specific sectors, something that necessarily escapes law-makers, at least in the short term. VDEs, in particular, require an extremely wide interdisciplinary competence, ranging from business organisation to information technology to law.

Summary

In spite of the rapid technological advancements in ICT, the associated legal aspects have been largely overlooked. This has been always a major inhibitor to the uptake of ICT to its full potential. nD modelling as an emerging innovative concept in modelling, in general, and Virtual Design Environments (VDE) in particular, is mainly driven and influenced by innovative ICT solutions. nD modelling technical solutions should be based on the building information model, and various technologies have demonstrated the benefits of the concept. However, these technologies are based on different standards that are not compatible to each other, hence the big advantage of using IFC as an open and neutral data format. This chapter investigated the legal issues associated with IFCs and VDEs which constitute the main environment for nD modelling. This includes access rights, ownership, liability issues, IPRs and protection of know-how throughout the life cycle of the VDE. The sources of law including EU directives and common practice informal rules for addressing legal issues are also presented. Future work should include the implementation of contractual frameworks to regulate the use of nD models in order to dismantle these legal barriers and hence encourage and facilitate the uptake of nD modelling to its full potential.

References

Carter, C.D., Hassan, T.M., Merz, M. and White, E. (2001) The eLEGAL Project: Specifying Legal Terms of Contract in ICT Environment, *International Journal of Information Technology in Construction*, 6, 163–174, ISSN 1400–6529.

Carter, C.D., White, E., Hassan, T.M., Shelbourn, M.A. and Baldwin, A.N. (2002) 'Legal Issues of Collaborative IT Working', Proceedings of the Institute of Civil Engineers (ICE), November, pp. 10–16.

CIRSFID (2001) ALIVE (Advanced Legal Issues in Virtual Enterprises) Project Deliverable (IST – 2000 – 25459) D11: ICT Specific for Virtual Enterprises.

Day, A. (1997) *Digital Building*. Butterworth-Heinemann, Oxford. ISBN 0 7506 1897 3.

Garrigues, CIRSFID, Loughborough University, VEA, QMW, LAY, KSW (2002) ALIVE (Advanced Legal Issues in Virtual Enterprises) Project Deliverable (IST – 2000 – 25459) D13: Intellectual and Industrial Property Rights Legal Issues.

Hassan, T.M. and McCaffer, R. (2003) Virtual enterprises in construction: ICT, social and legal dimensions, *Journal of Construction Procurement*, 91, 31–46.

Hassan, T.M., McCaffer, R. and Thorpe, A. (1991) 'Emerging Client Needs for Large Scale Engineering Projects', Engineering Construction and Architectural Management (ECAM), March, pp. 21–29, ISSN 0969 9988.

Hassan, T.M., Carter, C.D., Seddon, C. and Mangini, M. (2001) 'eLEGAL: Dismantling the Legal Barriers to ICT Uptake in Virtual Enterprises', Proceedings of the eBusiness-eWork Conference, Venice, Italy, October, pp. 602–608.

Lee, A., Marshall-Ponting, A.J., Aouad, G., Wu, S., Koh, W.W.I., Fu, C., Cooper, R., Betts, M., Kagioglou, M. and Fisher, M. (2003) Developing a vision of nD-enabled construction, Construct IT, University of Salford, UK.

Lee, A. Wu, S., Marshall-Ponting, A.J., Aouad, G., Cooper, R., Tah, J.H.M., Abbott, C. and Barrrett, P.S. (2005) nD modelling roadmap, a vision for nD-enabled construction, Construct IT, University of Salford, UK.

Merz, M., Tesei, G., Tanzi, G. and Hassan, T.M. (2001) 'Electronic Contracting in the Construction Industry'. Proceedings of the eBusiness-eWork Conference, Venice: 595–601.

Shelbourn, M.A. and Hassan, T.M. (2004) The Legal and Contractual Aspects of Networked Cooperation for the Building and Construction Industry', *International Journal of Design Sciences and Technology, EUROPIA*, 112, 141–152, ISSN 1630–7267.

URL1: http://www.law.upenn.edu/bll/ulc/fnact99/1990s/ueta99.htm

URL2: http://www.arl.org/info/frn/copy/ucitapg.html

Chapter 16

Interactive experimenting of nD models for decision making

Kalle Kähkönen and Jukka Rönkkö

Introduction

Information and communication technologies (ICTs) are around us in versatile forms, and are essential tools for everyday work and savoury applications for spare time. The number and variety of different essential ICT tools we are using is continuously increasing and our daily routines are equipped with planning, design, recording, monitoring, analysis and informing solutions. In practice, it is very difficult or just impossible to name those in a comprehensive way.

Several technological innovations and developments are significantly changing the business of construction engineering. In particular, the speed and power of computing are opening new possibilities compared with traditional engineering analysis software. Sawyer (2004) has particularly named multi-physical simulations as enabling sophisticated analyses for engineering companies of various sizes. Even small-size engineering enterprises can develop mega-size building projects and carry out comprehensive what-if analyses and scenario studies than previously where the field once stood only for rare specialists. This is not paramount currently but the use of computers for engineering analyses and the availability to apply these tools have entered a new era where engineering practice is equipped with new possibilities.

The named multi-physical simulations are also gradually developing towards the 'softer' sides of construction engineering. This means testing viewpoints such as human comfort, safety, ergonomic issues, health issues and ecological consequences.

Modern construction engineering analysis solutions are developing towards interactive communication environments. This seems to be an overall trend in the development of ICT (Figure 16.1). Interactive graphics are increasingly affecting all ICT applications:

- how the applications are used;
- for what purposes they can be used; and
- by whom they can be used.

Interactive graphics is both a user-interface to analysis software and a self-explaining media that provides platforms for effective communication. The traditional paradigm

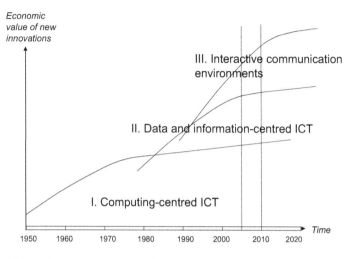

Figure 16.1 Development trends of ICT.

for running computer analysis software can be characterized with the chain fill-data-table, run-analysis and see-results. It seems that now the users want to have something different. With the modern analysis software based on interactive graphics, the user is all the time interacting with the graphical model, changing it, seeing affects of changes and running variety on analysis at the same time.

In this chapter, exploration of some chances for new types of construction analyses based on interactive graphics is undertaken. In particular, avatar technology meaning artificial creatures with human characteristics is covered in the chapter with respect to new possibilities for building engineering analyses.

Virtual testing for decision-making needs in building construction

Traditionally, building construction has been understood as comprising of several sequential phases from early draft until building commissioning. Additionally, it was often assumed that the previous phase is fully completed before the next phase is started. This can still be sometimes the case nowadays, for example in public construction, but in most cases the different project phases are highly overlapping and the process as a whole is very dynamic including parallel design and construction, incomplete or changing user requirements and networked contracting (Figure 16.2).

Management of the kind of process described earlier requires different skills, procedures and tools compared with traditional building construction. Effective communication plays here perhaps the most important role. Results from recent industrial construction projects have revealed that lacking communication skills

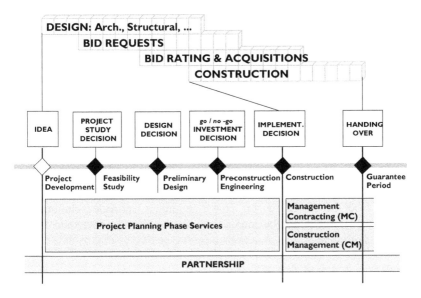

Figure 16.2 Typical decision points in modern construction process.

of project managers can cause big problems and even situations where project managers must be changed.

Continuous need for client's decision making is something that characterizes the modern building construction process. First, one needs to put to attention that all decision-making points are known and understood. Second, one needs to facilitate each decision-making situation in a way that the client can make a confident decision. This requires sufficient information and understanding of all aspects having an impact on the decision. This is where we see the big potential for interactive graphics and technologies.

Clients' decision making is often on the critical path of construction projects. Starting with the manufacturing of building components or the fixing of walls according to the floor plan, the decisions require final approval. The client must be able to make this decision and is so often required to do this in a short time frame. These are situations where interactive graphics and animations are a means for decision makers to explore the designs by themselves. The explorations can be directed, semi-directed or fully interactive meaning that the user has full control of, viewing directions, viewing aspects and details. Directed or semi-directed viewing of design provides a means for controlled or partly controlled design studies. These can sometimes be more effective than fully interactive viewing which may turn out to be indiscriminate.

The heart of modern construction engineering and management practice is in the interactive knowledge. This means visual communication plus engineers' enhancement of that visual knowledge with explanatory noting that can be produced with various media in relation to the particular situation. Henderson (1999) has approached this practice regarding the work by design engineers where the sketches in the margin and hand gestures on the shop floor are examples of breakthrough communication that finally gets the job done. Engineering documentation, and likewise the computer-based building models, can only partially capture the professional knowledge that is behind the proposed solution.

The latter part of the discussion in the previous paragraphs has led us to the following proposition: the rationale of building engineering models is very difficult or impossible to understand without sufficient means for accessing and communicating the interlocking knowledge that is missing from the model but essential for comprehending the engineering logic behind the model. Our research is about joining specialists' knowledge with interactive graphics for virtual building model-testing purposes. This is aiming to provide solutions where building performance can be interactively explored and understood from the viewpoints of interest.

Avatar technology

This section introduces avatar technology from an interactive animation point of view. An overview of technologies that allow a programmer to incorporate animation control to his or her application in the field of construction industry is provided. It becomes apparent that it is possible to deal with animation control at various levels: individual animated character level as well as crowd level. These technologies from a single-character level, going from simple to more advanced avatar presentations, are discussed. An insight into crowd simulation is also described. The simulation of characters can be only visual computer graphics presentations or it can be physically based with collision detection and dynamics calculations. Finally, the possible future in the nD modelling field: ergonomic analysis is illustrated.

Mention was made earlier of software tools as examples in the context of specific techniques. It is noteworthy however that no tool exists today, which would allow a programmer to easily access individual character, crowd or ergonomic simulation features and utilize these features in a very straightforward manner with construction industry product models. Therefore, this chapter is forward looking in nature.

Avatar structure for animation

Avatars can be human or non-human computer-animated figures. They need to be animated in real time to offer interactive experience to the users. The feeling of real-time simulation requires sufficient frame rates (25 frames/sec or more) in rendering the 3D scenes. The avatar movement within virtual spaces is either controlled by programmed algorithms or by human operators. However, even human-controlled avatar movement can be assisted algorithmically, so that human input only controls where the avatar moves, not how it moves. Also crowds of avatars can be simulated for instance for evacuation simulation from a building. Thus there exist different levels of abstraction of control mechanisms for avatar animation. Higher-level systems allow the animator to define intentions of single or multiple avatars, low-level systems require the animator to specify the animation parameters in detail – such as individual character joint position and orientation (Watt and Watt, 1992).

Animated avatars are represented as hierarchical control structures. In commonly used skeletal animation paradigm the animated body is represented by visible skin geometry and a control structure called skeleton. The skeleton consists of bones that represent transformations in 3D space relative to their parent bone, thus forming a control hierarchy (see Figure 16.3). The bones are not to be taken literally as visually existing items in an animated character, even though 3D-modelling programmes tend to give some visual presentations to them to help in the animation construction process.

The visible geometry of an avatar, the skin geometry, is deformed by the transformations of bones of a skeleton. All the vertices in the skin geometry are assigned to the corresponding bones. This process is called skinning. When the animated figure moves, each vertex is transformed by a function that produces values relative to the bone positions that are assigned to affect the particular vertex position. The skinning procedure can be done manually in a 3D-modelling programme, which is the most common method nowadays.

One can buy libraries of animal, human and machinery geometries from the Internet. Some libraries have skinned models with skeleton definitions, but most of the existing resources offer only geometries that can be attached to skeletons in a 3D-modelling package. Some dedicated character animation software

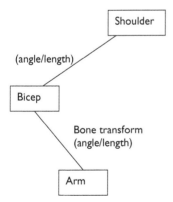

Figure 16.3 Hierarchical skeleton structure; bones are transformations relative to their parent bone; they can be described with angle and length.

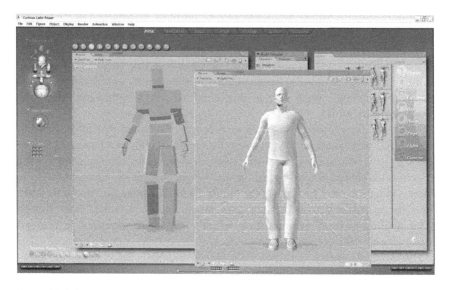

Figure 16.4 A human skeletal animation structure of a human character and its corresponding skin geometry shown in Poser6 – animation package (Poser6, 2005).

(see Figure 16.4) provide tools for creating animations as well as offer pre-made, customizable figures for human, animal and machine avatars with skeletal definitions. It is also possible to acquire geometrical avatar data by 3D body scanners. In this case a real person or object is scanned and the resulting geometry can be used as an avatar model after fitting the skeletal structure to the geometry and skinning. Currently it is under research how the skeleton fitting and skinning could be automated (James and Twigg, 2005) reducing the time-consuming manual work phases.

From an application programmers' point of view it is worth noting that 3D-graphics programming libraries can have features that enable the programmer to load a modelled character via an API (Application Programming Interface). Additionally it is possible to move the character and play skinned animations as well as to attach items relative to its skeleton parts. The current situation seems to be that game-oriented graphics libraries (e.g. Cipher (Cipher, 2005), Ogre3D (Ogre3D, 2005) and RenderWare (RenderWare, 2005), see Figure 16.5) tend to have character animation features while engineering and simulation-oriented graphics libraries often don't have these features.

Avatar movement

There are several ways to achieve animated avatar movement. For instance in motion capture, a real person performs actions such as walking. His or her movements

Figure 16.5 Cipher graphics library used in Martela office building visualization. Some of the characters walk along predefined paths while some of them are directed by users. The use of animated characters gives the sense of dimensions of the building to the users.

are tracked by a motion capture device. This movement can be mapped to avatar skeleton movement. The animations generated this way are very realistic, but strictly pre-defined. Animation blending alleviates this problem a little. In animation blending different animations can overlap: for example at a certain point in time a character can walk and wave his or her hand at the same time. These pre-defined animations can also be created with 3D-modelling packages. To gain more flexibility to react realistically to events in the virtual space, statistical methods for animation control have been investigated. Avatar movement languages represent a higher-level control system for avatars (Kshirsagar *et al.*, 2002). These language systems can utilize a database of pre-recorded sequences controlled statistically to provide parametric animations that differentiate from the recorded animations. For example it can be possible to use a control language to direct an avatar to point to an arbitrary location by providing x, y, z coordinates of the target point.

Usually some form of kinematics is required to produce the desired movements. Kinematics is the study of motion independent of underlying forces that produce the motion. It covers position, velocity and acceleration. Articulated structures are structures composed of rigid links connected by joints. The joints can have various degrees of freedom (DOF) of movement. For example in human avatar the shoulder joint can be seen as a rotational joint only without translational movement ability. Therefore the skeleton described earlier is a form of an articulated structure, if its parts are connected by joints. Kinematical animation of articulated structures can be categorized to forward kinematics animation and inverse kinematics animation. In forward kinematics the motion of an open end of an articulated structure – such as fingers in a human skeleton – is determined as an accumulation of transformations that lead from a central, parent link to the open-ended link via joints.

In inverse kinematics the position and orientation of each link can be determined by traversing the structure from the open end of the structure to the core link. The degrees of freedom of the connecting joints set constrains on orientations and positions the joints can take. An example of the inverse kinematics is that if one wants the avatar to touch a button with its finger, the finger position and orientation as well as the joint constraints are used to calculate the positions and orientations of the palms, forearm, bicep and the orientation of the shoulder joint. To produce human-like or animal-like motion, kinematics calculations require additional constraints and movement parameters in addition to joint constraints. This information can come from an user of a modelling package or more automatically from pre-recorded animation libraries. The information can also come from physics calculations. 3D-modelling software that has character-animation capabilities usually has kinematics features incorporated to simplify animation design. Kinematics packages for controlling avatar movement at run-time in an application are also available; for example IKAN (IKAN, 2005) is and inverse kinematics toolkit for software developers.

If complex interactivity with physical surroundings of an avatar is required, the avatar has to be modelled physically for collision detection and response. Thus we enter from the purely kinematical world to dynamics-based world. Physical avatar modelling may be required for evaluating the usability of building constructs for handicapped people, for example. In a simple case the physical model of an avatar consists of simple rigid collision bodies with inertia properties that are connected together with joints. To simplify the task, only the major joints of a body can be used in the skeletal model. The collision bodies can be, for example, spheres, boxes and cylinders. These collision-detection rigid body objects are connected to each other by physically defined joints, such as hinges and prismatic joints. Rigid body dynamics calculations together with collision-detection algorithms are used to determine the positions and orientations of the collision bodies. This is called rag-doll physics and is used in games to simulate avatar behaviour when they are falling for example. However this physical model is not sufficient to produce more active, goal-directed motion. In practice the motion of an avatar can be a combination of pre-recorded motions, user directed movement and physics-based calculations. Many of today's game-oriented physics libraries provide rag-doll physical entities out of the box. Examples of such physics packages include Newton (Newton, 2005), PhysX (PhysX, 2005) and Tokamak (Tokamak, 2005).

Some of these physics toolkits offer also more extensive, active physics-based character animation control like Havok (Havok, 2005).

A more advanced, purely physics-based approach to animation is based on motor-driven motion. Motors are physical units that provide torque or forces to rotate or to move objects. In motor-driven motion one approach is to determine the desired trajectory of an end bone (e.g. a foot or finger) of a skeleton and then use motor objects to drive the joints of the bones leading to the end bone to the corresponding angles. Inverse kinematics is used to calculate the joint angles together with data that describe the particular joint behaviour. Producing realistic motion in a physically simulated and relatively accurate way is very complex especially when the skeletons of moving characters have many degrees of freedom. Many simulations and games can already present motor-based, simple actors like moving cars quite convincingly. Naturally the level of detail of the models varies from implementation to implementation.

Crowd simulation

Many applications in construction process or final-product-usage design can benefit from crowd simulation. The concept of crowd simulation can be thought to include simulation of humans operating in their everyday environment in such a way that the individual behaviour is not important, rather how people operate as a crowd: how they constantly enter public buildings, cross roads, wait for busses and so on. Similar principles can be applied to human-controlled vehicles for traffic simulation near planned construction sites and buildings.

There is naturally a trade-off between highly detailed behaviour models and computational cost: very detailed models are more difficult to compute efficiently and may require offline operations thus disabling more interactive experimentation of different solutions. The algorithms that steer individuals of a crowd (or a vehicle in a group of vehicles) can be simple scripts or more complex AI (Artificial Intelligence)-based methods. Sung *et al.* (2004) present a noteworthy method for structuring crowd behaviour in complex environments: certain actions are applicable only to certain areas of the environment and the behaviours can be composed of the situation-related higher-level actions. The higher-level actions are composed of lower-level actions that are selected probabilistically among possible choices. In the situation-based approach, when a character enters a situation it is equipped with machinery to deal with the particular situation. The machinery is removed when the character leaves the situation. The simulated individuals don't need to know how to deal with situations that took place earlier or that will take place later.

Programming tools for crowd behaviour include tools that are part of integrated game engine platforms as well as separate libraries. RenderWare (RenderWare, 2005) products are an example of the former and include tools from graphics rendering and physics to AI components. OpenSteer (OpenSteer, 2005) is an example of a separate C++ library that is meant be included into other systems.

Ergonomic simulation and its visualization

Ergonomic analysis includes the study of human-operator performance in a particular environment in terms of parameters like posture, repetition, reach, visibility and muscular stress. This type of analysis can give useful information to improve the quality of human working environment and also to improve the efficiency of working procedures.

Ergonomic simulation related to nD modelling can take various approaches. The simulation can be related to the building construction phase: how workers are able to perform the required tasks at the construction site. This can give insight on how to improve working efficiency. On the other hand ergonomics simulation can be used to simulate the usage of various spaces of a building from the viewpoint of handicapped users in terms of accessibility and reach. Neither field has yet received much attention, but both can offer interesting possibilities in the future. Naturally the usage of the finished building is easier to simulate due to the fact that it is a more controlled and less varied environment than a construction site. The results of these simulations can be visualized and experienced in 3D via avatar presentations of the simulated humans. There are tools for ergonomics simulation like JACK (JACK, 2005). Imtiyaz Shaikh *et al.* (2004) describe how to connect JACK to external, live VR (Virtual Reality) simulation that enables the users to interactively analyse procedures in VR.

Requirements for nD models

To utilize the technologies described so far in this chapter, the models composing the environment should have structural information that can be used in guidance for avatar movement and in case of autonomous characters, character behaviour. For instance classifications of geometrical objects to roads and floors can be used to determine walking or driving surfaces. Objects like doors can be used to construct avatar movement paths from one space to another. On the other hand, objects classified as walls are naturally used for impenetrable objects for collision detection in the environment.

Some of the objects, like roads and active building blocks like elevators, can carry other useful design information, for instance about their capacity. This information can be readily used when constructing the simulations.

Level of confidence in decision making with avatars

With avatars we are trying to approach the field of human perception of the physical world where our tools are now the artificial humans (avatars) together with virtual worlds (building information models plus interactive visualization environments). Regarding human perception we have earlier addressed physical quantities in relation to ergonomic simulations. Here we have an example of reliable and objective source of data which after interactive visualization can turn into sufficient amount of information for confident decision making.

However, for understanding the final satisfaction of building tenants and other building end users we need to approach softer, and more challenging, areas (Ruck, 1989). These are human factors and perception in relation to comfort. Concerning buildings and building models, interest is being stepped up in the indoor comfort domain where the main elements are thermal comfort, visual comfort and indoor air quality. It is worth emphasizing that each of these three subjects is already an important research field of its own and has well-established knowledge for practical solutions. Obviously, taking careful advantage of this knowledge and building the virtual building testing applications based on the established and acknowledged principles, we can reach credible solutions.

Thermal comfort and visual comfort are both good examples of areas where the experienced level of comfort depends both on indoor (in our case) environment variables and personal variables (Hershenson, 2000). Regarding thermal comfort the environment variables consist of air temperature, radiant temperature, air velocity and air humidity. When addressing visual comfort, it is the lighting intensity and direction that are the most important environment variables. In addition to these, either the level of thermal comfort or the level of visual comfort cannot be estimated without personal variables where we need to take into account clothing (when estimating thermal comfort), human activity (in both cases) and personal preferences (in lighting comfort cases).

Conclusions

This chapter has covered aspects from decision making in modern building construction and the use of interactive animation as a special set of interactive graphics for facilitating client's decision making. Avatar technology is providing a new extension to building engineering analysis solutions. With avatars, the building performance can be analysed interactively from the human viewpoint. In particular, this mean comfort (e.g. thermal comfort) that earlier had been rather difficult in practice to approach with engineering analysis solutions.

References

Cipher (2005) Cipher Game Engine. Retrieved on Dec. 14, 2005 from www. cipherengine.com

Havok (2005) Havok Physics Product Family. Retrieved on Dec. 14, 2005 from www.havok.com

Henderson, K. (1999) *On Line and On Paper*. The MIT Press, Massachusetts Institute of Technology, Cambridge, Massachusetts, USA.

Hershenson, M. (2000) *Visual Space Perception*. Massachusetts Institute if Technology, Cambridge, Massachusetts, USA.

IKAN (2005) *Inverse Kinematics Using Analytical Methods*. Retrieved on Dec. 10, 2005 from http://hms.upenn.edu/software/ik/ik.html

JACK (2005) *Jack Ergonomics and Human Factors Product*. Retrieved on Dec. 14, 2005 from http://www.ugs.com/products/tecnomatix/human_performance/jack/

James, D.L.J. and Twigg, C.D. (2005) Skinning mesh animations. *ACM Transactions on Graphics (SIGGRAPH 2005)*, 24(3), August 2005.

Kshirsagar, S., Magnenat-Thalmann, N., Guye-Vuilleme, A., Thalmann, D., Kamyab, K. and Mamdani, E. (2002) Avatar Markup Language. Eight Eurographics Workshop on Virtual Environments 2002.

Moore, R. (2003) The Extent of Under-heating in Dwellings, Workshop on Comfort and Temperatures in UK Dwellings, University of Oxford, October.

Newton (2005) Newton physics toolkit. Retrieved on Dec. 14, 2005 from http://newton-dynamics.com

Ogre3D (2005) Ogre3D graphics toolkit. Retrieved on Dec. 14, 2005 from www.ogre3d.com

OpenSteer (2005) OpenSteer C++ library for constructing steering behaviors. Retrieved on Dec. 14, 2005 from http://opensteer.sourceforge.net/

PhysX (2005) PhysX physics toolkit. Retrieved on Dec. 14, 2005 from www.ageia.com.

Poser6 (2005) Poser6 avatar animation package. Retrieved on Dec. 14, 2005 from www.curiouslabs.com

RenderWare (2005) Renderware game development platform. Retrieved on Dec. 14, 2005 from www.renderware.com

Ruck, C. (1989) *Building Design and Human Performance*. Van Nostrand Reinhold, New York.

Sawyer, T. (2004) 10 Electronic Technologies that Changed Construction, Engineering News Record, June 21, 2004, ISSN 0891–9526, pp. 24–33.

Shaikh, I., Jayaram, U., Jayaram, S. and Palmer, C. (2004) Participatory ergonomics using VR integrated with analysis tools. Simulation Conference, 2004. Proceedings of the 2004 Winter Simulation Conference. Volume 2, Dec. 5–8, 2004, pp. 1745–1746.

Sung, M., Gleicher, M. and Chenney, S. (2004) Scalable behaviors for crowd simulation. *Computer Graphics Forum (EUROGRAPHICS '04)*, 23(3), 519–528.

Tokamak (2005) Tokamak Game Physics SDK. Retrieved on Dec. 14, 2005 from www.tokamakphysics.com

Watt, A. and Watt, M. (1992) Advanced Animation and Rendering Techniques, *Theory and Practice*, ACM Press. 455 s.

Chapter 17

Technology transfer

Martin Sexton

Introduction

nD modelling technology ushers in novel possibilities to transform the design, production and use of buildings by providing a '. . . multi-dimensional model . . . [which] . . . enable true "what if" analyses to be performed to demonstrate the real cost in terms of the design issues . . . [and] . . . the trade-offs between the parameters . . . clearly envisaged' (Lee *et al.*, 2003: 5). The espoused benefits of a new technology such as nD modelling, however, are not sufficient in itself to ensure its widespread adoption and use within the construction industry. Technology transfer is widely considered to be a potentially powerful mechanism to provide the construction industry with new technologies that can, where appropriate, transform and complement current technologies to create and sustain better levels of performance (DETR, 1998; Mitropoulos and Tatum, 2002; DTI, 2002). Commentators, however, have noted that the construction sector often lags behind other industries in technology adoption (e.g. see Sauer *et al.*, 1999; Fairclough, 2002; Sexton and Barrett, 2004). We understand effective technology transfer as being the 'movement of know-how, technical knowledge, or technology from one organizational setting to another' (Roessner in Bozeman, 2000: 629). This 'movement' is driven by a combination of actors' unique needs (Klien and Crandell, 1991) and tacit knowledge (Howells, 1996; Teece, 1977). The barriers to the diffusion of new construction technologies are reinforced by, among other issues, lack of information for relevant construction actors regarding technologies' characteristics and appropriate uses (Anumba, 1998; NAHB, 2001; Sexton and Barrett, 2004).

The aim of this chapter is to identify the key enablers and obstacles to the effective adoption and use of nD modelling technology. The structure of the chapter is as follows: first, a technology-transfer framework will be presented and supported by a review and synthesis of the relevant literature. Second, the methodology of the primary data will be briefly described. Third, using the technology-transfer framework, principal findings will be reported. Finally, conclusions and implications will be drawn.

The 'technology-transfer system' model

Performance improvement based on construction research and innovation absorbed into companies through technology transfer, can and does, occur successfully. However, present construction-industry technology-transfer endeavours are being severely hampered by a lack of proper understanding of technology-transfer processes and their interrelationships to both company capabilities and processes, and the knowledge characteristics of the technologies being transferred:

- First, current approaches tend to view technology transfer as a mechanistic 'pick-and-mix' exercise – identifying new technologies, and trying to insert them in their existing form into (unsurprisingly) unreceptive construction companies (Barrett and Sexton, 1999).
- Second, current technology transfer mechanisms are not sufficiently informed by, or engaged with, company strategic direction and organisational capabilities and processes necessary to enable them to absorb technologies and to turn them into appropriate innovations. Experience from the manufacturing sector, for example, has stressed that the capacity of companies to understand and effectively use new technologies from external sources is heavily influenced by the level of prior-related knowledge and expertise (e.g. see Adler and Shenhar, 1993).
- Finally, current technology transfer mechanisms do not fully appreciate that both the ability and motivation for companies to absorb and use new technology is significantly influenced by the knowledge characteristics of the technology. Grant and Gregory (1997), for example, concluded from an investigation of manufacturing technology transfer that the extent of the required tacit knowledge to absorb and use manufacturing technology had a significant impact on the success of the transfer. 'Hard' technologies which are characterised by explicit knowledge require very different diffusion mechanisms and organisational capabilities and processes than those required for 'soft' technologies which are tacit in nature.

The interactive nature of these elements stresses the systemic nature of technology transfer, and can be fruitfully viewed as a 'technology-transfer system' made up of inter-organisational networks, organisational direction and capability, and knowledge characteristics of technology. This more fluid, dynamic view of technology transfer is consistent with the move away from traditional, sequential models of innovation and technology transfer (Van de Ven et al., 1999). This position reconceptualises the technology-transfer phenomenon from a simple cumulative sequence of stages or phases, to multiple progression of divergent, parallel and convergent paths, some of which are related and cumulative, and others not. The argument is that technology transfer will only be effective if all three elements are appropriately focussed and integrated to achieve a specific aim.

The three elements of the 'technology-transfer system' are discussed in more detail in the following pages.

Organisational direction and capability

The motivation and ability of companies to absorb and innovate from new technologies has to come from *within* the company: through envisioning technology strategies and supporting organisational capabilities. Technology is a crucial strategic variable in creating competitiveness; indeed, the connection between technology and corporate strategy is confirmed as follows: 'increasingly managers have discovered that technology and strategy are inseparable' (Kantrow, 1980: 4). The technology strategy of a company should address the issues of acquiring, managing and exploiting technologies which progress corporate aspirations (Clarke et al., 1989). An appropriate focus for a company's technology strategy can be usefully considered as the generation of a balanced portfolio of enabling, critical and strategic technologies. Enabling technologies are classified as those technologies which are necessary at a functional level for the company's survival, and are widely available to all competitors in the construction industry. Critical technologies are viewed as those technologies which are unique to the company, and which give the company a differentiated technology over its existing and potential competitors. The acquisition and exploitation of enabling and critical technologies are in the domain of the company's short- to medium-term technology strategy. Strategic technologies are those technological trajectories pursued by a company that they anticipate (through technological and market forecasting) to become the critical technologies of the future. Strategic technologies, by their intrinsic nature, are in the domain of the company's long-term technology strategy.

The motivation and capacity of firms to innovate and to absorb and use new technology are very much shaped by general business and project environments. Research in innovation in small construction companies, for example, concludes that there are two principal modes of innovation: mode 1 and mode 2 (Sexton and Barrett, 2003). Mode 1 innovation focuses on progressing single-project, cost-orientated relationships between the client and the firm – this mode of innovation is more driven by rapid change and uncertainty in the interaction environment. (The interaction environment is that part of the business environment which firms can interact with and influence.) Mode 2 innovation concentrates on progressing multiple-project, value-orientated relationships between the client and the firm – this mode of innovation is more aligned to improving the effectiveness of a firm's relationship with its clients. The mode of innovation is substantially determined by the nature of the *interaction environment*: an enabling interaction environment encourages Mode 2 innovation, and a constraining environment is conducive to Mode 1 innovation. An enabling interaction environment is one which the firm can influence to a significant extent, *enabling* the firm to innovate within a longer-term and more secure context. A constraining interaction environment is one which a small construction firm can only influence to a limited extent, *constraining* the firm to innovation activity undertaken within a shorter and more insecure context.

Coupled with strategic direction and business context, companies need the organisational capability to absorb and use new technology. This capability is influenced by the level of prior-related knowledge and expertise (i.e. basic skills, shared language, technological acumen) that exist in the organisation (Cohen and Levinthal, 1990). Thus, it is clear that organisations will not be able to accomplish many of their key strategic and operational goals for technology transfer without adequate complementary capabilities (e.g. see Adler and Shenhar, 1993; Bröchner *et al.*, 2004).

Inter-organisational networks

Companies do not operate in a vacuum; rather, they are situated in a number of fluctuating inter-organisational networks of varying complexity (Betts and Wood-Harper, 1994; Bresnen and Marshall, 2000a,b,c). Inter-organisational networks promote and facilitate the development and exchange of knowledge and resources needed to encourage learning and innovation in participating companies (e.g. see Grandori and Soda, 1995; Ebers, 1997; Barlow and Jashapara, 1998). Indeed, it has been argued that the greater the number of inter-organisational networks that a company is involved in, the greater the likelihood of generating and supporting successful innovation (e.g. see Porter, 1990; Ahuja, 2000).

There are two main types of inter-organisational networks. First, companies are exposed to 'business networks' through their normal client and supply-chain interaction. These networks can encourage innovation because the companies involved are able to share needed expertise and resources (Hauschlidt, 1992). Indeed, the knowledge necessary for innovation processes is often created at the interfaces of business network technology transfer. The innovation process is therefore in part a knowledge and technology mobilisation process, based on intensive social and economic interaction processes (Hakansson, 1987). Second, companies are embedded, to varying degrees, in 'institutional networks', such as educational institutions, government bodies, research institutions and professional associations. Such networks are also potentially useful in providing companies with the knowledge and expertise needed for innovation (e.g. see Hauschlidt, 1992; Abbott *et al.*, 2004). Professional associations, for example, disseminate a particular body of knowledge to industry via their members, and thus act as vehicles for the diffusion and translation of knowledge needed for innovation (Allen, 1977; Constrinnonet, 2004).

Knowledge characteristics of technology

Technology is not transferred as a self-contained artefact; rather, for success, the technology *and* the knowledge of its use and application must be transferred (e.g. see Sahal, 1981). The extent to which new technology can be effectively absorbed by construction companies is thus substantially influenced by the knowledge characteristics of the technology being transferred. Two characteristics are especially important. The first is the extent to which the knowledge embodied in the technology is explicit or tacit. Tacit knowledge is hard to formalise, making it difficult to communicate or share with others. Tacit knowledge involves intangible factors embedded in personal beliefs, experiences and values. Explicit knowledge is systematic and easily communicated in the form of hard data or codified procedures. Often there will be a strong tacit dimension with how to use or implement explicit knowledge (e.g. see Nonaka and Takeuchi, 1995). The second characteristic is complexity. Whether based on explicit or tacit knowledge, some technologies are just more complex than others. The more complex a technology, the more difficult it is to unravel (e.g. see Gibson and Smilor, 1991).

In summary, the knowledge conversion concept argues that technology transfer is a social process of interactive learning within and between inter-organisational networks, from which a shared language of tacit and explicit knowledge can be developed.

Research methodology

The research methodology was designed to develop a greater understanding of the absorption and use of new nD modelling technologies into construction companies. The primary data was from semi-structured interviews with key staff from six construction companies comprising of architects and contractors. A semi-structured interview approach was taken for the data collection to provide both the stability to follow a predetermined route of enquiry, and the flexibility to probe further where the interviewee felt that supplementary information was valuable to bring new insights into the discussion, or to amplify answers in response to preset questions.

The semi-structured interview was shaped by the following questions: What are the benefits of the technology? Who would use the technology? Where would they use the technology? and When in the design process would the nD modelling technology be used?

Results and analysis

Organisational direction

nD modelling technology was generally seen as potentiality useful by architects but, at present, not sufficiently proven to risk disrupting current ways of working and existing technology infrastructures. nD modelling technology was described by one architect, for example, as having the potential to 'improve upon the exchange of construction information', with the benefit being articulated by another architect as being 'the idea that the drawing is only ever produced once, and so mistakes made by copying the drawing is only ever produced once – this can happen with the mechanical and electrical engineers at times.' This potential benefit was conveyed by the contractors, with one commenting that

> nD seems to be a great prospect, especially if contractors can be involved in the design process. The design plans and elevations may be checked and checked, but there will always be problems with it. It is human nature that mistakes is made. We as contractors only find all of these mistakes once the building has been built. If there can be a checking mechanism, and a way in which we can get involved and see the designs first, I think problems will be minimised.

The risk dimension, however, was stressed by both the contractors and the architects, with concern that nD modelling technology needed to be demonstrably proven to offer greater added value, compared to existing arrangements, before firms would be sufficiently motivated to adopt and use it. This view was captured in the argument that 'change takes time and costs a lot of money, so you would need to demonstrate the benefits of nD in terms of time, cost and quality before it would be widely adopted.' This generic view of risk was brought to life by an architect who asked, 'Who will own the drawings? How will risk be assigned? How does this impact upon professional indemnity? Unless the whole structure and legal system of the industry is changed, I cannot see how nD can be implemented soon.'

This position was reinforced by the view that the incumbent ways of working were widely understood and appreciated and that, according to one architect, 'we have a document management system in place, and it is ideal. It has speeded up the process of sending drawings and we, as a company, are happy with this. I think document management systems are more aligned to the way a construction project team works than the nD tool.'

Inter-organisational networks

There was consensus across the contractor and architect interviewees that the nD modelling tool needed to be appropriately integrated and mobilised through the supply chain if it was to generate the sustained impact to encourage widespread adoption and diffusion. The architects interviewed argued that the principal role of the nD modelling tool was in the design phase, with involvement from a number of project actors:

> this tool can only be used during the detailed design stage. It is during the early design stage where the architect experiments with the design of the building, and nothing should interrupt that process. Besides, the tool would be more useful to check conformity of regulations after the engineers have started to add to the design. We as architects are clear about regulations, and

usually our designs conform to the majority of the regulations. It is when other engineers start to move things in the drawing around, and someone else wants to move something somewhere else, do things get complicated. Actually, in this instance I think the tool would really benefit our work.

The need for project supply chains as a whole to adopt the technology was stressed, with one contractor articulating that

the tool will only be successful if everyone in the project supports and fully adopts it. It can only really work if it was an online tool, so that everyone could access the latest version of the model. I think that the client should own the design, as at the end of the day, it is the client who is paying for the project. The client should own the design and not the architect. In this way, there can exist a shared product model.

Knowledge characteristics of the nD modelling technology

The nD modelling tool's human computer interface was seen as important, with a contractor remarking that 'the tool seems to be user friendly and pretty self-explanatory'. The need for compatibility with existing design packages was noted, with one architect observing that

we can still use our own design packages and do not have to learn new ones. This is a great advantage and must not be overlooked. Architects really have to be specialists on a particular design package. If a new package is used, drawings take a lot longer which costs more money to us, and there is more chance of mistakes being made. And this is a bigger problem if you have to use another design package on another project – this can be very costly.'

In addition, potential adopters need to experiment with the new technology. One architect, for example, emphasised, 'I need to have the opportunity to play about with the tool on my own, that is when you can assess how practical something is'. The theme of 'getting to know the technology' was continued into the required capability development after adoption, with a contractor remarking that, 'there needs to be training sessions delivered on the tool, so that people will adopt it without training, adoption will be sporadic'.

Conclusions

The results indicate that architects and contractors appreciate the potential significant benefits of nD modelling technology. The findings further reveal, however, that construction firms will only be motivated to absorb and use nD technology when the technology has sufficiently proven that it will contribute to the business in a quick, tangible fashion, and which can be dovetailed into organisational and technological capabilities they already possess. At present, nD modelling technology is seen as too embryonic, too far removed from construction firms' 'comfort zones', requiring too much investment and containing too many risks.

nD modelling addresses a real need to identify and appropriately reconcile competing design criteria. The challenge for nD modelling technology, along with any new technology, is to shift from its 'technology push' emphasis to a more balanced 'market orientated' stance which allows the technology to be shaped by both strategic design concerns, and the rough and tumble of day-to-day operational needs. If this trajectory is pursued, nD modelling technology has a positive future.

References

Abbott, C., Ong, H.C. and Swan, W. (2004) A regional model for innovation and cultural change in the construction industry. Proceedings of Clients Driving Innovation International Conference Surfers Paradise, Australia, 25–27 October.

Adler, P.S. and Shenhar, A. (1993) Adapting your technological base: the organizational challenge. *Sloan Management Review*, 32, 25–37.

Ahuja, G. (2000) Collaborative networks, structural holes, and innovation: a longitudinal case study. *Administrative Science Quarterly*, 45, 425–455.

Allen, T.J. (1977) *Managing the Flow of Technology*, MIT Press, Cambridge, MA.

Anumba, C.J. (1998) 'Industry uptake of construction IT innovations – key elements of a proactive strategy', Proceedings of the CIB W78 Conference, Stockholm, Sweden: 3rd–5th June.

Barlow, J. and Jashapara, A. (1998) Organisational learning and inter-firm partnering in the UK construction industry. *Learning Organisation*, 5, 86–98.

Barrett, P. and Sexton, M.G. (1999) The transformation of 'out-of-industry' knowledge into construction industry wisdom. *Linking Construction Research and Innovation to Research and Innovation in Other Sectors*, Construction Research and Innovation Strategy Panel, London, 24th June.

Betts, M. and Wood-Harper, T. (1994) Reengineering construction: a new management research agenda. *Construction Management and Economics*, 12, 551–556.

Bozeman, B. (2000) Technology transfer and public policy: a review of research and theory. *Research Policy*, 29, 627–655.

Bresnen, M. and Marshall, N. (2000a) Partnering in construction: a critical review of issues, problems and dilemmas. *Construction Management and Economics*, 18, 229–237.

Bresnen, M. and Marshall, N. (2000b) Motivation, commitment and the use of incentives in partnerships and alliances. *Construction Management and Economics*, 18, 587–598.

Bresnen, M. and Marshall, N. (2000c) Building partnerships: case studies of client-contractor collaboration in the UK construction industry. *Construction Management and Economics*, 18, 819–832.

Bröchner, J., Rosander, S. and Waara, F. (2004) Cross-border post-acquisition knowledge transfer among construction consultants. *Construction Management and Economics*, 22, 421–427.

Clarke, K., Ford, D. and Saren, M. (1989) Company technology strategy. *R&D Management*, 19, 3.

Cohen, W.M. and Levinthal, D.A. (1990) Absorptive capacity: a new perspective on learning and innovation. *Administrative Science Quarterly*, 35, 128–152.

Constrinnonet project (Promoting Innovation in Construction Industry SMEs) (2004) *Constrinnonet Project Final Report: Innovation Issues, Successful Practice and Improvement*, European Commission, Brussels.

Department of the Environment, Transport and the Regions (1998) *Rethinking Construction*, DETR, London.

Department of Trade and Industry (2002) *Rethinking Construction Innovation and Research*, The Stationery Office, London.

Ebers, M. (ed.) (1997) *The Formation of Inter-organisational Networks*, Oxford University Press, Oxford.

Fairclough, J. (2002) *Rethinking Construction Innovation and Research: a Review of Government R&D Policies*, DTI: London.

Gibson, D.V. and Smilor, R. (1991) Key variables in technology transfer: a field-study based empirical analysis. *Journal of Engineering and Technology Management*, 8, 287–312.

Grandori, A. and Soda, G. (1995) Inter-firm networks: antecedents, mechanism and forms. *Organization Studies*, 16, 183–214.

Grant, E.B. and Gregory, M.J. (1997) Tacit knowledge, the life cycle and international manufacturing transfer. *Technology Analysis and Strategic Management*, 9, 149–161.

Hakansson, H. (ed.) (1987) *Industrial Technological Development: A Network Approach*, Croom Helm, London.

Hauschlidt, J. (1992) External acquisition of knowledge for innovations – a research agenda, *R&D Management*, 22, 105–110.

Howells, J. (1996) Tacit knowledge, innovation and technology transfer. *Technology Analysis and Strategic Management*, 8, 91–106.

Kantrow, A.M. (1980) The strategy-technology connection. *Harvard Business Review*, 58, 4.

Klien, G.A. and Crandall, B. (1991) Finding and using technology-specific expertise. *Journal of Technology Transfer*, 16, 23.

Lee, A., Marshall-Ponting, A., Wu, S., Koh, I., Fu, C., Cooper, R., Betts, M., Kagioglou, M. and Fischer, M. (2003) *Developing a Vision of nD-enabled Construction*, University of Salford: Salford.

Mitropoulos, P. and Tatum, C.B. (2002) Technology adoption decisions in construction organizations. *Journal of Construction Engineering and Management*, 125(5), 330–338.

NAHB Research Centre (2001) *Commercialization of Innovations: Lessons Learned*, NAHB Research Centre: Upper Marlborough.

Nonaka, I. and Takeuchi, H. (1995) *The Knowledge Creating Company: How Japanese Companies Create the Dynamics of Innovation*, Oxford University Press, New York.

Porter, M. (1990) *The Competitive Advantage of Nations*, Free Press, New York.

Sahal, D. (1981) Alternative conceptions of technology. *Research Policy*, 10, 2–24.

Sauer, C., Johnson, K., Karim, K., Marosszeky, M. and Yetton, P. (1999) 'Reengineering the supply chain using collaborative technology: Opportunities and barriers to change in the building and construction industry', *IFIP WG 8.2 New Information Technologies in Organisational Processes: Field Studies and Theoretical Reflections on the Future of Work*, International Federation for Information Processing, Kluwer Publishers: University of Missouri, St. Louis: 21–22 August.

Sexton, M.G. and Barrett, P. (2004) The role of technology transfer in innovation within small construction firms, *Engineering, Construction and Architectural Management*, 11(5), 342–348.

Sexton, M.G. and Barrett, P.S. (2003) Appropriate innovation in small construction firms. *Construction Management and Economics*, 21(6), 623–633.

Teece, D. (1977) Technology transfer by multinational firms: the resource cost of transferring technological know-how. *The Economic Journal*, 87, 242–261.

Van de Ven, A.H., Polley, D.E., Garud, R. and Venkataraman, S. (1999) *The Innovation Journey*, Oxford University Press, New York.

Chapter 18

The role of higher education in nD modelling implementation

Margaret Horne

Introduction

University education is no stranger to change, but the rapid development of today's information technology (IT) is posing great challenges to academics who have to consider its appropriate integration into carefully designed curricula to meet the expectation of students and the requirements of industry. Since the advent of computer aided design (CAD) into academic programmes in the early 1990s, there has been ongoing debate and concern relating to IT implementation within students' education and whilst there have been many advances over recent years, CAD for many still means electronic drafting. The majority of construction organisations throughout the world are still working with 2D CAD (CIRIA, 2005), and although there is growing evidence of 3D CAD, many companies use 3D primarily for presentation and marketing purposes. Nonetheless, leading researchers and executives, attempting to predict likely developments in the twenty-first century, forecast more widespread acceptance of 3D modelling and the ability to describe the look, sound and feel of an artificial world, down to the smallest detail (Gates, 1996). Built environment higher education has an important role to play in enabling this vision. The emergence of today's easier-to-use 3D modelling software, rapidly advancing computer hardware combined with an increasingly computer literate student population is beginning to result in an increase in the adoption of 3D CAD for built environment applications. This increase is likely to continue alongside industry's adoption of new tools and the emergence of a generation of students with the IT skills required by their profession. A more widespread acceptance of 3D modelling within organisations is seen as a significant milestone in advancing the adoption of nD modelling technology and furthering the vision for a single integrated project model shared by the key participants in the design and construction process.

The need for a strategic approach

In order to successfully implement and evaluate any new technologies into the academic curriculum of students, several strategic approaches have to work alongside each other. IT implementation cannot work in isolation of teaching and learning strategies and the overall business objectives of a School. Embedding new technologies into built environment programmes involves preliminary discussions with employers and professional institutions. There is a need to raise awareness of the importance of incorporating new technologies into undergraduate programmes. Processes of software selection and evaluation have to be set up alongside training workshops to inform academic staff about the availability and development of new materials. Programmes of evaluation and feedback should enable reflection on decisions taken and plans for future direction.

Traditional approach to introducing IT into the academic curriculum

The initial inclusion of computer-related subjects as stand-alone modules in the structure of academic programmes is an established technique within the School of the Built Environment, Northumbria University. The School endeavours to introduce students to IT applications appropriate to their subject discipline. Undergraduate students are introduced to industry-standard, commercially available computer programs and not to computer programming. Academic staff feel it is important that students understand the theory that lies behind computer applications and not just the use of the software. Students' education still promotes the theory of architecture, materials and construction, planning, costing as well as traditional representation techniques. Thus, students learn about real buildings and current practices alongside the appropriate application of IT for their profession. It could be argued that students are now producing some of the highest quality designs, and some of the most interesting projects, ever to come from university schools.

Modular approach

When a new technology emerges that is of importance to built environment students, an adopted approach for integration has been to develop a stand-alone IT module to be included within the modular structure of an academic programme. The potential of IT to become an integrating factor in the curriculum has been discussed by Asanowicz in the context of computer implementation in the architectural curriculum (see Asanowicz, 1998). A stand-alone module provides opportunities to assess the stability of new technologies, raise staff awareness and appraise potential applications for other subject areas. Lead users (or 'champions') of software liaise with academic staff, software suppliers and technical support services to ensure efficient management of such modules. However IT modules, such as 2D CAD, are one of several subjects competing for time in the academic curriculum. Students are taught many subjects within a programme and often these subjects are not connected with each other. Figure 18.1 illustrates how an IT module may initially sit alongside other different subjects.

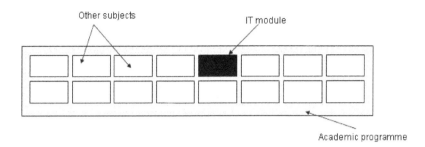

Figure 18.1 Traditional approach of IT as a separate subject of an academic programme.

Software selection and progression strategy

In addition to placing an IT module alongside other taught subjects within a programme, some consideration is necessary to place new technologies into the appropriate year of study. Students are entering universities increasingly computer literate and continual adjustment is being made to the IT components of programmes to acknowledge this. The chronological approach, adopted by the School of the Built Environment, introduces CAD/visualisation as follows:

- Year 1   2D CAD (AutoCAD) – to provide accurate 2D representation of project design.
- Year 2   3D CAD (Revit) – building information modelling used for purposes of design, spatial assessment, visualisation, energy analysis and costing.
- Year 3   Year in industry – to experience current practice.
- Year 4   Virtual Reality (3dsMax, VR4Max, Stereoworks) – to enable interaction and depth perception when viewing design.

One of the criteria for software selection in the School of the Built Environment is that software is commercially available and used by industry. Software recently selected to introduce the factors of interactivity and immersion into 3D modelling is 3ds Max and VR4Max, selected as being suitable for VR for AEC applications. This strategic approach ensures that students are introduced to the underlying theory in each subject area, as well as developing knowledge and skills in applying the software. Having acquired this IT knowledge, students are encouraged to apply it in other subject areas, and demonstrate appropriate integration into their modular programme.

IT integration

Once students and staff develop familiarisation and confidence with a computer application, further integration into other subject areas evolves, and, to date,

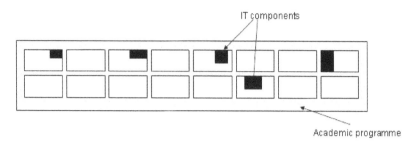

Figure 18.2 IT integrated into other modules.

elements of IT appear in many modules throughout the School. Hence many IT applications develop from being introduced as subjects in their own right into component parts of other subjects in the curriculum (Figure 18.2).

This phased approach to the adoption of IT applications can lead to a change in the teaching methods of subjects, but it does not lead to a change in the curriculum as a whole, because in both of these models there still may not be any connection between separate subjects of the curriculum.

An increasing demand for 3D computer modelling is one example of an application of IT being integrated across several disciplines in the School and some interesting applications are emerging.

Case studies

Case study 1: 3D modelling for design

Students on the final year of the BSc Architectural Design and Management programme are given an opportunity to apply 3D computer modelling to a design project. Figure 18.3 illustrates a student project with a chosen location of Forth Bank, Newcastle upon Tyne. This site has witnessed, over the centuries, a continuing process of industrial renewal and decay.

The project was a physical narrative of this process: two post-industrial monuments, clad in a thick external crust of devices and services, stand defiantly in the face of the area's latest phase of decay and pending gratification. The crust preserves pure interiors from any external forces; the galleries only exhibit progress, in the form of exemplar products displaying technological and design advancement from around the world. The student visualised this project using traditional 2D representation as well as a 3D computer model developed in 3ds Max. The model was then imported into VR4Max to enable interactivity and further design exploration.

Case study 2: 3D modelling for construction

The role of the architectural technologist is emerging as the profession which can 'bridge the gap' between design and construction. Figure 18.4 illustrates a student

Figure 18.3 Forth Bank Gallery (student: Oliver Jones).

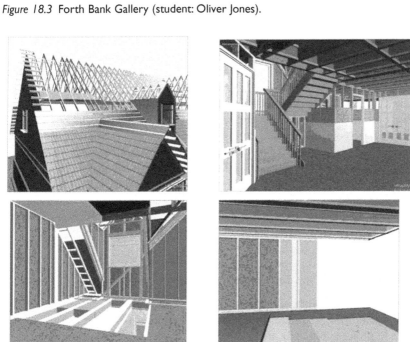

Figure 18.4 Fully interactive VR model of timber frame house showing construction details (student: Peter Jones).

project which created a fully interactive 3D model of a timber frame house. The end results communicated the technology of modern platform floor methods of timber frame construction. Students of architectural technology need to be able to 'analyse, synthesise and evaluate design factors in order to produce design solutions, which will satisfy performance, production and procurement criteria' (CIAT, 2005). CAD and VR are key subjects for the architectural technologist who needs to visualise and analyse new technologies and communicate options to various design teams and clients. The student imported files from AutoCAD into 3ds Max and then imported these into VR4Max to enable interactivity within the model. This model was also viewed in a semi-immersive passive stereo facility (Section 4 of Figure 18.4) which enabled a feeling of being immersed in the building.

Case study 3: 3D modelling for building services analysis

Figure 18.5 illustrates work from a final year design project and is the culmination of the course of B.Eng. (Building Services Engineering). The project provided an opportunity for development of a major design, involving consideration of building environmental performance and building services systems design. The project was based upon simulated brief and architectural sketch scheme design ideas. The project provided the opportunity to apply the knowledge gained throughout the four-year course and allowed for individual development of personal interests within a range of building services engineering topics.

This study developed the requirements of a media centre through the initial stages of design, to establish effective building performance. In this case the student aimed for the building to achieve an Excellent BREEAM rating and he adopted a holistic approach and an understanding of 'what the building was about' before considering the selection of building services systems. Renewable opportunities were identified and high-efficiency low/zero carbon systems were matched to the buildings' characteristics making for a sustainable centre incorporating energy cogeneration where practicable. He considered the relationship between natural and artificial lighting and produced detailed visual and numerical simulations to support his design ideas. Energy analysis simulations involved data transfer between AutoCAD, Revit and IES software.

Case study 4: 3D modelling for quantity surveying

Figure 18.6 illustrates how students of quantity surveying are learning to understand traditional cost control procedures through the design and construction of building projects. Students are encouraged to make extensive use of appropriate information and communication technologies (ICT) to search for, synthesise and abstract building cost data and to effectively manage the cost information on

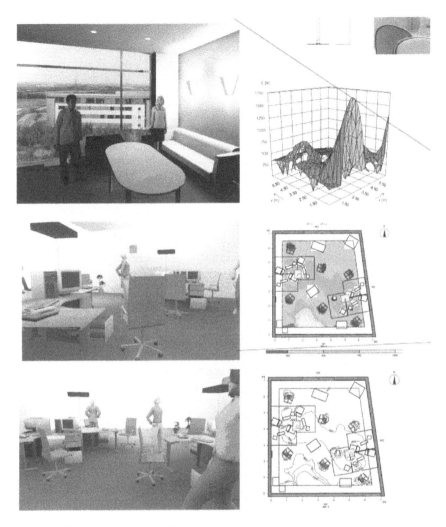

Figure 18.5 Lighting studies detailing the factors contributing to good lighting design for a successfully serviced media centre (student: Edward Miley).

a construction project. 3D object-oriented software is used to integrate design and cost information.

Revit is used with Year 2 surveying students as a tool for providing quantitative information which can be fed into the cost model for a building project. Students use the scheduling tool to generate quantities for significant elements of a project and import this information into a spreadsheet. Students are then tasked with

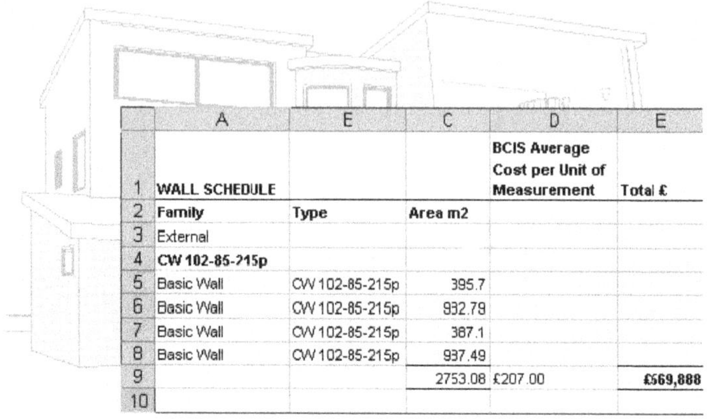

	A	E	C	D	E
1	WALL SCHEDULE			BCIS Average Cost per Unit of Measurement	Total £
2	Family	Type	Area m2		
3	External				
4	CW 102-85-215p				
5	Basic Wall	CW 102-85-215p	395.7		
6	Basic Wall	CW 102-85-215p	932.79		
7	Basic Wall	CW 102-85-215p	387.1		
8	Basic Wall	CW 102-85-215p	937.49		
9			2753.08	£207.00	£569,888
10					

Figure 18.6 External wall schedule converted into an estimate (sudent: Mark Janney).

sourcing matching cost information from an industry database in order to produce estimates for the building elements. The outcome involves an appreciation that the building model contains information which can be used for purposes beyond the drawn representation of the building, in this case the derivation of quantities for estimating. The process requires the students to examine the specification of the building element and therefore what the schedule quantities actually represent in order to be able to apply appropriate costs from another source.

Potential contribution of IT to further multi-discipline integration

The case studies outlined earlier are encouraging as they illustrate the interest and uptake of computer modelling applications in a multi-disciplinary school. Building Information Modelling, using Autodesk Revit, has been incorporated into modules which introduce design, visualisation and costing and energy analysis. Hence, a new model is beginning to emerge (Figure 18.7) in which IT can play an integration role. However, IT itself doesn't enable this communication if it is attached to separate modules and is not treated as a whole. This model can also lead to problems in some subject areas if students, having gained an appreciation of the multi-purpose capabilities of the software, apply it to other subjects prior to any underlying theory being covered. The classic ways of teaching some subjects need to be examined in order to manage and exploit the capabilities of the emerging IT technologies of interest to industry.

As long as IT applications are attached to separate modules and not treated as a whole, this situation will continue. However, technologies such as 3D CAD and

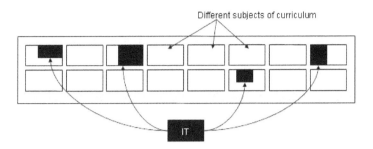

Figure 18.7 IT as an axis for academic curriculum integration.

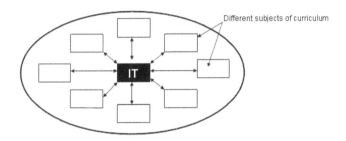

Figure 18.8 IT facilitating academic integration.

VR are offering a chance for a more collaborative way of working and for students to work on the type of integrated projects they will meet in industry. The lessons to be learned from adopting a more integrated approach may lend themselves also to the progression from 3D to nD technologies, which have application throughout the complete building process, from design, management, construction and maintenance.

Figure 18.8 shows a model which brings all subjects into contact and enables exchange of information. Such a model will only be possible if we start to think of IT applications in a different way. Virtual Reality is one such technology which is beginning to demonstrate this, and should provide lessons for the introduction of nD modelling into the academic curriculum.

Virtual Reality for collaborative working

VR adds the dimensions of *immersion* and *interactivity* to 3D computer generated models. It enables the exploration of designs and the consideration of alternative options not offered by any other form of traditional representation. However the technology is still perceived by many as being inaccessible and cost prohibitive, but this is beginning to change and current lessons being learned in the implementation of VR for higher education students could be applied for the subsequent implementation of nD-enabled construction.

VR systems

VR can be of a type known as '*full sensory immersion*' where the user wears headsets and maybe gloves to gain a total immersive feeling of 'being' in a simulated environment. Applications of immersive VR have been seen in the fields of aeronautics, medicine and military applications – developed often at high cost by those whose need for a prototype was justified. To enable more participants to experience a simulated environment, a type of '*semi-immersive VR*' describes small cinema-like studios where audiences can share the feeling of being in a scene, although the navigation is usually in the hands of an experienced operator. CAVE installations can also be described as '*semi-immersive*' as the participants can be surrounded on three sides by screens onto which are projected images of a scene. A third type of VR is becoming known as '*desktop VR*' and the increasing power of PCs and graphics cards is making the technology accessible to those with computers typical of those found in many offices.

The performance of the computer hardware selected for VR has to be such that simulated scenes can be navigated in real-time and that users' exploration is not hampered by hardware limitations. Hardware needs to be compatible with software already in use and software planned to be used, and configurable to display the scene on a large screen system.

Considerable advances have been made in CAD software and architectural design programs such as AutoCAD, ArchiCAD, Revit, MicroStation and SketchUP. Commercially available VR software is capable of producing visual interpretations in real time which can enable the exploration of ideas and scenarios. There are have been a number of PC-based VR systems, including Superscape, VRML and World Tool Kit, tried for their suitability for use in the AEC sector. Some products are also being developed which offer VR as an 'add on' to CAD packages already widely used by the industry. As in the early days of CAD, software selection and implementation is far from straightforward as users are faced with many choices. A key criterion for selection is usually 'ease of use' which can be facilitated by choosing familiar computing environments and tried and tested commercial applications wherever possible (Otto *et al.*, 2002).

Users of VR systems

Successful implementation of VR begins by considering *who* will be the users of the facility and *how* it will be used. Various design decisions have to be made when considering a VR facility. The financial and human resource investment in implementing VR has to be balanced against the returns on any investment in computer hardware, software and space requirements for the VR facility.

The requirements of the users will also determine the location of a VR facility. Whilst a single mounted headset display may be the most appropriate type of VR simulation for flight simulation, in the AEC industry it is likely that VR will be used for collaborative purposes and the space and location should be designed with this in mind. The integration of hardware and software within the VR facility still remains in the hands of the VR specialists, but advances in computer PC hardware and software are resulting in more affordable systems which can be used by non-computer specialists.

Integration of VR into the academic curriculum

Increasingly, it is becoming acknowledged that visualisation can assist all AEC professionals, from the presentation of an initial concept to the effective planning, management and maintenance of a project (Bouchlaghem *et al.*, 2005).

However, the relatively new profession of the architectural technologist has a specific need to be able to simulate and interact with a model in order to resolve both design and technical issues and to ensure optimum building performance and efficiency. Students of architectural technology have an immediate requirement to apply VR to enhance buildability and performance in their design projects. A stand-alone module entitled VR for the Built Environment was designed to offer progression from the introduction of CAD (Year 1), Building Information Modelling (Year 2), with VR strategically placed to be introduced to final year students. This module was designed to introduce students to the theoretical aspects of VR, including the planning, management, documentation and archiving of VR projects, as well as hands-on use of the software.

An approach was adopted to implement the integration of the knowledge and skills acquired in this VR module with the undergraduate architectural technology students' final year design project (Figure 18.9). This integrated approach was adopted to minimise the time pressures and risks on students, increase their motivation and encourage to them to apply the software to their discipline in a focused way.

The first group of twenty undergraduate architectural technology students in the School of the Built Environment to use VR did so from October 2004 to April 2005. Their degree is a four year modular programme structured around several core areas. IT provision had been introduced strategically alongside other modules, culminating in the integration of knowledge and skills within a design project.

Desktop VR facilities were introduced to provide access for up to 30 students. VR software (VR4Max) was chosen that could interface with the commercially available CAD and Visualisation software (AutoCAD and 3ds Max) already used in the School, enabling the direct transfer of data from traditional CAD systems into a VR application.

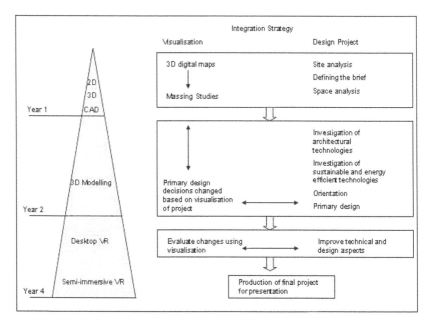

Figure 18.9 Integration strategy of visualisation into architectural technology curriculum.

Collaborative Virtual Environment

In addition to desktop VR software installed in computer labs, the School has also implemented a semi-immersive Virtual Environment, located in a well-situated, central position to allow convenient access for students and staff (Figure 18.10). The facility can be used by groups of up to 30 participants and allows students to view their designs in stereoscopic format, from multiple viewpoints, and to navigate through space in real time. A cordless mouse allows user interaction with the system and a wireless control panel enables projectors, mono/stereo display, lights and sound to be controlled from one convenient source. The familiar Windows desktop environment resulted in an immediate confidence by users to operate the system and encouraged other academics in the School to consider that the technology was not as 'out of reach' as they had perceived. The implementation of the semi-immersive Virtual Environment included the additional installation of six workstations to be used by students and staff specifically for 3D modelling and visualisation. This was seen as an important inclusion in order to maximise the use of the facility from the outset, to promote possibilities of collaboration and to demonstrate potential applications of VR to projects.

The architectural technology students who had linked VR to their design project used the Virtual Environment for their final presentation and students presented their VR models alongside paper plans, sections, elevations and 2D

Figure 18.10 Collaborative Virtual Environment.

renderings, supporting the view that VR, if used appropriately, supplements and extends other forms of traditional representation (Giddings and Horne, 2001). The process, end results and feedback from the students has provided valuable lessons which can be used in determining a correct integration methodology between two subject areas (Horne and Hamza, 2006). The VR module and Design Project module were assessed separately to minimise the risk to students at this stage of investigation. Tutors noted that some students spent a longer time learning how to use the software, which affected their focus on design and choice of architectural technologies. Future integration for architectural technology will encourage the students to focus more on visualising the interface between various constructions and finishing materials. Also the development of further subject-specific VR tutorial material and workshops will speed up the process of skill acquisition and enhance the appropriate application of the technology. Student feedback encouraged continuation of this integration with more emphasis on using visualisation for communicating the interface between materials and related operational and maintenance issues.

3D to nD

The approach outlined in this chapter has been focused on selecting and implementing appropriate technologies alongside the strategic integration of IT into the academic curriculum of built environment students. In the case of 3D CAD and VR it has shown the following:

- 3D technologies must be selected first and foremost to be appropriate for the needs of the users.
- 3D technology is becoming less technically demanding and cost prohibitive.
- Fairly sophisticated 3D models can now be created by non-computer specialists using commercially available software.

- Increased student engagement in using 3D will enhance design and construction knowledge and extend traditional communication techniques.
- The careful selection of 3D CAD and VR systems may effect greater collaboration between AEC disciplines as visualisation develops to provide substantial integrative capability.

Further integration

The Virtual Environment was used initially only by students of architectural technology, but is already finding a wider role, with 390 students from across the School having currently been introduced to its capabilities. Academic staff, initially apprehensive about the supposed complexity of VR, are showing acceptance of a technology which employs the familiar Windows operating system, uses wireless, unthreatening peripherals and is conveniently located in an accessible location (Figure 18.11). Staff concerns centre on the constant need for training, necessary to keep abreast of what advancing technologies can offer. Nonetheless, after a series of staff-awareness events, 3D applications are being proposed for development which could aid teaching and learning programmes throughout the School. It is anticipated that the facility will enable the exploration of possible benefits of 3D across a wide range of projects and disciplines.

As students and staff gain more confidence in the use of 3D we should see more widespread and earlier application.

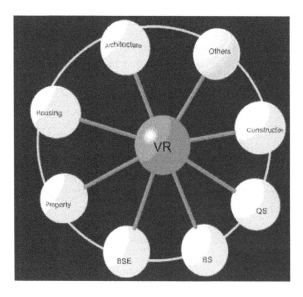

Figure 18.11 Integrative role of VR for the Built Environment.

VR earlier in the design process

This chapter has illustrated how VR is beginning to be integrated in the design process of built environment students, yet it is often used after traditional visualisation techniques have taken place. However, the growing interest in 3D modelling technologies is resulting in emerging strategies for earlier adoption, subsequent evaluation and potential for integration. The advent of parametric building information modelling software which considers massing elements and building elements (walls, columns, doors, windows, etc.) rather than geometric lines, points, faces and surfaces (as in traditional CAD software) heralds a new way of working. Evidence is emerging that adoption of building information modelling software is resulting in productivity gains of 40–100 per cent during the first year (Khemlani, 2005). Students and staff who are being introduced to such software, such as Autodesk Revit, are beginning to appreciate its capabilities and potential to be used throughout a project (Figure 18.12). Whilst the 3D models currently created with such software may still be too detailed to be used as VR models, some researchers believe that because the data is object oriented, it is easier to simplify or replace certain elements when used in a VR application (Roupé and Johansson, 2004).

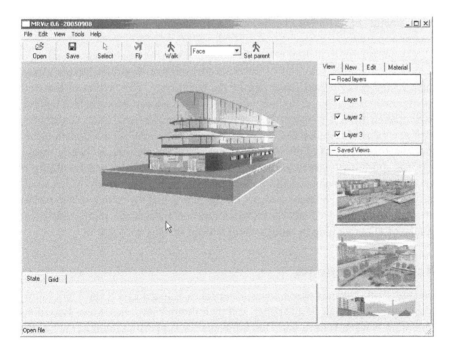

Figure 18.12 Student work transferred directly from Revit to VR application MrViz.

This could extend the *nature of interaction* within VR models from that of navigation and ability to decide *what* to look at, to that of intuitive modification of objects (facades, trees, roads, etc.) at run-time whilst viewing the model. This extension would lend support for the belief that VR could be successfully applied to the earlier stages of design (Leigh *et al.*, 1996) and offer the user greater control over the Virtual Environment, as is the case with CALVIN – an immersive approach to applying VR in architectural design. As building information modelling is further adopted into the design process, and a new generation of students emerge into practice with 3D modelling skills, the demand for a seamless integration between building information modelling and VR will increase. The established divisions between software will become less distinct as technologies develop and blend into one another (Whyte, 2002).

Future challenges

This chapter has outlined the classic modular approach in introducing new technologies into an academic curriculum. It has identified that it will be necessary to think of IT in a different way if we are to maximise the benefits that nD modelling can bring. For 2D–3D–4D–nD technologies to have a wider uptake will require a 'centralised' approach to their management and implementation throughout academic programmes and curricula. The introduction of nD modelling technology will require a change in the classic ways of teaching if it is to support the requirements of different disciplines, and facilitate and encourage a multi-discipline culture.

A systematic introduction of nD technologies into the academic curriculum of students across all disciplines of the built environment would enable the assessment of the benefits and feasibility of the technology. The School of the Built Environment is well placed to integrate a multi-disciplinary project which involves participation from a range of built environment professions, each contributing their specialist knowledge and skills. This could help establish a foundation for a more detailed analysis of the implementation of nD modelling, including organisational and human issues, in higher education. The benefits and feasibility of nD modelling technology could be assessed via such a pilot study which would have the potential to prepare students for the future, and for an industry in pursuit of more widespread collaborative environments. Indeed, the more widespread uptake of nD modelling in industry will depend on how we teach it.

Acknowledgements

Undergraduate students Oliver Jones, Peter Jones, Edward Myers and Mark Janney whose projects appear in this paper, and their tutors Paul Jones, Steve Baty, John Thornton, Neveen Hamza and Derek Lavelle for supporting them in their application of new technologies, and for permission to reproduce their work.

Senior Research Assistant Emine Thompson for furthering the application of VR in the School.

Funding for the Virtual Environment has been provided by Professor David Fleming, Dean of the School of the Built Environment, Northumbria University.

References

Asanowicz, A. (1998) Approach to Computer Implementation in Architectural Curriculum, *Computerised Craftsmanship, eCAADe Conference Proceedings*, Paris, pp. 4–8.

Bouchlaghem, D., Shang, H., Whyte, J. and Ganah, A. (2005) Visualisation in architecture, engineering and construction (AEC), *Automation in Construction*, 14, 287–295.

CIAT (2005) Chartered Institute of Architectural Technologist, www.ciat.org.uk, accessed 14 Nov. 2005.

CIRIA, Construction Productivity Network (2005) Virtual Reality and 4D Modelling in Construction, Members Report E5114.

Gates, W. (1996) *The Road Ahead*, Harmondsworth, Penguin Books.

Giddings, B. and Horne, M. (2001) *Artists' Impressions in Architectural Design*, Spon Press, Oxford.

Horne, M. and Hamza, N. (2006) Integration of Virtual Reality within the Built Environment Curriculum, *ITcon*, March 2006.

Khemlani, L. (2005) Autodesk Revit: Implementation in Practice, http://www.autodesk.com/bim, accessed 12 June 2005.

Leigh, J., Johnson, A. and DeFanti, T. (1996) CALVIN: An Immersimedia Design Environment Utilizing Heterogeneous Perspectives.

Otto, G., Kalisperis, L., Gundrum, J., Muramoto, K., Burris, G., Masters, R., Slobounov, E., Heilman, J. and Agarwala, V. (2002) The VR-desktop: an accessible approach to VR environments in teaching and research, *International Journal of Architectural Computing*, 01(02), 233–246.

Roupé, M. and Johansson, M. (2004) From CAD to VR – focusing on urban planning and building design. *AVR III Conference and Workshop*, Chalmers, Gothenburg, Sweden.

Whyte, J. (2002) *Virtual Reality and the Built Environment*, Architectural Press, London

Chapter 19

Designing fit-for-purpose schools

The nD game

Ghassan Aouad, Angela Lee, Song Wu, Jun Lee,
Wafaa Nadim, Joseph H.M. Tah and Rachel Cooper

Background

The concept of 'nD' is still emerging and will undoubtedly require a mind shift from the traditional approach to modelling in 3D. Education plays a major role in this and the next generation of professionals will be able to make the change if they are equipped with the right skills. School students, particularly at the secondary level, will need to be educated on how the new 'nD' approach will help them to understand the design process. This will ensure that the nD concept is targeting the larger community and future constructors, namely school students. This chapter proposes how to customise and exploit existing technology into a 3D:nD modelling game that enables school students to be involved in the design process of their new/refurbished school.

The Department for Education and Skills' (DfES) 'Building Schools for the Future' (BSF) programme in the UK, which aims to replace or renew all secondary schools over 10–15 years from 2005 to 2006, is currently embarking on several such projects worth illions of pounds sterling. Usual public participation methods (such as displays at local council offices, architectural plans, sections and elevations, written text, etc.) are not easily accessible or presented in an easy-to-understand format for young persons. Students, the users of the building, should be involved in this process, thus resulting in more cost-effective, user and environmentally friendly solutions to be deployed. The way in which the building fulfils the expectations of stakeholders is increasingly becoming the measure of its success (Lee *et al.*, 2003). However, it is not only during the design process that stakeholders voice their concerns; the completion of a building often brings about a multitude of additional censures that stakeholder groups did not 'see' in the building design (i.e. common amongst non-constructors). The scrutiny of the design of buildings is often the topic of everyday conversation. This is exemplified in the development of new schools, which is usually at the heart of a local community. The community usually is, at large, almost endeavouring to see the problems associated with the design rather than appreciating the complex decision-making process undertaken by construction stakeholders.

Playing the hand of the head teacher, a pupil or parent, school students will be able to create/build their ideal school and summon the subsequent design impact at will through immediate feedback. There is the need to incorporate enough space for student activities: playground, gymnasium, changing rooms, canteen, assembly hall and most importantly, classrooms. The layout of a building and its components will be assessed in terms of its crime deterrence, access facilities and egress. However, these factors may at times conflict with each other. For example, a window may be small in size, positioned high and open inwards to prevent intruders; a window may be positioned low and open outwards to enable operation by all users. Thus, the design has to be prioritised. The 3D:nD modelling game enables the user to assess these various priorities from the point of view of various stakeholders (i.e. head teacher, pupil or parent).

The 3D:nD modelling game will stretch the imagination of users to appreciate the complexity of the design and construction process, all centred on the creation of the 'best' school in the world! Not only will it provide hours of endless fun (it is proven that young persons learn quicker through play), the results will be used to inform the design process of the students' school, it will prove to be a valuable learning experience as it is aligned to the Design and Technology National Curriculum in the UK and it will also encourage students from challenged areas of Greater Manchester to consider careers in construction and research. For the past two decades there have been falling numbers entering the construction trade despite the rising demand for new buildings. The industry is often seen as dirty, fragmented, low paid, dangerous and with a lack of personal development. The 3D:nD modelling game, through assistance with CITB-ConstructionSkills, aims to encourage new constructors into the industry and be the mechanism for the promotion of science and technology research in general.

The nD game project aims to tailor existing nD technology. The main benefits of developing this nD game are as follows:

- To foster a two-way learning approach of the ongoing development of new and refurbished schools.
- To inspire future generations of construction professionals, and science and technology researchers in a seemingly challenged area.
- To highlight the complexity of the design and construction process and to challenge student perception of the image of design engineering.
- To demonstrate the impact of crime, accessibility and sustainability on the architectural design of (school) buildings.
- To demonstrate how our nD prototypes can be easily customised and used to enhance learning and understanding, thus further exploiting the concept.

Methodology

In order to develop the nD game, it is essential to follow a structured methodology. The following was undertaken:

Requirements capture for interactive 3D:nD modelling

- Ascertain plans for case study schools. Review school building design guidelines, including the following:
    - local government educational development proposals;
    - crime, accessibility and sustainable development design;
    - special educational needs and disabilities.
- Continually review the Design and Technology National Curriculum to ensure that the 3D:nD modelling game fits within its criteria.
- Conduct survey questionnaire of users/ students' perception of their school building design and understanding of the design process, and demonstrate the importance of nD modelling.

The work considers previous research done on Design Quality Indicators (DQI) and how these can be customised to fit the design of new schools. The results of this phase have been fully documented in Nadim (2005).

3D:nD modelling game development

The 3D:nD modelling game is developed using the case study schools and involves the development of a set of customisable school building components, such as classroom, playground, gymnasium, science lab and so on, which can be dragged and dropped into a 3D environment to build a scenario school building. Knowledge and technologies are deployed to demonstrate the knock-on effect on various design options (crime, accessibility and sustainability). Users (young persons) will be invited to test the game at various stages of the development to ensure friendliness of the user interface. Key activities in the development of a 3D:nD modelling game are as follows:

- develop 3D design environment based on existing nD modelling platform and game engine;
- develop 3D building components for school buildings;
- develop interactive user interface.

Demonstration and dissemination

The 3D:nD modelling game will be available to download from its dedicated website and on CD ROM, for those without internet access. It will be deployed initially at two case study schools. An evaluation survey questionnaire with the users, and semi-structured interviews with the teachers will be administered to ascertain the relevance or otherwise of the game.

Although the initial work concentrates on two schools in the first instance, once the game has been developed it can be used as a case study exemplar or can be easily customised and applied on any refurb/renew school nationally.

Why the use of computer games for nD education?

From the time the first computer game 'Space War' was developed by Steve Russell at MIT, games have achieved a remarkable growth (UbiSoft, 2005). The game industry currently brings in around $36 billion dollars annually world-wide. It became bigger than the world music and movie industries (Fullerton *et al.*, 2004). The influence of games over the children and youth of today is akin to the cultural influence of music, political movements and even religion on adults. Hence, it has become an increasingly important issue for game developers or educators to study the application of computer games in the filed of education. Gaming is one of the most exciting ways for pupils to learn new things. From this point of view, the nD game which is a multi-user simulation/building game, based on nD technology adopted a multi-level approach involving five levels starting with basic principles and culminating in level five which uses the nD engine. The nD game aims to provide an opportunity for the school students to learn basic and advanced building design concepts through playing the game.

Computer games are today the most popular pieces of software in the world. They have become so popular that blockbuster films are being made out of the plots of popular computer games. Today's teenagers live and swear by the cult of computer games. Computer games have gained unprecedented access to the minds and souls of young people. Hence educators have found a possibility by which computer games can be used as excellent educational tools. However, it must be understood that without proper application of the computer game in the educational field, the computer game cannot be recognised as a potential tool for education. Up to now, computer games have had very little educative features in general. This is slowly changing. Right from games for school-age kids (Where in the World is Carmen Sandiego) to games for would-be astronomers (Starry Night), today's computer games have more educative content than at any time (Bates, 2001).

Developing computer games is a complex and long-run process and it also involves the team effort of many individuals spanning dozens of professions all across the game industry and spilling into other industries (Rage, 2005). To understand a complex nature of game development, it is increasingly necessary to know what a game is made of. Generally, all games do not share the same exact structure (DNER, 2005). For example, a card game has a very different structure from that of a 3D action game. There is something, however, that they must share, because we clearly recognise them all as games. Figure 19.1 shows widely accepted game production parts as a member of game structure. On the whole, game structure contains other parts, not just game production parts, such as business and post-release parts. However, only game production parts are analysed at this time because we are placing emphasis on the

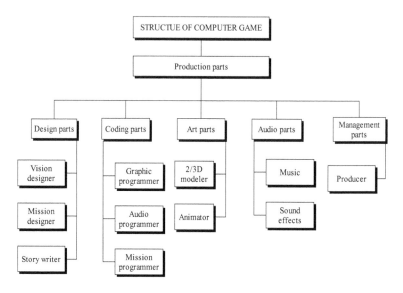

Figure 19.1 An overview of the game structure.

technical side:

- Design parts: Each game usually has a visionary game designer. The nature of the visionary game designer is typically to coordinate the design staff in the effort to create timely, thorough, compelling game design specifications that the rest of the team can readily use and which are readily understood by the game's publisher and other key stakeholders. Mission designers are sometimes programmers writing scripting codes for a mission. Sometimes these designers are artists laying out tiles of a map and designing triggers, and sometimes they work in pure text, describing to others how the game should be laid out.
- Coding parts: The designers can write documents and create specifications' but the game will not be anything other than what the programmers and artists create. The programmers' roles are to obviously create the code: the 3D engine, the networking library, art asset converter and such, to realise the vision for the game.
- Art parts: Games used to have a single artist or modeller drawing the characters for these electronic heroes to carry out their missions. In the earliest days the programmer, designer and modeller were one and the same person. Starting in the mid-1980s small teams of modeller, usually no more than three, would work on a project. Starting in the early 1990s game projects grew substantially in their art requirements and budgets. Famous examples of these are Wing Commander by Origin, where over $10 million was spent

(Bates, 2001). Modellers are now differentiated by their skill sets. It is interesting to know that many modellers can build 3D models of the most arcane objects quite accurately and swiftly without being able to sketch them.

- Audio parts: Audio assets come in three main flavours: sound effects, music and voiceover. In the beginning there were only crude sound effects performing buzzes, beeps and whistles. We now have full Dolby 5.1 3D sound. Music has come a long way from clever timing of beeps to compositions by film composers performed by 50-piece live orchestras. And voice acting is now an art form performed by many Hollywood stars.
- Management parts: As game project is getting bigger and complicated, the roles of manager or producer for the project become critical for the effective execution of the project.

Stages of nD game development

The development of computer games is a complex and long-run process. Hence, few games are profitable. In 2001, over 3,000 games were released for the PC platform; it is likely that only less than a 100 of the games turned a profit, and around 50 made significant money (ESA, 2005). To fight with these problems, the game industry has evolved some practices for the efficient development of computer games. A core aspect of the process is that a project is developed and approved in stages. Each stage is defined by milestones.

Figure 19.2 is a graphical representation of the stages of development. Notice that the five stages are drawn in a 'V' shape. This is to represent that in the beginning of a project, the creative possibilities are broad, open and changeable.

Early on, changes can be made with little financial consequence. As the process moves along, ideas must become more focused, and smaller and smaller changes can be made to the design without disrupting the production. By the time the project reach the middle point, it's virtually impossible to alter the broad vision of a game, but it is possible to think with some features or concepts within it. During later stages of production, it becomes increasingly difficult and more expensive to make any modifications to the game design beyond tweaking variables that have been set up to be flexible. As it is approaching the testing stage, only the completion of details is an open area of discussion. Any major or even moderate changes are usually out of the question.

- Concept stage: although the developer has a new concept for the game, it is a different matter to get a funding for it. The goal at this stage is for the developer to get a publisher to commit to funding at least the first milestone. The project plan is an important element in this stage. It includes a budget and schedule. The project plan shows the publisher that developers have thought through every element of the production and understand what it will take to implement.

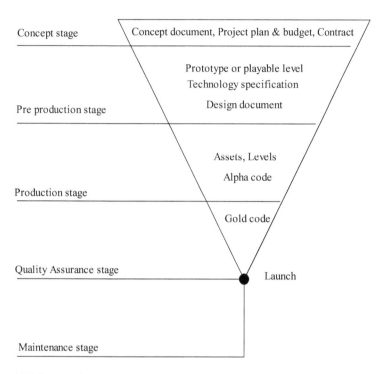

Figure 19.2 Stages of game development.

- Pre-production: during pre-production, a small team will be put on the project to verify the feasibility of the idea. This team will work to create one playable level or environment of the game, focusing on proving out differentiating features. The team at this point will be small because a small team is inexpensive. Until the publisher is certain that the game concept is going to work, they won't fund a full team. The job of the small team is to further refine the plan for the design and implementation of the entire production, from staffing and resource allocation through to schedules and deliverables. If a software prototype was not created in the concept stage, now is the time to make it and test it. This is also the time to write up a detailed design document and technical specification. When the full team is hired for the production stage, they will use these documents as a guide. The design document and technical specification impact the project plan and vice versa. The project plan cannot be fully fleshed out until the design is complete, and the limitations of schedule and budget will impact decisions that are made regarding the design. Typically, these documents are refined in parallel, each one influencing the other. In addition to refining the game design, pre-production is the time to build much of the risky technology and prove its feasibility. This helps to reduce the potential risk for both developer and publisher. At the end of pre-production, the publisher will evaluate the prototype or completed level, the progress on the technology, the design and technical documents and the fully developed project plan, in order to make a decision whether to finance the production or kill the project.

- Production stage: production is the longest and most expensive stage of development. The goal in this stage is to execute on the vision and plan established in the previous stage. In the process of improving and executing the design, some changes will inevitably be necessary that must be reflected in the design document. During this stage, programmers write the code that makes the game function and artists build all the necessary art files and animation. The sound designers create sound effects and music. Writers write dialogues and other in-game texts. The producer works to make sure everyone on the team is communicating and is aware of the overall progress. The ultimate object of this stage is to get to 'alpha' code. This means that all features are complete and no more features will be added. Sometimes the team has to cut ambitious features in order to meet the alpha milestone and stay on schedule. As the programmers work on the code, they will periodically assemble versions of the project, which are called 'builds'. Each build is given an incremental number, so that any issues or bugs can be referenced as to what build they were found in. When the developers achieve alpha code, they send a build to the publisher's QA (Quality Assurance) team.
- Quality Assurance stage: the QA stage is almost the final stage where the developers has a chance to truly see their game for what it is and to make sure it offers the best possible experience for the player. In order to achieve this, the QA team tries out the test and this test is carried out according to the test plan. The test plan is created by the QA team and it describes all the areas and features of the product. It also shows how the test should be carried out under the various conditions. The ultimate goal of this test is to detect the 'bug' that makes a game not behave according to plan. The final goal of this stage is to reach what's called 'gold code'. This means that all bugs have been resolved.
- Maintenance stage: nowadays, updating the game is an increasingly important issue because of easy accessibility to internet to most game players; game players generally have a tendency to look for very up-to-date and revolutionary games.

The nD game so far is in the production stage.

During the concept stage, the aim, target and application method of the nD game was established. From the technical point of view, proprietary game engines were investigated at this stage. A game engine called '3D rad net' was initially selected. The primary reasons behind this choice were because it was cheap and easy to programme. However, difficulties soon arose. First, the quality of image that can be handled by this game engine was very poor (Figure 19.3), and it did not allow customisation of own rules, which was fundamental to development.

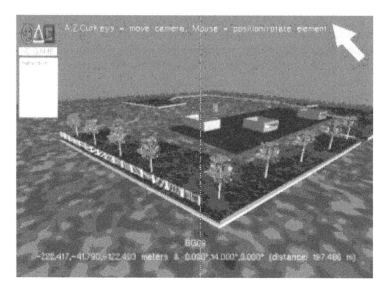

Figure 19.3 3D rad net prototype development.

Figure 19.4 Quest prototype development.

Subsequently, the game engine called 'Quest' was adopted. It enabled the use of high-quality images (Figure 19.4). However, it too had its own computer programming language, dissimilar to common languages such as C or Java. It had predefined commands and so it was difficult to create specific actions or rules.

As a result of this, the nD game was developed using the computer language called visual C^{++}, which allows the creation its own rules, and which can handle good quality images. Though it required us to work from scratch, it enabled freedom to create exactly what we wanted.

For the pre-production stage, all the rules for the nD game were defined. The tool will cover five key aspects of nD:

a   design internal lay-out, ensuring that the initial design fulfils the requirements set by the brief (cost, functions, time, etc.);
b   inclusivity, ensuring that the design conforms to accessibility regulations;
c   security, ensuring that the design conforms to crime regulations;
d   fire safety, ensuring that the design conforms to egress regulations;
e   nD modelling, ensuring that the design conforms to all of these requirements.

Users are required to select, drag and drop building components within a pre-defined outline school building (Figure 19.5). They select building components based on criteria (such as capacity, cost, inclusivity rating, security rating, fire rating etc.) and determine where best to position within the school. Accompanying notes will be provided to educate users in terms of design regulations and requirements. The design can be viewed three dimensionally. Once the requirements of the level are completed, the design is then assessed to ascertain whether it has passed the design critefia within the prescribed level. Scores are

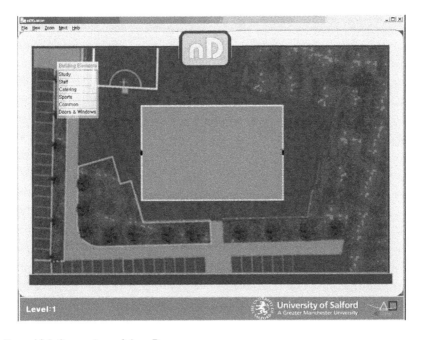

Figure 19.5 Screenshot of the nD game.

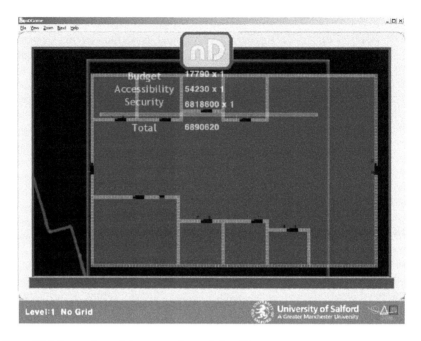

Figure 19.6 Screenshot of the score-sheet of the nD game.

Figure 19.7 Screenshot of the 'pass' certificate of the nD game.

allocated (Figure 19.6), and users are either requested to revisit the design so that they fulfil all the requirements, or are issued a certificate for passing the level (Figure 19.7). In the final nD level, users are requested to design a new school block that should pass all the design criteria, hence nD modelling. In this way, the game educates users of different design criteria and highlights that requirements may conflict with each other.

The nD game is currently in the production stage, and is due to be complete by March 2006. A series of testing and validation will follow before widespread dissemination across UK schools.

Conclusion

This paper reviewed the current condition of computer games being used for the purpose of education. Until now, computer games haven't had great amount of educative features in its contents. However, current tendency is different. Many people start to see the potential in the game as an educational tool. It is an increasingly important issue. If students can learn something through playing a game it is not so hard to image what kind of responses educators can get from students. Educators and game developers must realise the enormous opportunity, which computer games being used as a educational tools, has opened for students as well as themselves. This chapter presented the development of an nD game, targeted to teach school kids the process and purpose of design.

Acknowledgements

The authors would like to thank the Engineering and Physical Sciences Research Council for funding the development of the nD game, and to the project partners: CITB-ConstructionSkills, Setpoint, Enabling Concepts and Perfectly Normal Productions.

References

Bates, B. (2001) *Game Design: The Art & Business of Creating Games*. Stacy L. Hiquest, California, USA.
DfES (2004) http://www.teachernet.gov.uk, accessed 10 March.
DNER (2005) The Distributed National Electronic Resource. [online] Available from: <http://www.jisc.ac.uk/dner/>.
ESA (2005) The Entertainment Software Association. [online] Available from: <http://www.theesa.com/>.
Fullerton, T., Swain, C. and Hoffman, S. (2004) *Game Design Workshop*. Group west, San Francisco, USA.
Lee, A., Marshall-Ponting, A.J., Aouad, G., Wu, S., Koh, I., Fu, C., Cooper, R., Betts, M., Kagioglou, M. and Fischer, M. (2003) Developing a vision of nD-enabled construction, Construct IT Report, ISBN: 1-900491-92-3.

Nadim, W. (2005) Towards a virtual construction environment: designing fit for purpose schools using an nD based computer game. MSc unpublished dissertation, University of Salford.

Rage, R. (2005) Games Software Company. [online] Available from: <http://www. rage.co.uk/>.

UbiSoft (2005) UbiSoft: Games Software Company. [online] Available from: <http:// www.ubisoft.com/>.

Part 4

nD modelling
The future

This book has so far outlined the concept, scope and application of nD modelling. The final part of this book seeks to paint a vision of the future. It is anticipated that the chapters presented here will give readers a good overview of recent research, stimulate further research, as well as encourage those working in the construction industry to seriously think about applying nD modelling in their business environment.

Lee *et al.*'s chapter sets the scene for the future by illustrating 'where next' for nD modelling and research at Salford. A roadmap for future research is presented, which accumulates the findings from a number of national and international workshops involving leading researchers from both academe and industry – many of the authors of chapters in this book. The roadmap is further broken down into a research framework and technology framework, in an attempt to elucidate key action areas.

Hamilton *et al.*'s chapter extends and demonstrates how nD can be applied to a city/urban scale. The EU project IntelCities is used to describe how data can be comprehensively and coherently integrated. How the required data, in terms of building and geospatial information, can be obtained, integrated and extracted is detailed and presented using a scenario case. The concept of an urban nD model has the potential to support a wide range of regeneration services to a variety of stakeholders, and thus encroaches towards the utopia of a true 'intelligent city' of the future.

Continuing the theme of city/urban modelling, Kohlhaas and Mitchell's chapter illustrates the development and potential of creating computer-based city models. The need and current use of city modelling is described, before a chronology of how major city models such as Berlin, Stuttgart, Hamburg, Nottingham and Coburg have influenced their evolution. The potential and future drive for such an approach is highlighted with the introduction of IFC, BIM and IFG technologies. Application of nD city modelling at the University of New South Wales based on technologies developed in Singapore and Norway demonstrates how it can be incorporated in education. Finally, the potential benefits for local government of computer-based City Modelling based on open standards are canvassed.

Finally, but by no means least, Fukai presents the novel idea of how to present nD in 2D! He clearly and cleverly portrays how difficult and cumbersome it can be to read 2D and 3D architectural drawings, and uses a 'comical' way to guide the reader on how to understand nD in a 2D format. This approach is pertinent and timely as we move towards adopting nD modelling globally, and how complex and n-dimensional that it is, it needs to be simply laid out for clarity and understanding.

nD modelling

Where next?

Angela Lee, Song Wu and Ghassan Aouad

This chapter summarises the nD modelling roadmap, which was developed as part of the 3D to nD modelling project at the University of Salford (£0.443 million EPSRC Platform Grant funded). The roadmap draws its results from a number of research team workshops (academic), a Construct IT national workshop (industrial) and two international workshops (both industrial and academic) that took place over a period of 3 years:

- The first academic workshop sought to gain a consensus amongst the 20+ strong group of multi-disciplinary research teams of the boundaries of the vision; applied or blue-sky, short-or long-term industry implementation? An electronic voting tool was used to ascertain the consensus of the participants in their own grappling of what they perceived would be the nD model.
- The second academic workshop set about defining the need and scope of nD modelling using a case study exemplar. Research topics such as data quality/availability and decision-making mechanisms were identified.
- At the national workshop, the findings of the academic workshops were presented to enable a consensus with both industry and academia across the United Kingdon. Those who attended included contractors, clients, suppliers and architects, ranging from single persons to large multi-national organisations. The spectrum of participants was to reduce any inherent exclusion of industry players in the vision development, and to gain interest and acceptance of the work. The positioning of IT in the industry was established.
- The first international workshop, held at Mottram Hall February 2003, brought together world-leading experts (both from the industrial and academic communities) within the field of nD modelling (see Figure 20.1). There were 52 participants from 32 collaborating organisations, from 10 countries. The workshop developed a business process and IT vision for how integrated environments would allow future nD-enabled construction to be undertaken. A summary of all the four workshops is published in 'Developing a Vision of nD-Enabled Construction' (Lee *et al.*, 2003).
- The aim of the second international nD modelling workshop was to build upon the findings of the preceding events, concentrating on tackling the

Figure 20.1 First international nD modelling workshop participants.

strategic and operational issues confronted by the widespread application of nD-enabled construction, and defining the future research agenda. The event was held on the 13–14 September 2004. The workshop brought together 45 participants from 27 organisations and representing 12 countries: Australia, Canada, Chile, China, Denmark, Finland, The Netherlands, Norway, Malta, Qatar, Singapore and the United Kingdom.

The results of all five workshops are combined to produce a roadmap of the future. The wide range of participants from over 15 different countries has enabled the roadmap to be of a global outlook. This chapter presents the nD modelling roadmap for the future, which discursive itself into a:

- nD modelling framework
- and a nD modelling technology framework

nD modelling roadmap

The nD roadmap is illustrated in Figure 20.2. It highlights of research over the coming years to enable widespread nD implementation. The diagram helps

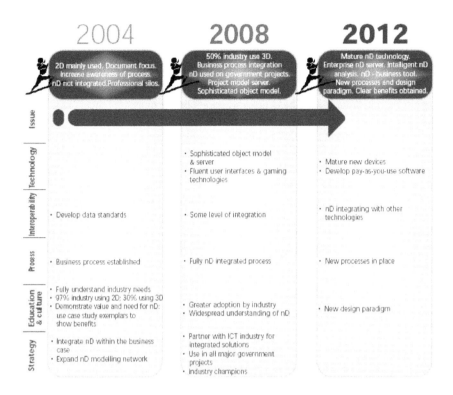

Figure 20.2 nD modelling roadmap.

illustrate and summarise the futures that were suggested during the workshops. The findings of the workshops were clustered into themes and are presented as repertory grids in the final roadmap. Whilst this report highlights the barriers and opportunities of implementing nD modelling in the construction industry, the roadmap presents a research agenda for global nD uptake.

The roadmap attempts to capture all the discussions that were generated during the workshops and shows that in order to reach the final nD destination landing place, notably the roadmap demonstrates that there are more obstacles that are human and organisational in nature than technological. Barriers of this kind were cited in all the workshops as most problematic in nature due to the unpredictability of the behaviour of groups and individuals, and the multiple effects that that behaviour can have. Whilst the development and uptake of technology is also unpredictable, the specific issues do not persist in the same way as for the social aspects. More specifically, whilst we now have increasingly more sophisticated ways of remaining in touch as technical barriers continue to be broken down – telephones, fax machines, mobile phones, video-conferencing, wireless laptops, multimedia tools – the social problems of their usage in the workplace for collaboration and decision making remain. Education/culture and performance measurement, and business case and process were established as being the biggest challenges.

Despite this though, there were a number of contradictions during the workshop discussions, and these are likely to result in slightly different pathways to, hopefully, the same goal. These contradictions had a more technical bias: some countries argued that lack of investment in technology is a problem, whilst others thought there was too much focus upon technology. Likewise, some argued for the strong belief in the need for change while, others stated that there was a lack of awareness of the importance of this. Whilst it was accepted that the barriers and enablers of nD are global and not regional, there are still important differences in the impacts that regional culture, politics and government can have.

In focussing upon the implementation problems and practical actions for change, it is hoped that the message of the requirement to understand and change the social and cultural underpinnings before technical implementation is driven home and a significant part of this involves making explicit what it is we need to know, for example establishing good practice exemplars. As the ideas and work related to nD modelling are developed in different contexts and on different scales, it may be that the practical benefits and added value of the concept becomes clear.

nD modelling research framework

In order to action the nD roadmap (Figure 20.2), the nD modelling research framework illustrates key focus areas. The nD modelling prototype so far (see Chapter 1) stands as a what-if analysis tool that enables the impact of various design perspectives to be highlighted. However, the roadmap and the development of the nD prototype identified a number issues that need to be accommodated, such as scalability, different actors/users and so on and should truly mimic the design process. Architects unconsciously think about buildings in several ways while they are designing (adpated from Lawson, 2004):

- a collection of spaces which may be indoors, outdoors or hybrids such as courtyards and atria;
- a collection of building elements such as walls, windows, doors and roofs;
- a collection of systems such as circulation, structure, skin and, service;
- a collection of voids and solids, as from an architectural perspective;
- a series of layers such as floor levels.

When designing, they oscillate without noticing between these descriptions of the building, thus adopting parallel lines of thought. In a similar way, if we are to fully adopt the nD concept we must first understand the human processes of

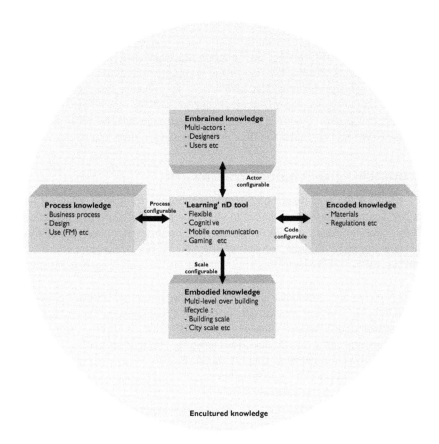

Figure 20.3 nD modelling research framework.

designing so that it can be mimicked in the technology that we develop. We must simultaneously look at (see Figure 20.3) the following:

- Embrained knowledge: encompassing the viewpoints of different stakehold-ers/users of nD such as the client, architect, access auditor and so on in terms of both feedforward and feedback of design information, thus being actor configurable.
- Process knowledge: so that it can harness and be harnessed within various operating schematics such as the business process, operation process and so on, thus being process configurable.
- Encoded knowledge: ensuring the design conforms to the respective design standards, thus being code configurable.

- Embodied knowledge: enabling the scalability of use, covering a single building to city and urban use, thus being scale configurable.

Framed in these terms, this curious relation and messy mapping between problems and solutions in design is one of the reasons why architecture is not only challenging for the designer but also often so impenetrable for the client. Moreover, in doing so, the nD tool should be self-learning. The embrained, process-encoded and embodied knowledge can each in turn be self-improving, more intuitive and pro-active and thus, formulates encultured knowledge.

nD modelling technology framework

As a result of the five workshops, a new technology framework (Figures 20.4 and 20.5) emerged to support the nD modelling roadmap (Figure 20.2). The technology framework that provides the architecture for different domain applications which are directed by business process is based on a service-oriented platform supported by various technologies such as visualisation, decision support and analysis:

- The technology framework is based on the service oriented architecture (SOA). The participants of the workshop recommended that a common framework and interface should be provided so that applications from different domains can work together. Furthermore, common technologies such as visualisation and decision support should be part of the platform. SOA is a technology which can be deployed to support this platform. According to the World Wide Web Consortium (W3C), SOA is 'a set of components which can be invoked, and whose interface descriptions can be published and discovered'. The nD modelling framework should provide the common technology components and data access as services for each domain application.

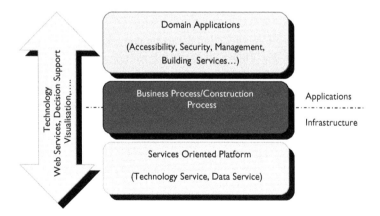

Figure 20.4 The overview of the technology framework.

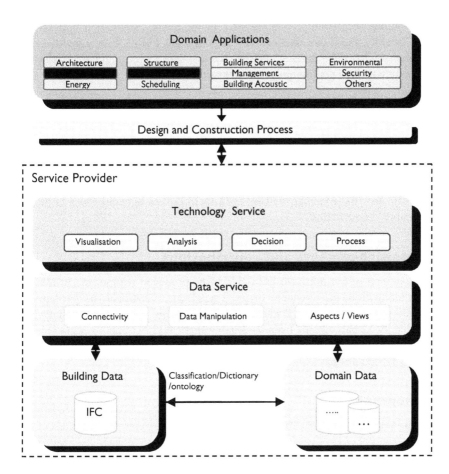

Figure 20.5 nD modelling technology framework.

- It was also recommended that the framework needs to be driven by the business process. Different applications will be used at different stages of the process and the data requirements will be different. The process control mechanism has to be in place to ensure that the right data can be served to the right application.
- The services include technology service and data service. The technology service provides common technologies, such as visualising the building data, multi-aspects decision support, data analysis for thermal, structure and so on and process-control mechanism. The data service provides the data access to two types of data sources, building data and domain data. Building data can be described as data definition and representation of the building model and it has to be supported by interoperable data standards such as IFCs.

Domain data is specific data related to each domain such as regulations for building accessibility, weather data is for energy simulation and so on. Currently, these two data sources are often not linked. It is suggested that research has to be done to integrate the two data sources through developing common concepts (ontology/classification/dictionary).

Summary

This chapter summarises the progress of the 3D to nD modelling project and cultivates the findings of five workshops to form an nD roadmap of the future. The roadmap has contributed to the ongoing debate on the value of nD modelling and as such, presents a research and a technology framework. The authors anticipate that the roadmap will be used to formulate agendas towards global nD-enabled construction.

Acknowledgements and contributions

The roadmap would not have been able to be produced without the invaluable contribution from the attendees of the 2 academic, 1 national (64 participants from 39 organisations) and 2 international workshops. (The first workshop was attended by 52 participants from 28 organisations, representing 10 countries; the second workshop was attended by 45 participants from 27 organisations, representing 12 countries.)

Bibliography

Adachi, Y. (2002) Overview of IFC Model Server Framework. In proceedings of the International ECPPM conference – eWork and eBusiness in Architecture, Engineering and Construction, Portoroz, Slovenia.

Alshawi, M. and Ingirige, B. (2003) Web-enabled project management: an emerging paradigm in construction. *Automation in Construction*, 12, 349–364.

Alshawi, M., Faraj, I., Aouad, G., Child, T. and Underwood, J. (1999) An IFC Web-based Collaborative Construction Computer Environment. Invited paper, in proceedings of the International Construction IT conference, Construction Industry Development Board, Malaysia, 8–33.

Aouad, G., Sun, M. and Faraj, I. (2002) Automatic generation of data representations for construction application. *Construction Innovation*, 2, 151–165.

Durst, R. and Kabel, K. (2001) Cross-functional teams in a concurrent engineering environment – principles, model, and methods, in M.M. Beyerlein, D.A. Johnson and S.T. Beyerlein (eds), *Virtual Teams*, JAI, Oxford, pp. 167–214.

Eastman, C. (1999) *Building Product Models: Computer Environments Support Design and Construction*, CRC Press LLC, Florida.

Graphisoft (2003) The Graphisoft Virtual Building: Bridging the Building Information Model from Concept into Reality. Graphisoft Whitepaper.

Halfawy, M. and Froese, T. (2002) Modeling and Implementation of Smart AEC Objects: An IFC Perspective, CIB W78 Conference 2002 – Distributing Knowledge in Building, Denmark, 2002, 45–52.

Karola, A., Lahtela, H., Hanninen, R., Hitchcock, R., Chen, Q., Dajka, S. and Hagstrom, K. (2002) BSPro COM-Server – interoperability between software tools using industry foundation classes, *Energy and Building*, 34, 901–907.

Kunz, J., Fischer, M., Haymaker, J. and Levitt, R. (2002) Integrated and Automated Project Processes in Civil Engineering: Experiences of the Centre for Integrated facility Engineering at Stanford University, Computing in Civil Engineering Proceedings, ASCE, Reston, VA, 96–105, January 2002.

Lawson, B. (2004) Oracle or draftsman. In proceedings of the International Conference on Construction Information Technology (INCITE) conference, Lankawrim Malaysia.

Lee, A., Marshall-Ponting, A.J., Aouad, G., Wu, S., Koh, I., Fu, C., Cooper, R., Betts, M., Kagioglou, M. and Fischer, M. (2003) Developing a vision of nD-enabled construction, Construct IT Report, ISBN: 1–900491–92–3.

Leibich, L., Wix, J., Forester, J. and Qi, Z. (2002) Speeding-up the building plan approval – The Singapore e-Plan checking project offers automatic plan checking based on IFC, The international conference of ECPPM 2002 – eWork and eBusiness in Architecture, Engineering and Construction, Portoroz, Slovenia, 2002.

Stahl, J., Luczak, H., Langen, R., Weck, M., Klonaris, P. and Pfeifer, T. (1997) Concurrent engineering of work and production systems, *European Journal of Operations Research*, 100, 379–398.

Chapter 21

Integration of building and urban nD data to support intelligent cities

Andy Hamilton, Hongxia Wang, Joseph H.M. Tah,
Ali Tanyer, Steve Curwell, Amanda Jane
Marshall-Ponting, Yonghui Song and Yusuf Arayici

Introduction

Cities are dynamic living organisms that are constantly evolving. Thus city planning has always been difficult. Today, our rapidly changing society makes the job of predicting future needs of city dwellers, and those who depend on the services cities provide, even more problematic. Particular problems include: transport, pollution, crime, conservation and economic regeneration. Thus in addressing the complex problems of city planning it is not sufficient just to be concerned with the physical structure of the city; the interplay of intangible economic, social and environmental factors needs to be considered as well (Gilbert, Stevenson *et al.*, 1996).

Those involved in the sophisticated art of city planning use a variety of tools. In particular, several forms of IT based models (graphical, analytical, hybrid) of cities have been in use for a number of years (Hamilton and Davies, 1998). Recently it has become clear that in order to provide a more efficient city management and administration, it is important to integrate information from city authorities, utility and transport system providers and so on (Wang and Hanilton, 2004a; Hamilton, Wang *et al.*, 2005).

Data integration based on nD models has significant potential for use in mapping the complex multi-dimensional nature of urban information. nD modelling research is already well established in the construction industry and recently this has been followed by increasing interest in urban issues. The concept of an nD urban information model was discussed in Wang and Hamilton (2004b), Hamilton, Wang *et al.* (2005). The nD urban information model can integrate multi-dimensional urban aspects, such as economic and environmental aspect, with 3D urban models plus the temporal dimension. An nD urban information model can provide comprehensive information support to a range of city services such as transport and urban planning.

This chapter discusses how innovative approaches to data integration, based on nD modelling and other techniques, played an essential role in the development of a conceptual model for Building Data Integration which was developed for use in the urban context. This development took place as part of the Intelligent Cities (IntelCities) Project. IntelCities was an integrated project funded (6.8 million Euro, Project No: 507860) by the EU Information Society technologies programme under framework 6. It ran from January 2004 until October 2005. It aims to help achieve the EU policy goal of the Knowledge Society in cities by 2010 through new forms of electronic governance of cities and greater social inclusion through enhanced access to services by citizens and businesses (please refer to http://www.intelcitiesproject.com/ for further details).

The objective of the IntelCities project is to integrate citywide information to support integrated services accessible to all stakeholders, thus forming intelligent cities. As shown in Figure 21.1, the concept of the intelligent city is implemented through the development of an e-City Platform (e-CP) which is being achieved through a number of Work Packages (WPs) and tested across Europe as part of the research project.

The e-CP supports new services for e-Governance as well as the integration of legacy systems. It has been developed across a set of European pilot cities, which are as follows: Marseille, Siena, Helsinki, Rome, Leicester, Dresden and Manchester. The e-service tasks are respectively implemented in WP1-WP5 and corresponding cities. These work packages are as follows: WP1 e-Administration, WP2a e-Inclusion (wired), WP2b e-Participation (wireless), WP3 e-Mobility and Transport, WP4 e-Land Use Information and WP 5 e-Regeneration. The core e-CP with service framework to integrate all the services was mainly developed in WP6 (Integration and Interoperability) and WP7 (Simulation and Visualisation).

This paper focuses on one task of WP5 e-Regeneration (Hamilton, Burns *et al.*, 2005). A key aspect of WP5 is the development of systems for specific use in urban regeneration and their integration with other city systems to get the

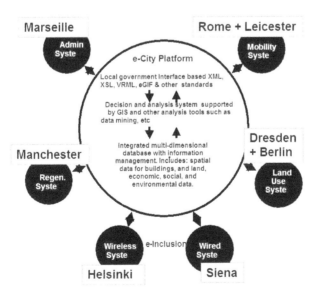

Figure 21.1   e-City platform and e-Services architecture of IntelCities.

benefits of working in a holistic environment. The Building Data Integration System is one of the two main tasks in this work package. It aims to illustrate how digitised historic building data can be integrated with other types of city data, using nD modelling, to support integrated intelligent city systems. This has been related with the nD modelling database developed in WP7 of IntelCities. The three main tasks of the Building Data Integration System are as follows:

- to capture building data using 3D laser scanning system;
- to integrate digitised building data with geospatial information and other types of city data;
- to build a conceptual system to illustrate how nD modelling supports intelligent city systems.

In order to illustrate the objective of this research work and give a sense of purpose to the system developed based on the building project database, the Gaudi Bank scenario was proposed.

A situation is imagined where the Gaudi Bank in Barcelona, with branches in the United Kingdom, wants to set up a new training centre in Manchester. Gaudi Bank has located a building in Manchester, Jactin House, as a potential base. However, they have to make a number of checks and assessments about the building and its location before making any detailed plans. Furthermore, once a suitable building is selected, a plan for its refurbishment needs to be developed. Gaudi employs property developers to select and purchase buildings and construction professionals, such as architects and quantity surveyors, to carry out the refurbishment work.

In order to simplify the situation, the conceptual Building Data Integration System only considers the design phase of the regeneration process. The design phase is the initial stage of project. It is necessary to identify the actors involved in the design process. The main actors involved were identified as

- property developers/finance sector and
- professionals (architects, planners, construction managers, etc.).

The main duty of the property developer is to find a building in a suitable location that is appropriate for the intended use of the building. Thus they are concerned with issues such as the following: transport, pollution, building status and open space around the building. The construction professionals need to ensure the selected building can be converted to a training centre within the budget. They need to assess the construction work needed to convert Jactin House into a training centre. Thus they are concerned with the physical structure of the building, the amount of space in it, its general condition and schedules of building elements, such as windows and doors.

The scenario described earlier set a challenge to the IntelCities partners. The authors of the paper all worked on the development of the system for the lead

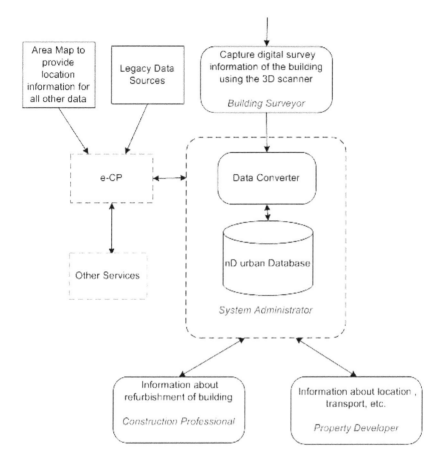

Figure 21.2 e-City Platform and Building Data Integration system.

partner, the University of Salford. However it was very much a co-operative effort and without the essential work of other partners, the system would not have been the same. Details of the partners involved are given in the acknowledgements section at the end of this chapter.

Figure 21.2 shows the basic structure of the system. It should be noted that the current conceptual model is only integrated with the e-CP at the data level in that common data structures have been used. Service level integration has not been developed in the conceptual system. This is discussed later in this chapter.

In order to support the property development and building refurbishment application, it was fundamental to have detailed and integrated comprehensive information as well as the tools and technologies that allow an effective manipulation and operation of the integrated data. The following section details the data needed and collecting the required information. The next section focuses on the integration of geospatial information, building information and other kinds of city information. The development of a conceptual prototype system is then presented, and how the Building Data Integration System and nD modelling can support an intelligent city is discussed. A summary and assessment of the research work is given at the end of this chapter.

Data requirement and acquisition

Data requirement

The evaluation and decision-making process of urban design and planning needs to consider a wide range of information. An integrated approach is necessary to link all the information gathered for this process. In an IntelCities workshop to determine the design of the conceptual Building Data Integration System, using the expertise of the Bilbao-based research and consultancy institute Labein who specialise in information for re-use of buildings, we developed the following list of information needed in pre-construction stages of building re-use projects:

- availability, multiple ownership;
- demographic, economic and social data;
- local plan (use, potential changes of uses, designated use for the site/building, development briefs and so on);
- transport, communication, environmental impact assessment;
- structural survey of building (areas, volumes, diagnosis and so on).

For the Gaudi Bank scenario described previously, the property developers need to assess general building information and surrounding environment information, such as transport information, pollution, parking facility, safety and open spaces around the building, in order to select an appropriate location and property. The architects need to consider the building structure, material and damage to the building to assess the suitability of the building from a construction point of the view. The bank directorate will balance and assess the criteria of location, the condition of building and budget for purchasing and refurbishment. Therefore, this scenario actually comprises two aspects: property development and building refurbishment.

The information required includes not only data about the building itself but also about the surrounding environment. Here the required datasets are cate-gorised into two groups according to the general semantic consideration of the data sources:

- Spatial data mainly describes the physical structure of the urban environment (3D city model). It includes geometrical and topological information like Geospatial Information (GI), CAD model of buildings and so on.

- Non-spatial data describes different themes like transport, environmental, crime and so on. These kinds of data can be census data like population or survey data like traffic volume on a road.

Data acquisition

One of the most time-consuming and difficult tasks is data acquisition in order to support a particular project or development (Nedovic, 2000). It is important to investigate what data is available and how to obtain the required data.

For the Building Data Integration System, it is impractical to collect all data from scratch. The first task is to find available data sources and relevant data holders. The data holders can be at every level of an authority according to the administrative differences. They also can be commercial data suppliers. Some main data holders are identified as government organisations, commercial data suppliers, local agencies, survey organisations, private organisations and groups, and other public bodies. The following table (Table 21.1) lists some of the related data sources in the United Kingdom.

In the conceptual Building Data Integration System, the surrounding building information like transport, safety, pollution and so on come from the local authority or other public or private organisations. Geospatial information comes from Ordnance Survey's Land-Line datasets (http://www.ordnancesurvey.co.uk/oswebsite/products/landline/). Ordnance Survey Land-Line is a comprehensive dataset, depicting man-made and natural features ranging from houses, factories, roads and rivers to marshland and administrative boundaries. There are almost 229,000 Land-Line tiles covering Great Britain, surveyed and digitised at three different scales according to location.

However, it is often the case that not all the required data is available. For our scenario, there were no available sources for the required building structural information. Effective data acquisition methods to capture building information were deployed. Survey and remote sensing have been the two most usual methods for primary data acquisition (Laurini, 2001). Due to the resolution requirement and performance-cost consideration, a new technique, 3D laser scanning (ground-based LiDAR), was employed to capture the building survey information for this research and development. How to use 3D laser scanner technology to capture building information and make the information usable is discussed later in this chapter.

For the conceptual system, only a subset of the location information was developed and installed. The purpose was to demonstrate the potential of the system not to build a fully working system. However data acquisition issues have to be considered when fully developed systems providing a business service is developed in this area. Data acquisition is difficult; there are often administrative and political problems in acquiring data. The e-CP, when adopted, would greatly alleviate these problems. A fully developed Building Data Integration System would rely on the e-CP, and is an example of the benefits of the e-CP.

Table 21.1 Related data sources (UK)

Types	Sources	Owner	Formats
Land and property	National or local land and property gazetteer (http://www.nlpg.org.uk/)	Local authorities	BS7666 (British Standard for Spatial datasets for geo-referencing)
Digital map	Ordnance Survey (OS) (http://www.ordnancesurvey.co.uk)	Ordnance survey	OS MasterMap (GML), OS Land Line (e.g. Shp)
Terrain data	Ordnance Survey (http://www.ordnancesurvey.co.uk)	Ordnance survey	Grid (raster data format)
UK Geospatial metadata	UK Standard Geographic Base (UKSGB) aims to provide access to commonly used geographical units in the United Kingdom. (http://www.gigateway.org.uk/)	Government	UK Geospatial Metadata Interoperability Initiative (GEMINI)
Development plan	Local governments	Local governments	Document
Census Data	http://www.statistics.gov.uk/census2001/default.asp	Government	Database

Notes
These data sources are still limited to one country due to political, economic and technical reasons. EU-scope datasets are not well standardised and unified at present although there are current initiatives such as INSPIRE.

Transforming the LiDAR building model into the IFC building model

The spatial data can be obtained by using 3D scanner. By post-processing the captured spatial data, outputs for different purposes can be obtained, such as CAD modelling, which can be converted to IFC model. The use of this technology allows the physical structure of the building to be represented in an nD database.

For the efficient creation of the database, it is useful if building elements (windows, doors, etc.) can be automatically and directly inputted to the appropriate part of the data model. The IntelCities research to date on object recognition from laser scan data is related to the pattern-matching approach. Pattern matching works by searching and matching strings and more complicated patterns such as trees, regular expressions, graphs, point sets and arrays (Arayici *et al.*, 2005). Using this approach capture of building elements is mostly automated. Figure 21.3 illustrates outlined building elements.

Data integration

Generally speaking, data integration is the problem of combining data residing at different sources, and providing the user with a unified view of these datasets (Lenzerini, 2002). Various datasets are useable by actors at the appropriate stage of the property development and refurbishment process. It is important to integrate all the required information in order to significantly improve the quality and efficiency of work.

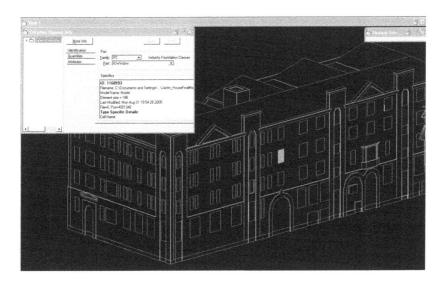

Figure 21.3 IFC building model of Jactin House.

Comprehensive spatial information and 3D modelling can greatly enhance the awareness and understanding of development plans and options. For the Gaudi Bank scenario, spatial information like building information (CAD) and Geospatial Information System (GIS) are important for all the stakeholders. Current spatial data integration is problematic due to the heterogeneity of these data types and data formats. The discussion will focus on the spatial data integration, that is building information and geospatial information integration.

Methods of data integration

As mentioned earlier, this research work aims to integrate building data and other types of city data to support city systems. In order to implement this integration, the central problem is to choose a suitable method. There are many issues on data integration, such as database-based integration, application-level integration and data standardisation (Taylor, 2003). These issues are discussed in the following sections.

Application-level integration

The application-level integration method is to develop an application system to support data exchange among different applications. For example, GIS application can import CAD data format or directly connect CAD layer in its application. However, application-level integration could not solve building and geospatial information integration problem because this method relies on either CAD or GIS application but cannot be used to tightly integrate the whole range of data.

Data standardisation

Data standardisation is a process of normalising data for easier information representation and exchange between different systems (Taylor, 2003). Many international and industrial standards have been developed or are under development. For example, the Open Geospatial Consortium (OGC)'s Geographical Mark-up Language (GML) and the International Alliance for Interoperability (IAI)'s Industry Foundation Classes (IFC) have been widely recognised as the GIS and building information standard respectively. City Geographical Mark-up Language (CityGML) and IFC for GIS (IFG) are two standards under development for urban environment:

- IFC (ISO/PAS 16739) is a comprehensive data representation of the building model. IFC is developed by IAI and is a universal model to be a basis for collaborative work in the building industry and consequently to improve communication, productivity, delivery time, cost and quality throughout the design, construction, operation and maintenance life cycle of buildings.
- GML (http://www.opengeospatial.org/) is an XML-based schema for the modelling, transport and storage of geospatial information. It was developed

as a data exchange standards interface by OGC to achieve data interoperability and reduce costly geographic data conversions between different systems.

- CityGML has been developed in the ongoing work of the Special Interest Group SIG3D of the Initiative for a Geodata Infrastructure in Northrhine-Westphalia (Germany) (http://www.gdi-nrw.de). It is a GML application profile for the generation of cities, the geometry and topology is based on ISO 19107.

- IFG project (http://www.iai.no/ifg/) is trying to provide a demonstration of the concept of using the IFC model as the specification for the exchange of limited but meaningful information between GIS and CAD systems and vice versa.

Data standardisation aims to provide single unified data standards for both CAD and GIS. However, for a holistic urban model, there is no appropriate data standard so far despite the ongoing efforts of IFG, CityGML. Thus it is not possible to build systems that are both interoperable and that conform to international standards at this level.

Database-based integration

There are two common database-integration approaches, lazy and eager approaches, to integrate data sources (Tatbul, Karpenko *et al.*, 2001). For the lazy approach, the data is accessed from the sources on-demand when a user submits a query. Federated databases and wrapper/mediator systems belong to this approach. For the eager approach, some filtered information from data sources is pre-stored in a repository and can be queried later by users. Data warehouse is a typical example of eager-approach integration.

For the Building Data Integration System, the eager database integration method was employed. A centralised database is used to tightly integrate the required information and provide the support of data accessing and manipulation. The integration can be enhanced with sophisticated, customisable data conversion. Tightly integrating all the information required for property development and refurbishment could significantly improve the quality and efficiency of work carried out. Every user only needs to connect to this centralised database interface, and then is able to share the rich information. For example, construction professionals could work more flexibly because they can access building structural information and also transport and other related information as well. Furthermore, a unified database could provide a consistent interface of service provision.

Designing the data model for nD modelling database

Building Data Integration System is reliant on the e-CP, and therefore is needed to be integrated with WP7 of IntelCities where a conceptual urban model is

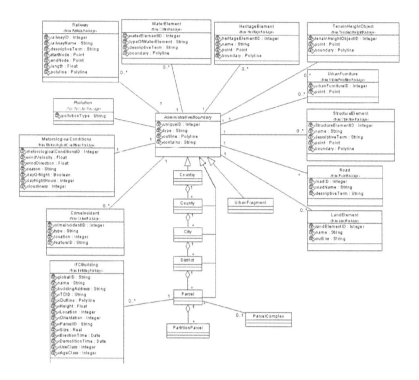

Figure 21.4 Conceptual urban data model (Tanyer, 2005).

established in order to integrate multiple sources of urban data, as shown in Figure 21.4 (Tanyer *et al.*, 2005). Based on Gaudi Bank scenario's requirement, the model is extended and developed into the project database to support property development and building refurbishment.

The relational model is chosen for this data model due to its popularity with all actors and its mature technologies (SQL and main stream database products). It is adopted by relevant cities in this project. In order to design the relational data model, the spatial relation between geospatial information and other kinds of useful information is used as the main linkage data to implement the data integration.

The important element of the relational data model is the linkage data. A common geospatial location is used as such a linkage that can bring various datasets together. In geographical science, a location is a position on the Earth's surface or in geographical space definable by coordinates or some other referencing systems, such as a street address or space indexing system (Clarke, 1999). Specifically, a location can be any geo-referenced information such as address, x/y coordinate, postcode, place name, route location or other. It can be represented by some simple geometry like a point, line or area.

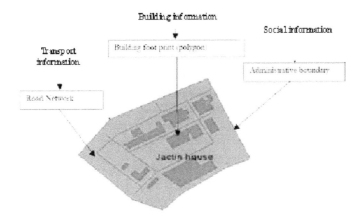

Figure 21.5 Location-based data linkage to form geo-referenced database.

Location-based mapping has been used to integrate various data sources as shown in Figure 21.5. IFC building information is linked with building footprint location information. Similarly, transport information is mapped to road network in geospatial information. Census information is mapped to administrative boundary.

Spatial information is described by geometry models like point, line and area. Non-spatial information includes attribute information and various thematic information like transport, crime datasets and so on. Every theme needs to be depicted by some entities. The attribute information of the entities like identification and description can be organised into a relational data model.

Non-spatial information has a close relationship with spatial information. For example, bus entity has attributes like bus stop locations and bus routes that are described by spatial information, while some binary document and multimedia data like the photos of the building are organised by file-based data structure. The hyperlinks between the attributes and these binary files were setup in relevant entities.

For the Building Data Integration System, nD model database is used as a platform. On the other hand, Building Data Integration system could contribute to the design and contents of this nD urban model database. The database is extended to describe the required information for the Gaudi Bank scenario. The property developer needs to know the ownership, availability and development permit for the building, so these attributes are added to the building model. Considering surrounding information, public transport data model is designed. For example, bus information is organised into the following entities:

- bus (BusID, Number, Serving time, Bus route);
- busstop (Name, StopID, RoadID, Location);

- busandstop (BusID, StopID);
- roadandstop (RoadID, StopID);
- road (RoadID, Road name Road line, Roads type, Traffic density).

Integrating building information

Comprehensive 3D models of buildings are important for a variety of tasks including simulation, urban planning and disaster recovery. Building modelling in the construction industry has experienced several changes. How to employ building models developed for the construction industry in the urban domain is a problem that we need to investigate.

Building model and IFC

As discussed in Chapter 1 of this book, the CAD Building model has experienced the shift from 2D drafting to 3D Building Information Model (BIM) in the construction industry. Besides realistic representation, 3D BIM employs object-oriented concept to describe building objects with attributes and association (Cyon, 2003). The recent research work on 4D modelling (Alyazjee, 2002) and nD modelling (Lee *et al.*, 2003) will result in efficient project process and management.

CAD-based building models from different vendors use their own data formats such as DWG from Autodesk, DGN from MicroStation, Active-CGM from Intergraph and so on. DXF is probably one of the most widely supported vector formats in the world today. However DXF suffers as a cross-system standard through having been designed purely for AutoCAD system (Harrod, 2004).

IFC is a comprehensive data representation of the building model. The IFC object descriptions deal with not only full 3D geometry but also relationships, process, material, cost and other behaviour data. Integrating a CAD model with IFC enables the accurate geometric representation to be integrated with structural and behaviour elements and facilitates linking with external applications (Ding *et al.*, 2003).

Extraction of building information

Buildings are one of the most important elements of the urban built environment. In a physical sense, a city is a collection of buildings. Currently there are many methods to describe 3D building models in the urban environment. Digitised building models have been employed to improve public awareness and visual impact. Five popular 3D modelling methods are identified as CAD modelling, GIS-based block models, LiDAR modelling, Image-based modelling and Panorama photographs model method (Wang and Hamilton, 2004a). Among these, the emerging LiDAR technology is becoming the easy and effective building modelling method especially for historical and cultural heritage.

This chapter investigated earlier the transformation of LiDAR data into IFC models. As discussed, IFC does not only preserve the full geometric description in 3D, but also its properties and characteristics. However, some IFC elements of facility management, architecture and construction are too detailed and complex to be employed wholly in urban application, for example furniture, construction equipment and so on. It is necessary to simplify the IFC of construction industry for urban application and extract the structural information of building.

In the Gaudi Bank scenario, building structural information is required by architects to assess and finish the refurbishing task. The structural information contains information about the geometry and attributes data of the building used in the IFC model. It includes the building elements like the walls, the roof, the floors and the windows. Extracting the structural information from an IFC model is the core and also the difficult problem for data processing. There are two methods to implement this information extraction.

The first method, which we used in our Build Data integration System, is using some commercially available tools to help extract information from the IFC model. This method has been tested in WP7 using Express Data Manager (EDM). EDM is an object-oriented database which supports modelling, application development and database management (EPM, 2002). It is used as a tool to transfer the IFC building in three stages. First loading an IFC building model to an EDM database, then extracting the required structural information from the database based on the EDM application interface and finally, mapping the extracted information into a user-defined data model.

Another potential method to implement the building information extraction is using XML technology. IFC is defined initially by EXPRESS. With the development of XML technology, some XML-based IFC schemas are developed like IFCXML, BLIS XML. This solution implements the data extraction from IFC in three steps: First converting the CAD IFC objects into XML instances based on a XML schema like IFCXML, then using XML parser and IFCXML schema driven programs to extract the required information from IFCXML model and finally, mapping the extracted information into user-defined data model.

Due to the flexibility and popularity of XML, the second method looks very promising. It does not rely on commercial software. The method can be extended into the conversion of other XML-based information like GML. Future system would be developed into common converter to convert XML-based information.

Integrating geospatial information

Geographical information is the core of information integration as geographical information provides the structure for collecting, processing, storing and aggregating various datasets. Research shows that 80 per cent of municipal information is geo-referenced (Curwell and Hamilton, 2002). Indeed, geospatial information is often the only factor different datasets have in common.

Geospatial information and GML

Historically, geographic information has been treated in terms of two fundamental types called vector data and raster data (Longley *et al.*, 2001).

'Raster data' deals with real-world phenomena that vary continuously over space. It contains a set of values, each associated with one of the elements in a regular array of points or cells. The raster data model uses the techniques of imagery, photos, grids and so on, to present spatial information. An aerial photograph stored in a computer file is a good example to show this method of presenting spatial information, where these kinds of images look more realistic than an ordinary map. Two raster data format examples are Grid and LAS. A Grid is a rectangular mesh of square cells. LAS file format is a public file format for the interchange of LIDAR data between vendors and customers (Campos-Marquetti, 2003).

'Vector data' deals with discrete phenomena, each of which is conceived of as a feature. The spatial characteristics of a discrete real world phenomenon are represented by a set of one or more geometric primitives (points, curves, surfaces or solids). Two popular vector data formats are SHP file and Triangulated irregular network (TIN). The shape file (.shp) format by ESRI is widely used in different GIS applications though it has some major drawbacks against other database-oriented data repositories used in GIS. And it is popularly adopted for data exchange between different systems. A TIN is a data structure that defines geographic space as a set of contiguous, non-overlapping triangles, which vary in size and angular proportion. Like grids, TINs are used to represent surfaces such as elevation, and can be created directly from files of sample points.

GML, as a geospatial data exchange standard, is based on the abstract model of geography developed by the OGC. This describes the world in terms of geographic entities called features. Essentially a feature is nothing more than a list of properties and geometries. Properties have the usual name, type, value description. Geometries are composed of basic geometry building blocks such as points, lines, curves, surfaces and polygons.

Geospatial information extraction

Building and urban-related planning and design processes are geographic-related processes. Thus the geo-spatial information-based technologies are used by most of the research consultants in this field, as well as by planning professionals. In the Gaudi Bank scenario, building, road and land parcel layer of the geospatial information are required.

The Building Data Integration System currently extracts the information from .shp files. The data converter has been developed to abstract geospatial data through a GIS application server (in the current system, GIS software from the company 'ESRI' and .shp file geospatial data are used), meanwhile, receiving tabular attribute data from the relational DB. The converter then uses ODBC to link to the nD modelling database and uses SQL to store the geospatial data into the database.

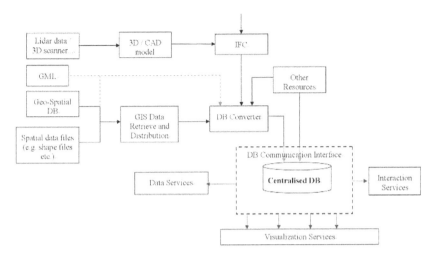

Figure 21.6 Structure of the data integration system.

Another method is to use an XML parser and GML schema driven program to extract the required information from GML-based geospatial information like Ordnance Survey's Master Map. This process of data extraction from GML are as follows: using XML parser and GML schema driven program to extract the required information from GML data source and then mapping the extracted information into user-defined data model. This method has the potential to support the geospatial information extraction in an open and extensible way.

An overall data integration system structure is shown in Figure 21.6. The centralised nD database gathers various data sources and data formats through a DB Converter or a series of converters. Currently, geospatial data (e.g. .shp file) and IFC data can be converted and saved in the centralised database. The conversion of GML has not been implemented yet. Many common application services can be built upon this nD database to provide support to applications like property development and building refurbishment of Gaudi Bank scenario.

Conceptual prototype system development

The nD modelling database was designed for various stakeholders to contribute to and share integrated data sources. The conceptual prototype system is developed based on this nD modelling database to support the property development and refurbishment for Gaudi Bank scenario. The system function is defined as follows:

- For property developers:
    - Search for a suitable building in an appropriate price and size range and view the building.

–   Invoke the map of the area where the building is located.
–   Conduct queries about the area (e.g. transport facilities) using the map for urban scale inspection.

• For construction professionals:

–   Create a dimensionally accurate 3D model of the building from which construction information can be abstracted.
–   Retrieve information about building elements such as doors and windows.

In order to support the system functions, the system administrator (see Figure 21.2) needs to select the relevant data from sources and set up the project database. The administer will also set up access to the system. With the initialising of the database, all system users and professionals would be able to query, represent the data and do maintenance thereafter with regard to their different planning/ construction tasks. For example, the system administrator can load the GIS data of the Jactin House area, and put it into the database. Also they can convert the IFC building model into the database and map them with the building ID and co-ordinates.

The system interface is shown in Figure 21.7. Property developers need to know general building conditions. Travel information, like bus, train and accommodation, is important in the application case because it is very important to make sure people can easily get to the place and find a suitable place to stay.

After the property developer has selected a suitable building on account of size, transport, hotels and so on, the construction professionals can then connect to the database and preview the area map. If he clicks on Jactin house, he can query the information about the building and generate a building elements survey report and the illustration of the windows and doors as shown in Figures 21.8 and 21.9.

It must be clear that this is a conceptual system development. The system can only support the initial design stage of the property development and building

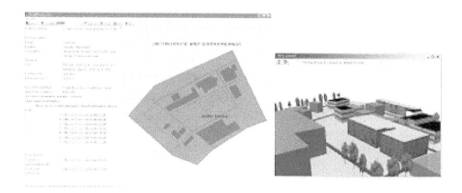

Figure 21.7 Application interface of property development system.

Scanning-Survey Report of Jactin House Elements - Windows and doors

Generated from Centralized nD Database

There are together 199 windows, in 4 sides of Jactin House, located in 5 floors and in different size

- **In each elevation of the Jactin House, the number of windows are:**
  - Front elevation window-number: 82
  - Left elevation window-number: 40
  - Right elevation window-number: 22
  - Rear elevation window-number: 55
- **The 199 windows are located in 5 floors (from the ground to the 4th floor)**
  - The ground floor window-number: 27
  - The 1st floor window-number: 47
  - The 2nd floor window-number: 56
  - The 3rd floor window-number: 43
  - The 4th floor window-number: 26
- **There are 6 different types among all these windows**
  - 0) There are 34 windows of type 1: 1.7 (w) x 2.5 (h)
  - 1) There are 41 windows of type 2: 1.0 (w) x 2.5 (h)
  - 2) There are 29 windows of type 3: 0.5 (w) x 1.7 (h)
  - 3) There are 33 windows of type 4: 0.60 (w) x 1.3 (h), with a curved shape on top (radius: 0.3)
  - 4) There are 31 windows of type 5: 2.0 (w) x 1.8 (h)
  - 5) There are 31 windows of type 6: 1.35 (w) x 1.6 (h)
- **Windows size illustration:**
  - http://127.0.0.1/jhmodel/Win&Doors/Windows.gif

There are together 5 doors, in the front side and left side of Jactin House, in different size

- 4 of the doors are at the front of the house
- 1 of the doors is at the left of the house
- **Doors size illustration:**
  - http://127.0.0.1/jhmodel/Win&Doors/Doors.gif

From the survey, it is suggested that all windows and doors need to be replaced.

A VRML model is generated following this report to show just the windows and doors around Jactin house

http://127.0.0.1/jhmodel/Win&Doors/Jactin_Housewindows.wrl

Figure 21.8 Digital survey report.

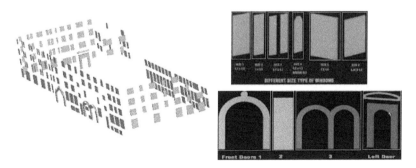

Figure 21.9 Jactin house building elements (left) windows/doors illustration (right).

refurbishment. Some data processing is not automated and does not cover all the required information yet. Also a good 'service-level' interface needs to be developed. However, the purpose of 'proof of concept' has been achieved.

Building Data Integration System and intelligent city

Currently the conceptual Building Data Integration System produced is not interoperable with the e-CP at the service level. However all the functionality in the form of Data Converters and so on is provided to create service-level integration. Only a service-level interface and a standard 'wrapper' for the system are needed.

The Building Data Integration System has been integrated with e-CP in two ways: first, building data integration system can provide data sources for e-CP service and also can get city datasets from e-CP services. Second, it has the potential to be developed into a business service with the capability of significantly improved work. This is shown in Figure 21.10.

The e-CP is built layer after layer, to ensure full operation of the Service Oriented Architecture (SOA) framework and a continuous development of services in this framework. The e-CP follows a multi-technology approach, able to deal with the different strategic and technological approaches of the various cities that could be part of some overall distributed collaborative process. (Please refer to WP6 of IntelCities for further information.)

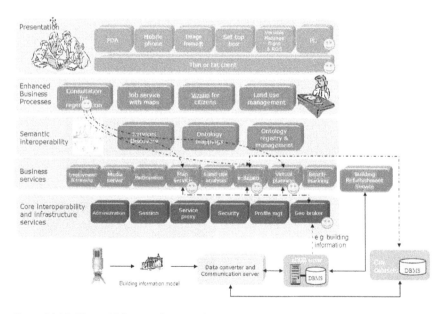

Figure 21.10 The e-CP Service Oriented Architecture and Building Data Integration System.

For the stakeholders from the average citizen to professionals working in the urban area a future when business services are underpinned by well-integrated databases and linked into the e-CP like systems provides many advantages. It will be possible to relate one city service to another to provide genuine 'joined up government'. For instance a property developer concerned about the location of Gaudi Bank training centre may want to look at library provisions for trainees. Information about library services, both online and through physical buildings could be accessed through the building data information system and it should be possible to show the location of the libraries on the area map.

For people developing business services, standardised connection to other services will greatly decrease development costs. Also when data is shared the continuous regeneration of data will be reduced. It will be in the interest of all service providers to produce high-quality data that other service providers will want to re-use and pay for. Whilst the cost model for re-use of data needs development, it will be in the interests of all for this to happen. Thus, in our scenario, when the nD model of Jactin house is produced it will immediately be of value to a wide range of potential users who require accurate city models. The incremental payments from many users will not put a burden on the individual.

Conclusion

The main aim of the research work was to develop a building data integration system 'to illustrate how digitised historic building data can be integrated with other types of city data, using nD modelling, to support integrated intelligent city systems'. This required the integration of data from a variety of different sources including spatial data such as laser scanning, CAD, digital maps and thematic data such as environment data and building data. As an example of an intelligent city system, we developed the Gaudi Bank Scenario in which a project team of property developers and construction professionals was to be supported in a development and construction project by the system we built.

Although the prototype system is implemented with the Gaudi Bank scenario data, it could also be used for any generic building development project. The system could help different stakeholders to get easy access to different data sets, customised analysis tools and so on. It has been shown that urban data structures can be constructed to integrate a variety of information into an nD model. This has potential to support a wide range of regeneration services to a variety of stakeholders for the intelligent city when it is fully linked to the e-CP.

Acknowledgement

The authors would like to thank European Commission for funding of the IntelCitites Project. For details of this project, please access its web site http://www.intelcitiesproject.com/

This work could not be achieved without the following partners' collaboration and support:

The Centre de Política de Sòl i Valoracions, Universitat Politècnica de Catalunya, Spain, provided expertise in use of the scanner as well as planning and construction knowledge. Labein, a Spanish research institute based in Bilbao, contributed their knowledge of commercial procedures for the procurement and restoration of historic buildings, on which the construction of the system was based. The Department of Construction Economics and Property Management, Vilnius Gediminas Technical University, worked with the University of Salford on the implementation of the system, building the section for property procurement.

Whilst not a partner in the project, the contribution of Dr Pedro Gamito, Visiting Professor in Real Estate Management ESAI, Lisbon, Portugal, was of great value to us in the development of a methodology for the use of the laser scanner for built environment rehabilitation purposes and with technical issues concerning use of the scanner and post processing of data.

References

Alyazjee, R.K. (2002) *Developments in Building Modelling*. School of Engineering, Leeds, University of Leeds.

Arayici, Y., Hamilton, A., Gamito, P. and Albergaria, G. (2005) Using the 3D Laser Scanner Data Capture and Modelling to Visualise the Built Environment: Data Capture and Modelling. Ninth International Conference of Information Visualisation, London, UK.

Campos-Marquetti, R.A.R. (2003) LIDAR Mapping and Analysis Systems: A GIS-based Software Tool for LIDAR Data Processing. Map Asia Conference 2003.

Clarke, K.C. (1999) *Getting Started with Geographic Information Systems*, Upper Saddle River: Prentice Hall.

Curwell, S. and Hamilton, A. (2002) ICT Challenges in Supporting Sustainable Urban Development. University of Salford, Salford.

Cyon (2003) *The Building Information Model: A Look at Graphisoft's Virtual Building Concept*. http://www.cyon.com

Ding, L., Liew, P., Maher, M.L., Gero, J.S. and Drogemuller, R. (2003) Integrating CAD and 3D Virtual Worlds Using Agents and EDM. CAAD Futures.

EPM (2002) http://www.epmtech.jotne.com/, EPM Technology. 2005.

Gilbert, R., Stevenson, D. and Jack, B. (1996) *Making Cities Work: The Role of Local Authorities in the Urban Environment*. London, Earthscan.

Hamilton, A., Burns, M.C. and Curwell, S. (2005) *IntelCities D5.4c Building Data Integration System*. BuHu, University of Salford.

Hamilton, A., Wang, H. and Arayici, Y. (2005) 'Urban information model for city planning.' *ITcon* 10 (Special Issue From 3D to nD modelling), 55–67.

Hamilton, A.C.S. and Davies, T. (1998) *A Simulation of the Urban Environment in Salford*. CIB World Building Congress, Gåvle, Sweden.

Harrod, G. (2004) A Standard Word Processor Format at last! . . . Now How About for CAD? Oxford University Press, Oxford.

Laurini, R. (2001) *Information Systems for Urban Planning. A Hypermedia Cooperative Approach*, Taylor & Francis, London.

Lee, A., Marshall_Ponting, A., Aouad, G., Song, W., Fu, C., Cooper, R., Betts, M., Kagioglou, M. and Fischer, M. (2003) 'Developing a Vision of nD-Enabled Construction.' University of Salford, Salford.

Lenzerini, M. (2002) *Data Integration: A Theoretical Perspective*. ACM PODS, Madison, Wisconsin, USA.

Longley, P.A., Goodchild, Michael, F., Maguire, D.J. and David, W.R. (2001) *Geographic Information Systems and Science*. John Wiley & Sons, Toronto.

Nedovic, Z. (2000) 'Geographic Information Science Implications for Urban and Regional Planning.' *Urban and Regional Information Systems Association (URISA) Journal*, 12(2), 126–132.

Tanyer, A.M., Tah, J.H.M. and Aouad, G. (2005) Towards n-Dimensional Modelling in Urban Planning. Innovation in Architecture, Engineering & Construction (AEC), Rotterdam.

Tatbul, N., Karpenko, O. and Odudu, O. (2001) *Data Integration Services*. Wiley, New Jersey.

Taylor, M.J.A.G. (2003) 'Metadata: Spatial Data Handling and Integration Issues.' School of Computing Technical Report.

Wang, H. and Hamilton, A. (2004a). Integrating Multiple Datasets to Support Urban Analysis and Visualization. Fourth International Postgraduate Research Conference, Salford, Blackwell Publishing.

Wang, H. and Hamilton, A. (2004b). The conceptual framework of ND urban information model. Second CIB Student Chapters International Symposium, Beijing, China.

Chapter 22

Modelling cities

Andreas Kohlhaas and John Mitchell

Introduction

This chapter examines the potential of creating computer-based city models. The context for this not only includes the new commercial services that are emerging that require 3D models such as for wireless network analysis for example, but a wider urban planning context that needs new solutions for the understanding of city environments, new demands of sustainability and so on and new methods that allow citizens and industry to understand and contribute to the operations of local government.

The use of wooden city models is described, and how they have remained unchanged from very early times until relatively recently. Following such, several recent city model examples are examined incorporating two different approaches to modern city modelling that derive from computer-based approaches. The first approach uses Geographic Information Systems (GIS) based on the disciplines of surveying, land ownership and related aerial photogrammetry. The second uses CAD tools from the construction industry and the new technology of building information models (BIM), a radical change from 2D drawings to intelligent accurate 3D representations of a building and its context. An example is described of the academic use of IFG (Industry Foundation Classes for GIS) tools which then canvasses the opportunity for integration of the building construction domain with the GIS/local government domain. Finally, the benefits arising from an emerging harmonization of data standards from both approaches are highlighted, and the increasing importance of open data and the converging activities of the GIS and building approaches in such standards as IFG, openGIS and CityGML (City Geographical Mark-up Language) are discussed.

A short history of city models

For a very long time city models made of wood were the standard in city planning. In Germany for example, almost all larger cities had their own wooden models in 50 cm tiles that were stored in drawers and assembled for specific purposes. When it came to a new planning proposal, a model maker hand-crafted the new building in wood at the appropriate scale (mostly 1:500 or 1:1000) and inserted it into the model. Expert opinion then decided, from different variants, the new skyline and outlook of a city. Though this technique is very old, it still contributes to architectural decisions with its tactile aspects and craftwork. The successor is Rapid Prototyping using modern materials with coloured surfaces where the models can be produced directly from Computer Aided Design software (CAD). Wooden city models inherit some interesting aspects that still are topical for today's Digital 3D City Models. The models have to be continuously updated; however, they can only illustrate reality up to a certain point of richness of detail, which means that there must be rules for abstraction and qualification.

Along with the change to digital construction of buildings from 2 to 3 dimensions at the end of the 1980s, computer science at last provided technologies that made 3D City Models possible. The basis for all 3D City Models is cadastral data that started to be collected in digital format with the entering of computer technology in public authorities in the 1960s. But the process of generation and distribution of 3D City Models 40 years later still is not obligatory and is characterized by a lively, if at times, inconclusive discussion of standards and emergence of new technologies.

Initial motivation for 3D City Models

In the case of wooden city models those who needed this tool, that is the departments responsible for urban development and planning, carried out the fabrication and maintenance. Digital 3D City Models were mostly assigned to public survey and cadastral organizations, because the sophisticated methods and techniques involved come from photogrammetry and Geographical Information Systems (GIS). There has been a lack of driving force from stakeholders of the 2D cadastral world and GIS who consider 3D Models as a 'nice to have' vision, but as commercially unrealizable.

Wave propagation analysis for mobile (cellular) radio

This situation changed dramatically with the introduction of cellular telephone. The standardization of the Global System for Mobile Communications (GSM) was completed in the 1980s. Both public and private providers were faced with a challenging task to efficiently plan their radio networks wherever they wanted to deploy their service. Due to the frequencies allocated for this new service former methods of wave propagation analysis had to be refined. Radio waves of the Giga Hertz frequency spectrum propagate almost like visible light and absorb rapidly within solid organic or inorganic material. Reflection on surfaces of buildings is also an issue that had to be taken into consideration. 3D City Models and Elevation-Models had to be created to carry out wave propagation analysis to find appropriate locations for mobile phone base stations. Public providers demanded full coverage to reach every possible customer. Other circumstances such as the opening of the former GDR with an almost nonexistent Telco infrastructure fostered new approaches. The public survey organizations could supply 2D data but mostly had no ready answer for the emerging 3D demand. In consequence every mobile network provider acquired their own digital surface model comprising the necessary 3D building shells with appropriate absorption and reflection coefficients for building surfaces and vegetation.

2D and 3D car navigation data and the Asian house numbering problem

Another strong emerging market also showed interest in 3D City Models. Car Navigation for many years has had a steady increase of 10–20 per cent annually (Figure 22.1). In those areas of the world where house numbering is set by the date of the building (and not by street order) drivers who do not know their destination are highly dependent on orientation from known points. Scientific investigations identified that so-called landmarks, that is buildings that are easy

Figure 22.1 Typical landmark for 3D-car-navigation with courtesy of Harman/Becker Automotive Systems innovative systems GmbH, Hamburg © 2005.

to remember by their elevation, construction style or beauty, are the cornerstones in intuitive navigation. The Asian market especially is adopting 3D navigational systems, to be introduced very soon, but they still lack the large amount of geographic and building data that once again has to be acquired. The crux of the problem is that the data for navigation and wave propagation analysis have very different requirements in terms of accuracy and detail. Unfortunately there is almost no way to use data that was once collected for a special purpose and supports use for different application fields.

3D City Models generated with Common Modelling and Virtual Reality Technologies

Early attempts for City-Modelling have been made in different cities all over Europe. Consequently companies have been looking for standard software and data formats; one outcome was the Virtual Reality Modelling Language (VRML, an ISO standard for 3D-Modelling) and available rendering modeller software packages for multimedia purposes. The workflow was to take as much from existing cadastral and elevation data as possible and to model the building shells with the aid of survey or laser scanning. Facade textures were equalized by image-processing methods and assigned to surfaces. These techniques produce nice 3D models that have been recently used for computer game animation with game engines (Schildwächter *et al.*, 2005). Due to the VRML definition that can only handle single precision floating point figures, the coordinates of projected coordinate systems can not be preserved fully or only by workarounds. This circumstance makes these models inadequate for spatial geo-referenced analysis and queries. Another issue is that most current modelling software produces highly tessellated surfaces, which are a perfect description of organic shapes (like morphing and avatars), but mostly produce no planar surface polygons, which are effective in data storage and can be reused for CAD.

A quite different approach uses Virtual Reality (VR) technology of 360° photographs that combined with camera standpoint mapping information creates colourful walkthroughs, but can be hard to update. The VR technology was first introduced to the public around 1994 by Apple Computer Inc. and still has a wide circulation because of the relatively easy data generation by photography. Despite the advantage of high Internet transfer rates this VR technology cannot claim to have an underlying model. Only photographic images of the reality can be captured and processed (Cyclomedia, 2005).

3D City Models generated with CAD software

Berlin 3D by Senatsverwaltung for Urban planning Berlin

The Berlin Senate Administration for Urban Development took a novel approach. In its search for an appropriate 3D City Model, the public office started around 1998

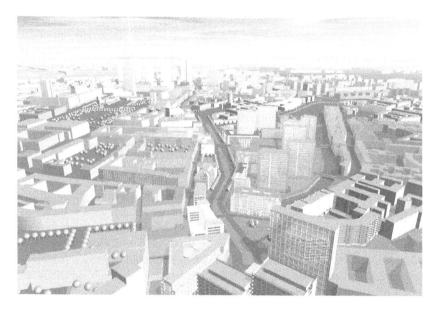

Figure 22.2 Perspective to the famous 'Alexanderplatz' TV-Tower, Berlin 3D (CAD-Model)
created by Senatsverwaltung für Stadtentwicklung Berlin – Architekturwerkstatt © 2004.

to create a model with its own resources. One of the results is illustrated in
Figure 22.2. Berlin's cadastral 2D plan was taken as the basis for all models, and
a master plan initiated, which provides coverage of about 100 km² of a total city
area of ca. 900 km². 50,000 buildings were modelled manually to obtain a complete coverage of the inner city. In Berlin, typical for many older cities, the architectural style of several historical periods creates a unique impression of the city
skyline in macro or micro scale for visitors and inhabitants. The architectural
styles embody standards of dimensions for rooms and storey heights. The cadastral information of Berlin stores also the date of construction and the number of
stories. It was therefore straightforward modelling block models in the CAD software in use as slabs with the estimated height as a product of both attributes.
Figure 22.2 also shows that this model is enriched by nicely detailed landmarks,
which are models that inherit higher geometric details and generic textures or
even real facade photos. Topologic surfaces stacked on a flat ground perform a
unique realistic impression with timeless abstraction. Complementing the former
wooden city models like the one shown in Figure 22.3 were typical representations of vegetation, for example trees made by balls and cones. This method is
still used in current CAD- and GIS-software.

Today the virtual 3D City Model (Stadtentwicklung, 2005) is kept up to date
periodically and enhanced on a project-by-project basis to visualize new planning
enterprises. In addition to the 3D model, a 2D master plan is also worked out on

Figure 22.3 One of the two former wooden city models of east and west Berlin (scales 1:1000 and 1:500) still existing but obsolete, Senatsverwaltung für Stadtentwicklung Berlin – Architekturwerkstatt © 2004.

the same data showing the impact for example on Berlin's urban fabric due to the Second World War and the resultant effects of new traffic planning. Another use is the simulation and analysis of reconstruction of certain parts of the city to reduce low occupancy and raise the quality of accommodation standards.

The European Community funded a research program to combine different data like this manually created CAD-Model with Photogrammetry Data (i.e. roof landscape) and make them available through a 3D database with specialized semantic clients.

Stuttgart City Information System (SIAS), City Survey of Regional Capital Stuttgart

Commercial products available in the middle 1990s for a city Information system were not acceptable in terms of price and functionality. Therefore, an invitation to tender was issued defining the exact requirements. Together with IBM a solution SIAS (Spatial Information and Access System, by Geobyte Software GmbH, Germany) was developed that is still in use. Today, SIAS is the graphical user interface for a SAP R/3 module for Plant Management and many applications followed for the Civil Engineering, Urban Development, Environmental Protection, Public Policy, Gardens and Cemetery and the City Survey Offices. In total about 1,280 users in 19 offices access SIAS 30,000 times a month (Bill *et al.*, 2003).

1998 was the starting point for 3D Models taken from aerial photographs, which became necessary for large urban projects like the development initiative Stuttgart 21 as well as measures for reducing noise and pollution emission. The approach for the city model was twofold, once it was realized that a realistic and precise roof model would take longer than expected. The detailed attributes of the cadastre (number and usage of stories, roof shape, year of construction, volume of converted space, as well heights of eaves and ridges) helped to produce a block model with full coverage of the city area rapidly. In contrast to a fully exploited photogrammetry model the amount of data is modest and this model serves most applications like noise calculations. As more accurate roof models are available from photogrammetry these have been used to enrich the block model (see Figure 22.4).

Data storage options using files or databases were examined. Databases would be used today, but no commercial products were available at that time of

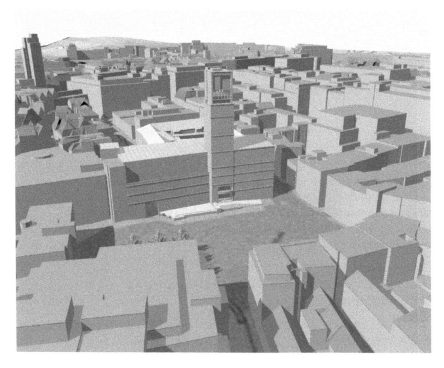

Figure 22.4 Stuttgart city hall in 3D City Model with courtesy of Stadtmessungsamt Stuttgart, © 2005.

the decision. The update process today is accomplished with ArchiCAD (by Graphisoft Deutschland GmbH) and initially other CAD and proprietary software like Microstation were used also for visualization.

Hamburg DSGK 3D (Digital City Base Map, Digitale Stadt-Grund-Karte)

A similar approach and almost at the time, the base model for the 3D City Model of Hamburg was designed and data collected. A study by independent consultants confirmed the technical direction. The first phase derived block models with flattened roofs from the cadastre; that is building footprints were extruded syntactically with an estimated storey height for the usage and multiplied by the number of stories. When software architecture and the process of generation were established (Welzel, 2002), the first model was available in 2000. The 2D cadastral information was exported from SICAD GIS and processed through an Add-on in ArchiCAD into the CAD model. The complete city model covers 750 km^2 and about 330,000 buildings and can be accessed by street and house number.

This model is continuously updated and enhanced by textures of streets, waterways and green areas.

The second step consists of buildings that come from photogrammetric evaluation of Digital Ortho Photos at a scale of 1:6000. This gives a resolution of approximately $+/-15$–20 cm in all coordinates. Almost one-third (230 km^2) with 120,000 building shells are available in this form. Interesting objects of this roof model were copied and moved from the elevated height to the flat ground model of the block model as Landmarks.

Since 2001 a precise Digital Surface Model (DSM) has been available of the whole city-state of Hamburg, derived from airborne laser scanning and consisting of 900 million points (Produkte und Dienstleistungen, 2005). It represents the basic ground surface for the roof model as well as for the block model. The building footprints of the block model were intersected with the DSM and the lowest z-coordinate along the footprint gives the baseline for each building block in height (see Figure 22.5).

Hamburg's Office for Geoinformation and Survey highlighted town planning and architecture as the most appropriate application for 3D City Models and consequently provide parts of both models to this discipline in various CAD-formats, but prefers now to use open IFC (Industry Foundation Classes) the only acceptable ISO standard for the Building Information Model or Virtual Building Model™. Both terms can be used synonymously to mean 3D Building Models, which inherits all information about materials and construction and component parts but are created by competing software vendors. IFC can bridge geometry and building semantics between the most popular and advanced software packages without loss of information. Other target groups are identified: environmental protection, civil engineering, traffic planning, emergency services, property market, conservation of historic monuments and buildings, mobile radio and meteorology.

Figure 22.5 Precise roof model (2. phase) of DSGK 3D Hamburg, with courtesy
of Freie und Hansestadt Hamburg, Landesbetrieb Geoinformation und
Vermessung, © 2005.

The roof model was provided mostly in DXF (Industry standard for CAD data
exchange of geometry and surfaces) format as building shells with roof geometry
and vertical walls. This Boundary Representation (BRep) of building entities are
converted from Constructive Solid Geometry (CSG) elements by extruding
surface to wall and roof volumes that suit the requirements for architectural
models. With the aid of CAD (ArchiCAD) modelling manipulations in the
geometry can easily be undertaken on photogrammetric raw data.

The model is under a continuous development and updating process. Next
steps include façade texture, vegetation model and more detailed street furniture.
Augmentation of the phase 2 (roof model) is also planned for the near future.

Nottingham 3D City Model

The 3D computer model of Nottingham was created for the City of Nottingham local
authority in 2004 (Figure 22.6). It was created to enable councillors to view planning
submissions in context. It is often difficult for those who are not used to reading
traditional 2D plans, sections and elevations to visualize a building proposal.
Perspective views produced by those submitting proposals are also of limited use

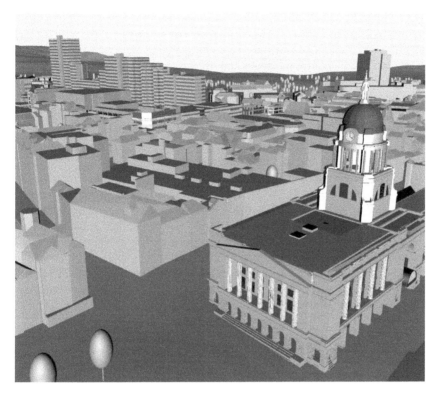

Figure 22.6 Nottingham 3D City Model with Points of Interest (Architectural Models) processed in ArchiCAD and visualized in NavisWorks Rendering software. With Courtesy of Bite Design Ltd., Nottingham © 2004.

since they will inevitably only be taken from the most advantageous position. Where there are multiple proposals for a development area, it is also particularly important to be able to view the various building proposals in context with each other, which can only be done with the help of a full computer model of the area.

The model itself was created more or less in the same manner as described for Hamburg. The Digital Surface Model consisted of a Triangulated Irregular Network (TIN) with break lines that could be easily imported from GIS software to CAD. A commercial photogrammetry company provided the roof landscape as BReps of the building shells. Landmarks were modelled directly in ArchiCAD and added to the city model.

Coburg 3D City Model

Smaller towns now a have the opportunity to compile and operate their own 3D-Town-Model for multiple purposes. Given a limited budget, a photogrammetry service provider and a CAD software vendor developed a concept that enables the city of Coburg to benefit from a 3D City Model from the first data deployment. Continuity, data standards in CAD and GIS were only one part of the wish list that was discussed. The main functions were city planning, data exchange of architectural models, tourism applications and flooding analysis. The challenge for the contractors was to balance expenditure between software development and supply of services.

At the end of 2004 a 3D City Model spanning the whole boundary of the city was delivered together with a set of software tools to maintain and operate this model. The model included a Digital Terrain Model (DTM), building shells in form of block models and roof models (Figure 22.7).

The DTM was evaluated from Digital Ortho-Photos (DOPs) forming a TIN with break lines covering the whole boundary of the City of Coburg with about 50 km². The 42,000 inhabitants live and work in ca. 20,000 buildings that form the entire block model. No information from the cadastre was available to help estimate the height of the buildings. The photogrammetric interpretations of roof heights were also taken from existing DOPs. In this case manual work seemed to be more efficient and superior to any other method. Finally Points of Interest (POI) like the town hall square, the castle and other relevant historic buildings were measured by terrestrial photogrammetry and furnished with façade textures, see Figure 22.8.

Initially three identical data sets were delivered. One is intended for a commercial standard GIS application with 3D capabilities in use by the city survey

Figure 22.7 Perspective of the 3D City Model of Coburg with courtesy of Stadt Coburg, Referat für Bauen & Umwelt, GIS-IT Koordination, © 2005.

and cadastre office. Second, a set of the building shells and the DTM was constructed for CAD applications. Finally, a fully detailed model with vegetation, buildings, textures and surface was provided as a VRML model for visualization purposes (see Figure 22.7). The VRML model can be derived from both CAD and GIS software packages, but cannot be transferred back. The bi-directional exchange between GIS and CAD is supported and guarantees the updating process.

Current motivation and driving forces

European directive for strategic noise mapping

One of the major driving forces in the immediate future will be the European directive for noise cadastres (Directive, 2002). Either complete measurement or calculations with the aid of 3D models and basic measurement can be used to produce required noise maps. The EU deadline is very close; at end of June 2007 the communities have to present the actual noise maps that show how severe the impact on living quality for their citizens actually is. A year later, July 2008, action plans for noise reduction have to be established.

Affected are 'all agglomerations with more than 250 000 inhabitants and for all major roads, which have more than six million vehicle passages a year, major railways which have more than 60 000 train passages per year and major airports within their territories'.

Figure 22.8 Town hall and square created by terrestrial photogrammetry as part of the 3D City Model of Coburg with courtesy of Stadt Coburg, Referat für Bauen & Umwelt, GIS-IT Koordination, © 2005.

One of the remedies to protect buildings near-by is by noise barriers. These have to be modelled and simulated in a 3D City Model environment to avoid costly aberrations. The technical requirements are a precise DTM and wall elevations. Roofs cannot contribute because in most cases they reflect noise vertically where it does not hit any obstacle. Obviously a lot of public and commercial institutions have to contribute their calculations (for example: airport, railroad companies and building authorities). They all need to rely on a standardised data exchange format. CityGML (CityGML is an application schema based on GML3, an XML definition enriched by geospatial concepts. See current results of Geo Data Infrastructure initiative of North Rhine Westphalia: www.citygml.org) is the only candidate, which hopefully will pass the acceptance of the OGC in early 2006.

3D-car-navigation still in perspective

As stated before, 3D-Town-Models are not only required for car-navigation, but also for Location-Based Services and other hand-held applications. Due to hardware restrictions and available data, combined block models of buildings are expected to be preferred solutions for the coming years in order to achieve full coverage. Galileo, the next generation of Europe's satellite-based navigation system, will enable the industry to find new applications.

Google earth opens markets

A first impression on the marketing of these new technologies is given by Google Earth. This easy-to-use internet application, which had been in existence for some years by Keyhole Inc, did not achieve the breakthrough then for widespread acceptance by the public as it has today.

One of the commercial objectives of Google Earth is to distribute 3D Data and make them widely available to architects and engineers as an environment to communicate their planning proposals to decision makers. The effect on industry will be very significant and it is difficult to anticipate its future impact reliably. It is however evident that no isolated 3D-Planning-Data will be adequate without a 3D- environment that has to be provided by Google itself.

A construction industry view

As the GIS world explores how to extend their technologies to encompass buildings, likewise the design and construction industry is also in the process of a fundamental change in paradigm for the representation of building information. The industry, by its nature, focuses largely on the construction of a single facility, or at least a set of buildings on a single development site. This fact, together with contractual arrangements that last only for that project, has meant that there have been little incentives to look at the lifecycle behaviour of a building nor examine the integration and or relationships to its immediate urban context. This has more often been the case in the English-speaking world, rather than in Europe where renovation and alteration is far more common.

Building construction authorities/urban city planning

Urban City Planning has a lot to do with visual impressions based on large-scale models. Public agencies currently seldom can afford models that suit their emerging digital needs. The history of architectural design technology, which has created separate approaches, that is, 2D construction drawings and 3D visualization models, initially did not encourage or support these new requirements. Today 3D BIM are accepted to serve facility development better and reduce costs in the construction industry. BIM has become a worldwide focus as a more effective alternative to the traditional use of 2D drawings to support the large and diverse processes and information needs in facility development and management.

BIM is an integrated digital description of a building and its site comprising objects, described by accurate 3D geometry, with attributes that define the detail of the building part or element, and relationships to other objects, for example this duct is-located-in the third storey of the building named Block B.

BIM is called a rich model because all objects in it have properties and relationships, and based on this, useful information can be derived by simulations or calculations using the model data. An example is the ability to perform automated code checking to confirm egress, fire ratings and so on, or a thermal load calculation. Whilst the concept of BIM – originally known as product modelling – has been in existence since the 1960s, it is only in the last 5–10 years that software vendors particularly in the architecture domain have delivered products that support this BIM approach. As computing capacity has been increased, the four major international CAD companies – Graphisoft and Nemetschek introduced product from the early 1980s in Europe and Bentley and Autodesk in the US – all now offer such products.

Complementing the BIM approach has been the efforts of the international construction business community represented by a group called the IAI (International Alliance for Interoperability) to develop a global information exchange protocol that enables not only an effective, and open standard for information sharing but also supported the many disciplines involved in construction, such as structural and service engineering, construction estimating, scheduling and procurement and asset and facility management. The IFC protocol was adopted as an ISO standard in 2002.

This new technology supports the incorporation of these models into 3D City Models and to visualize planned buildings in a digital urban context. The conjunction of IFCs and 3D City Models makes 3D City Models very attractive for widespread usage in urban planning scenarios as it was predicted some years ago (Laurini, 2001).

The IFC opportunity

As these two complementary and linked developments – an integrated building model representation BIM and an open data exchange protocol IFC – reached some maturity, a link was seen to be missing – how to integrate the building model with the 'site' meaning for example cadastre, terrain, utilities and transport/access systems. This is a crucial interface into local government which also needs information about developments for planning, regulation and increasingly important sustainability criteria. The capacity of local governments to plan, process and manage their own and their citizens' built environment in most developed countries is at a crisis point, putting enormous strain on human resources and delivery of timely services.

One solution to this issue has emerged through the work of the IAI, particularly the Singapore government's ePlanCheck building code checking system and the Norwegian government's Byggsøk Development approval system.

Singapore innovating regulatory services

The Building Construction Authority of Singapore – a member of the IAI – in 2003 released its first operational automated code checker accessed through a single web portal and newly integrated administration of some 13 agencies responsible for building assessment and regulatory approval.

The ePlanCheck system uses an expert system, based on an IFC model server, to interpret and check a building proposal submitted to it in IFC file format.

ePlanCheck has generated a lot of international attention – the automation, reuse of intelligent data, and time saving alone showed dramatic new potential; the IFC model data could not only be used as a better environment for multi-disciplinary design, it also was shared for building assessment and could act as reference data for local government and other government-agency uses.

An unexpected outcome of the project has been the realization that the rich model data now being delivered (in contrast to traditional drawing documentation, or GIS information about buildings – outlines and the like) could be used for local government purposes – city modelling, and other asset and facility management roles by government agencies and owners if during the construction process as-built information was updated in the model.

Norway integrating land and building data for local government

The Norwegian government visited Singapore in late 2003, and was impressed by the ePlanCheck system. The national building agency Statsbygg undertook the development of a Development Approval System – Byggsøk following the establishment of a Memorandum of Understanding (MOU) between the two countries. While Byggsøk shared a similar structure as the Singapore system, it needed land and planning information that was not yet supported by the IFC standard.

Convinced that an approach based on open standards, IFC and GIS, was necessary to remove a major impediment to the flow of information in the construction development industries, the IAI Nordic Chapter undertook the extension of the IFC model to include GIS objects so that the model can incorporate data usually manipulated by GIS analysis: cadastre, street centrelines, contour lines and aerial photography (Bjørkhaug *et al.*, 2005). The resulting version is called IFG (IFC and GIS). This was accomplished in just over 13 months, evidence of the rapidly improving base technology, expanding application support and widening knowledge of BIM, IFC and GIS.

Norway demonstrated this new environment in May 2005, to an international audience that included some 15 IFC compliant applications that all made use of IFC and IFG data. This development marks a major international milestone in the realization of open standards based on building models and GIS data to support a paradigm shift in the construction industry.

Design using BIM and IFG

In 2004, the Faculty of the Built Environment, University of New South Wales, Australia, commenced offering an elective of design collaboration based on a shared building model. The course takes advantage of the emerging maturity of BIM and diverse IFC compliant applications, with a focus on design decision making, as well as the process opportunities afforded by close multi-disciplinary collaboration and rapid feedback from design analysis. The class comprised 23 senior students from disciplines including architecture, interior architecture, landscape architecture, mechanical and services engineering, statutory planning, environmental sustainability, construction management and design computing.

Whilst the three teams that were formed had surprisingly good collaboration experiences using the technology despite technical and learning issues (Plume and Mitchell, 2005), an unexpected outcome was the failure of the town planning students to be able to access either planning context data or holistic building compliance and performance data required by regulations and so on. However part of this problem has been addressed when the elective was repeated in 2005, where students were able to use software (Powel Gemini is a GIS tool developed by Powel, Norway see http://www.powel.com) developed in Norway as part of the Byggsøk development.

The direct integration of cadastre, land zoning, terrain, aerial photography, roadways and water mains was achieved with data provided by the local council and a spatial data provider. Transformation of the data was carried out and the team design model merged with the IFG format land data. The results were very encouraging with the student planner gaining a unique introduction into planning and building software (Figure 22.9).

The design collaboration studio demonstrates that the technology is achieving its purpose – facilitating the accessibility of diverse land information in the development and design processes.

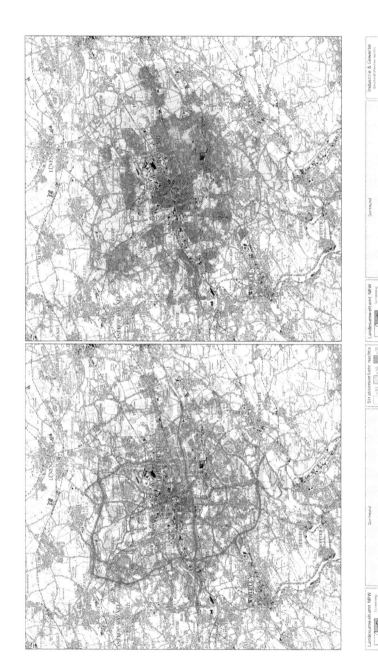

Figure 22.9 Plan projection of IFG model data showing aerial photo, terrain and building model insertion.

Planning at the urban level

Implementation by governments such as Singapore and Norway are underpinning innovative and much more efficiently delivered services by agencies responsible for the certification and management of built development. However, while this work marks an important achievement, the challenge for urban planning has many outstanding issues. As cities become more complex, it is becoming imperative that more powerful instruments be used to support decision making, and the quality of the outcomes hinges around the foundation of well-structured, detailed and efficient geospatial data. An evidence-based approach to policy development and planning in priority areas such as urban growth, sustainable development and infrastructure planning is now critical.

The University of New South Wales' City Futures Research Centre has adopted IFC/IFG technology in the belief that it is an optimal choice to address a core problem facing city and urban planners: the integration of the complex and diverse sources of data and information essential to successful management of the urban development decision process.

Traditional GIS tools provide for such systematic analysis but much of the data needed to support that analysis is held by a range of government instrumentalities and the information is typically organized around property boundaries, and not real building form. The diversity of models described earlier alone illustrates the plurality of techniques and approaches to 3D City Modelling currently in use and shows that, while they begin to address the fundamental need for integrated information by permitting visualization of that descriptive data, they fail to have a significant impact on urban planning because the data has no significant semantic richness. The data presented typically only describes the basic shape and location of objects, without explicitly referring to the objects as representations of the actual urban fabric (Figure 22.10). As shown earlier most current models are made for a specific purpose and are not transferable to another task. Beyond making visual judgements of the urban impact of specific developments or developing visualizations that depict future scenarios for urban growth, the models provide little opportunity for systematic social, economic and environmental analysis. Without this fidelity and breadth of detail, the capacity of these models to assist in complex decision making is limited by their lack of ability to harness more meaningful information about the geographic areas they are representing.

The Australian Government, for example has already established a robust spatial information infrastructure at the cadastral level through several data harmonization initiatives that maximize the utility of GIS data. At a sub-cadastral level, systems like BASIX (see http://www.basix.nsw.gov.au) and Nathers (see http://www.csiro.au/promos/ozadvances/Series4Houses.html) begin to serve as a strong base for complementary spatial extensions that operate at a finer urban scale. The University of New South Wales' City Futures Research Centre aims to build on this in order to provide a much-needed link between GIS analysis at the cadastral level, city modelling with its strong visualization capacity, the need for

Figure 22.10 Aerial view demonstrating the integrated 3D GIS and building model data.

more robust urban analysis at the sub-cadastral level and the plethora of existing datasets that could support such analysis.

Conclusion

In the current city model examples described, two key methods have been used to develop models for specific purposes. The GIS approach uses geographic data and related technologies to construct geometric models of the outer shells of buildings. In some cases the properties of the shells or their constituent surface-building elements have been articulated to analyse a problem, that is the behaviour of a wireless network.

In the second approach, where the models have been created using CAD tools, a more conscious 'building' model has been achieved, but apart from the under-lying cadastre in the Hamburg model for example, little real data about the build-ings is stored. In both these approaches the key value-added benefit is accurate visualization, providing a universally understood environment to show the scope and impact of proposals to lay and expert citizens alike.

Two inhibitors have been holding back building model development; the first has been the almost universal use of 2D drawings to model buildings, thus rendering their automated conversion to a 3D model cumbersome, low in semantics and error-prone; the second factor is the size and relevance of data implicit in a building model. In the worst case too much data is possible, and at this stage, use of BIM is in its infancy and few guidelines exist to guide the process for design and construction purposes, much less exploitation for city planning use.

However, IFG provides an integrating technology that can be used to interoperate effectively between traditional GIS applications and the full range of traditional CAD tools used by the built environment professions, by extending design analysis at the scale of a single building to an urban scale. Further, it creates an opportunity for the development of a whole suite of new computer tools that can access that meaningfull model data and undertake multiple analyses, ranging from issues of urban ecology, urban sustainability, transport planning, demographic change and economic development, as well as providing a platform for the efficient management of traditional planning and development processes such as development and building approvals. Since the information needed to be stored implies substantial orders of increase, IFC model server technology is being used to handle the large, multi-disciplinary, multi-facetted data.

These potentials highlight the need for open standards like CityGML and IFC/IFG, that are based on interdisciplinary cooperation between OGC and IAI. The SIG 3D GDI-NRW W3D Service and CityGML Working Group are other centres of work that will contribute to the effective and vendor-independent data exchange, necessary if comprehensive repositories are to be successfully created and maintained.

Google Earth provides a new 3D-GIS awareness and a commercial boost. BIM is recognized in the Building Construction Industry as the definitive way to reduce costs. The integration of both approaches offers a prospect of a new generation of tools that will serve communities' need to manage the built environment more equitably, more sustainably and more economically.

References

Bill, T., Ralf, C. and John, T. (eds) (2003) Kommunale Geo-informationssysteme: Basiswissen, Praxisberichte und Trends, Heidelberg, Wichmann, 2003, ISBN 3-87907-387-2.

Bjørkhaug, L., Bell, H., Krigsvoll, G. and Haagenrud, S.E. (2005) Providing Life Cycle Planning services on IFC/IFD/IFG platform-a practical example, 10DBMC International Conference on Durability of Building Materials and Components, LYON [France] 17–20 April 2005.

Cyclomedia (2005) http://www.cyclomedia.com

Directive (2002) DIRECTIVE 2002/49/EC OF THE EUROPEAN PARLIAMENT AND OF THE COUNCIL of 25 June, Official Journal of the European Communities, L 189/12.

Laurini, R. (2001) *Obert: Information Systems for Urban Planning: A Hypermedia Co-Operative Approach*, London, Taylor & Francis, 2001, ISBN 0-748-40964-5.

Plume, J. and Mitchell, J. (2005) 'Multi-Disciplinary Design Studio using a Shared IFC Building Model.' Computer Aided Architectural Design Futures 2005 Proceedings of the 10th International Conference on Computer Aided Architectural Design Futures/ISBN 1-4020-3460-1, Vienna (Austria) 20–22 June 2005, pp. 445–454.

Produkte und Dienstleistungen (2005) Landesbetrieb Geoinformation und Vermessung.

Schildwächter, B. Ralf, C. and Bill, T. (2005) 3D Stadtmodell Bamberg – Visualisierung mit3D-Game-Einiges in AVN Allgemeine Vermessungs-Nachrichten, Heidelberg, Hüthig, ISSN 002-5968.

Stadtentwicklung (2005) http://www.stadtentwicklung.berlin.de/planen/stadtmodelle/

Welzel, R.-W. (2002): From 2 to 3D in Hamburg, in Geoconnexion International, Godmanchester, UK, 9/2002 ISSN 1476-8941.

Chapter 23

nD in 2D

Dennis Fukai

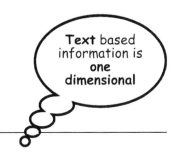

Look at this block of text: 12pt Times Roman, chapter title to match the others, Royal Octavo book format, margins and layout set by the publishers in order to keep the paragraphs and pages consistent. Ideas washed through editorial layers of protocol as we authors struggle to input letters and numbers, and periods and spaces. The result is linear, one dimensional text. Any hope of using the written word to communicate the potential of a multidimensional information system is mediated by the technologies used in its publication.

It's ironic that at the same time, our eyes drift to the thought bubble riding above this paragraph. The oval graphic and its caption pull a reader away from these regulated words. It's out of place, out of space, from another dimension. The natural visual attraction of a comic bubble suggests a way to shift the promise of multidimensional information systems off of a one dimensional page. In fact, the ability of graphic and visual explanations to support and clarify complex ideas is well known (Tufte). They are nonlinear forms of visual information that have been recognized to add a deeper understanding of complex systems (Guindi). Furthermore, diagrams and illustrations are known to be medium of construction communications from the earliest beginnings of human settlements (Tringham). As a practice, the construction industry has a long tradition of using illustrations and diagrams for communications.

We can also see and quickly visualize complex relationships in a graphic narrative. New works like *Persepolis* (Satrapi) and *The Yellow Jay* (Atangan), as well as older classics like *Comics & Sequential Art* and *Graphic Storytelling and Visual Narrative* (Eisner) communicate concepts that go well beyond the linear text of an ordinary book. They engage the reader with visual information and evoke the full potential of nD communications. Graphic narratives mix text with visual information relationships to carry their readers through time, process, motion, and description (McCloud).

For the construction industry, these graphic communication techniques suggest the integration of visual form, style, and content based on the multidimensional premise of nD modeling. Graphic narratives are quick to read, simple to understand, and widely used as cross cultural communication tools.

Though well within common standards of practice, this 2D drawing is **confusing** and **difficult** to "read." It takes time to interpret a 2D document and understand its **three-dimensional** relationships.*

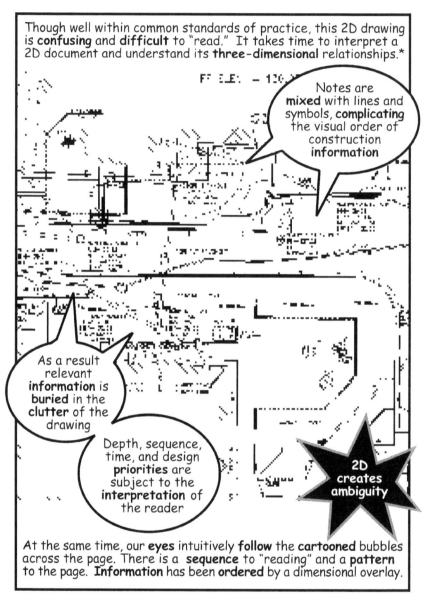

At the same time, our **eyes** intuitively **follow** the **cartooned** bubbles across the page. There is a **sequence** to "reading" and a **pattern** to the page. **Information** has been **ordered** by a dimensional overlay.

*Civil drawings and models are included here with the permission of the Office of Facilities Planning and Construction, University of Florida. Howie Furguson, PM, Holder Construction Inc and engineers Causseaux & Ellington Inc.

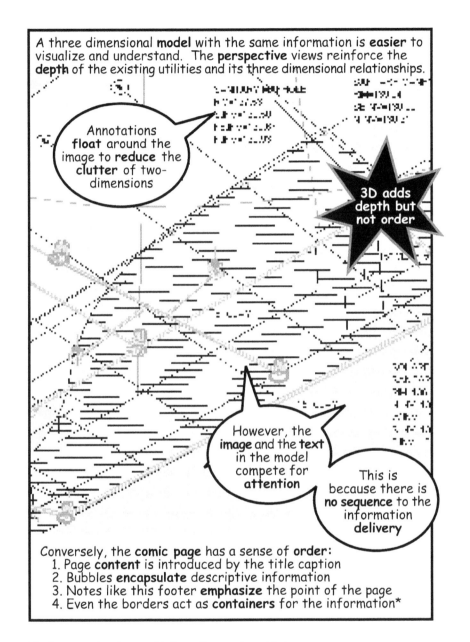

*In fact, another layer of information can be added
as a footnote to any word on the narrative page.

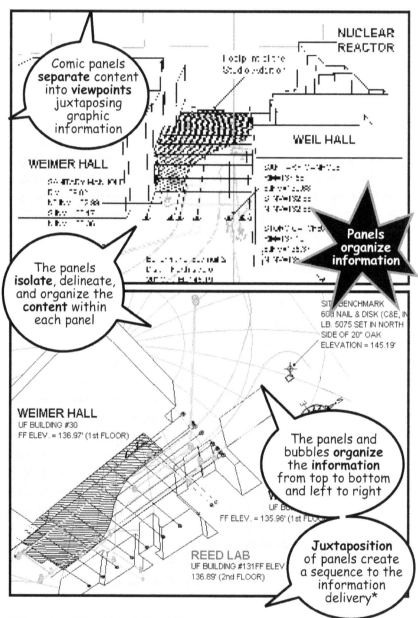

*These transitions hint at the ability to
control the sequence of information delivery

The **order and pace** of information delivery is **regulated** by graphic elements like the **shape** and lineweight of bubble and panel borders

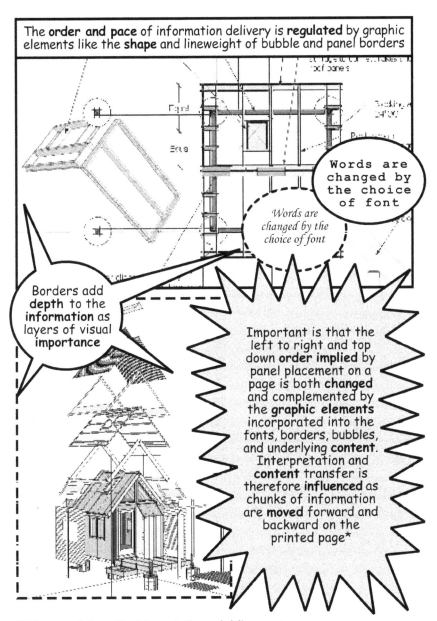

Words are changed by the choice of font

Words are changed by the choice of font

Borders add **depth** to the **information** as layers of visual **importance**

Important is that the left to right and top down **order implied** by panel placement on a page is both **changed** and complemented by the **graphic elements** incorporated into the fonts, borders, bubbles, and underlying **content**. Interpretation and **content** transfer is therefore **influenced** as chunks of information are **moved** forward and backward on the printed page*

*This means information interpretation and delivery can be dynamically controlled by simple graphic devices

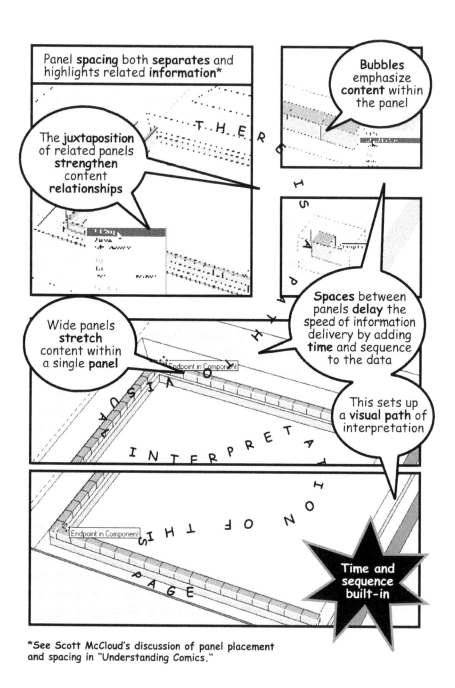

*See Scott McCloud's discussion of panel placement and spacing in "Understanding Comics."

Panel **sequences** can represent milestones, assemblies, sequences, phases, or paths in a CPM schedule. These **include** process animations, site utilization planning, team collaboration, and phased **submittals** & **reporting.**

The **construction** process is an engaging and interesting **narrative,** with a clear plot that moves to **capture** the reader's **interest** and imagination.

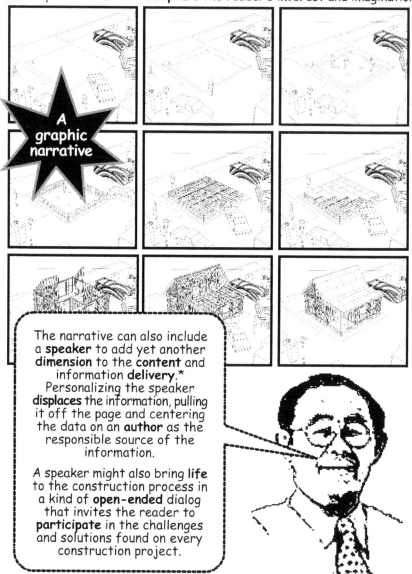

A graphic narrative

The narrative can also include a **speaker** to add yet another **dimension** to the **content** and information **delivery.*** Personalizing the speaker **displaces** the information, pulling it off the page and centering the data on an **author** as the responsible source of the information.

A speaker might also bring **life** to the construction process in a kind of **open-ended** dialog that invites the reader to **participate** in the challenges and solutions found on every construction project.

*Narrative "heroes" were first discussed at the 3D to nD Modeling International Conference in "VIRST: A Virtual Safety Trainer." 2003, Manchester, England.

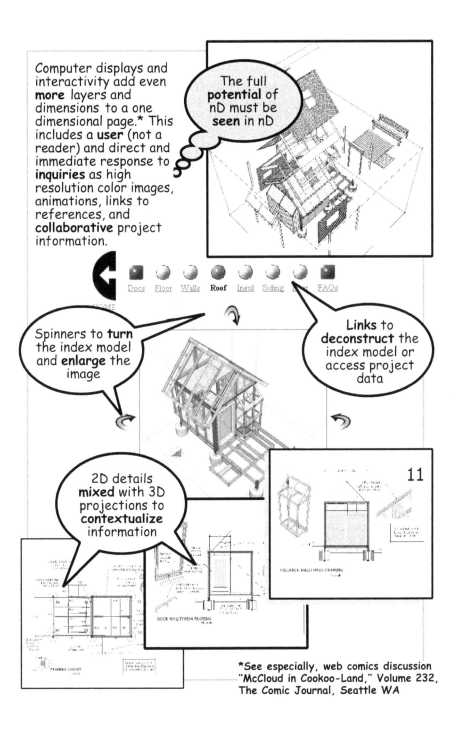

Computer displays and interactivity add even **more** layers and dimensions to a one dimensional page.* This includes a **user** (not a reader) and direct and immediate response to **inquiries** as high resolution color images, animations, links to references, and **collaborative** project information.

The full **potential** of nD must be **seen** in nD

Docs Floor Walls **Roof** Insul Siding ___ FAQs

HOME

Spinners to **turn** the index model and **enlarge** the image

Links to **deconstruct** the index model or access project data

2D details **mixed** with 3D projections to **contextualize** information

11

FULLBACK WALL TYPES FRAMING

DOOR WALLTYPE#4 FRAMING

FRAMING LAYOUT

*See especially, web comics discussion "McCloud in Cookoo-Land," Volume 232, The Comic Journal, Seattle WA

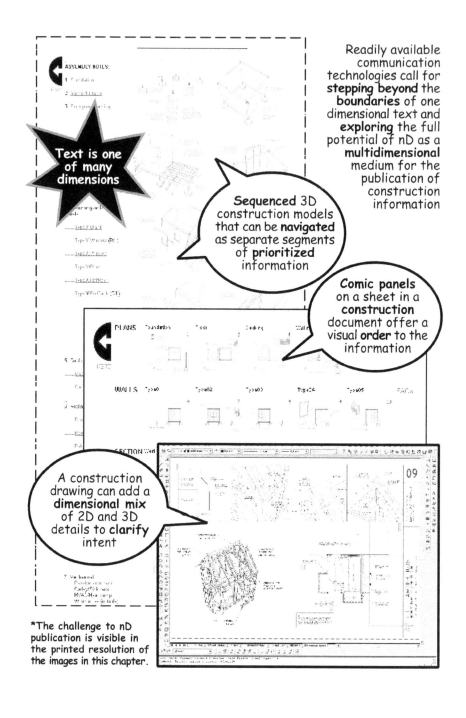

References

Atangan, Patrick. *The Yellow Jar: 2 Tales from Japanese Traditional.* Nanteir Beall Minonstchine Publishing, 2002.

Bloom, Howard. *Global Brain: The Evolution of Mass Mind from the Big Bang to the 21st Century.* Wiley, 2001.

Eisner, Will. *Comics & Sequential Art.* Poorhouse Press, 1985.

Eisner, Will. *Graphic Storytelling and Visual Narrative.* Poorhouse Press, 1996.

EL Guindi, Fadwa. *Visual Anthropology: Essential Method and Theory.* Alta Mira Press, 2004.

McCloud, Scott. *Understanding Comics: The Invisible Art.* Harper Paperbacks, 1994.

Satrapi, Marjane. *Persepolis: The Story of a childhood.* Panthron, 2004.

Tringham, Ruth. *Urban Settlements: The Process of Urbanization in Archaeological Settlements.* Warner Modular Publications, 1973

Tufte, Edward R. *Envisioning Information.* Graphics press, 2001.

Concluding remarks

Imagine this:

> A system that given an idea, illustrates alternatives, illustrates constraints, enables the understanding of both quantitative (time, cost, acoustics etc) and qualitative (aesthetics, usability etc) dimensions. A system that enables all stakeholders to participate in the design process in a medium conducive to their needs and ability; spreadsheets for the Quantity Surveyor, VR environment for the client perhaps. The system will determine the build specification, the manufacturing resources, the production process, provides drawings, the tool sets etc. A system that can adapt to a particular building type, or to the urban environment as a whole. A system where we can use our knowledge, in conjunction with the other stakeholders, to achieve the best solution, at the right cost in a faster time, and in a sustainable manner.

This is the utopia for nD modelling. The need will not go away, indeed as companies, systems, products and markets get even more complex we need an integrated product model to guide and help us make decisions. However, organisational behavior researchers dictate that we are not automatons – we will never work to a detailed and prescribed process and procedure, when situations demand innovation, creativity and constant change to enable us to compete. Yet, we do need systems to help us work through a complex world for the benefit of its inhabitants. The challenge is to understand what systems are the most appropriate, how we can best introduce them into organisations, and the impact that they will have on our work behavior and the future of the organisations who use them.

The Editors hope that this book has provided you with some insight into the potential of nD modelling. There is obviously much that needs to be progressed to fulfill the dreams of widespread nD-enabled construction, but the research covered in this book points a way into this future.

Index

Milton Keynes UK
Ingram Content Group UK Ltd.
UKHW031139141024
449569UK00024B/1211